Immune Mechanisms in Inflammatory Bowel Disease

Recent Volumes in this Series

Volume 568
HOT TOPICS IN INFECTION AND IMMUNITY IN CHILDREN II
Edited by Andrew J. Pollard and Adam Finn

Volume 569
EARLY NUTRITION AND ITS LATER CONSEQUENCES: NEW OPPORTUNITIES
Edited by Berthold Koletzko, Peter Dodds, Hans Akerbloom, and Margaret Ashwell

Volume 570
GENOME INSTABILITY IN CANCER DEVELOPMENT
Edited by Erich A. Nigg

Volume 571
ADVANCES IN MYCOLOGY
Edited by J.I. Pitts, A.D. Hocking, and U. Thrane

Volume 572
RETINAL DEGENERATIVE DISEASES
Edited by Joe Hollyfield, Robert Anderson, and Matthew LaVail

Volume 573
EARLY LIFE ORIGINS OF HEALTH AND DISEASE
Edited by Marelyn Wintour-Coghlan and Julie Owens

Volume 574
LIVER AND PANCREATIC DISEASES MANAGEMENT
Edited by Nagy A. Habib and Ruben Canelo

Volume 575
DIPEPTIDYL AMINOPEPTIDASES: BASIC SCIENCE AND CLINICAL APPLICATIONS
Edited by Uwe Lendeckel, Ute Bank, and Dirk Reinhold

Volume 576
N-ACETYLASPARTATE: A UNIQUE NEURONAL MOLECULE IN THE CENTRAL NERVOUS SYSTEM
Edited by John R. Moffett, Suzannah B. Tieman, Daniel R. Weinberger, Joseph T. Coyle and Aryan M.A. Namboodiri

Immune Mechanisms in Inflammatory Bowel Disease

Edited by

Richard S. Blumberg

Gastroenterology Division and Laboratory of Mucosal Immunology, Brigham and Women's Hospital, Harvard, Massachusetts, USA

Markus F. Neurath

1st Department of Medicine, Johannes-Gutenberg-University, Mainz, Germany

Springer Science+Business Media

Landes Bioscience / Eurekah.com

Springer Science+Business Media
Eurekah.com / Landes Bioscience

Printed in the U.S.A.

Springer Science+Business Media, 233 Spring Street, New York, New York 10013, U.S.A.

Please address all inquiries to the Publishers:
Eurekah.com / Landes Bioscience, 810 South Church Street, Georgetown, Texas, U.S.A. 78626
Phone: 512/ 863 7762; FAX: 512/ 863 0081
http://www.eurekah.com
http://www.landesbioscience.com

Immune Mechanisms in Inflammatory Bowel Disease, edited by Richard S. Blumberg and Markus F. Neurath, Landes Bioscience / Eurekah.com / Springer Science+Business Media dual imprint / Springer series: Advances in Experimental Medicine and Biology.

ISBN 978-1-4419-2157-4 e-ISBN 978-0-387-33778-4

Immune mechanisms in inflammatory bowel disease / edited by Richard S. Blumberg, Markus F. Neurath.
p. ; cm. -- (Advances in experimental medicine and biology ; v. 579)
Includes bibliographical references and index.

1. Inflammatory bowel diseases--Immunological aspects. 2. Inflammatory bowel diseases--Pathogenesis. 3. Gastrointestinal mucosa--Immunology.
[DNLM: 1. Inflammatory Bowel Diseases--immunology. 2. Immunity, Mucosal. WI 420 I33 2010] I. Blumberg, R. S. (Richard Steven) II. Neurath, M. (Markus) III. Series.

RC862.I53I44 2010
616.3'44--dc22

2005037056

PREFACE

Inflammatory bowel diseases (IBD: Crohn's disease, ulcerative colitis) are defined as inflammations of the gastrointestinal tract not due to specific pathogens. IBD were first described in clinical reports from the last half of the 19th century. Since then, major discoveries have been made in the field of IBD (particularly in the past decade) that allow a more detailed understanding of its pathogenesis. Specifically, recent key advances include the field of genetics and a better understanding of the role of the bacterial flora as well as of the mucosal immune system. This book aims at providing a timely overview of key concepts and recent progress in the field of IBD. First, the structure and function of the mucosal immune system are reviewed. Next, the book features new information on the genetic basis of IBD with special emphasis on the role of NOD2/CARD15 mutations in Crohn's disease. In subsequent chapters, environmental factors and bacterial antigens are discussed as triggering mechanisms of IBD pathogenesis followed by new findings on the immunologic and molecular basis of the chronic inflammatory process in IBD. In addition, the pathophysiologic role of individual cell populations in the inflamed gut in IBD is discussed including cells of the mucosal immune system, parenchymal cells and neuroimmune cells. Finally, the implications of the new findings for the design of novel diagnostic and therapeutic approaches for IBD are described.

A combined understanding of immunology, microbiology and genetics emerges as a prerequisite for a modern understanding of the pathophysiology of IBD. Hereby, IBD seem to develop as an immunologic dysfunction driven by bacterial antigens from the lumen in genetically susceptible individuals. A key area for future research will be to understand interactions between bacterial antigens and the mucosal immune system in health and disease and to determine the genetic and molecular basis of changes in the innate immune system in IBD patients. We hope that this book serves as a stimulus for modern understanding of IBD pathogenesis.

Markus F. Neurath
Richard S. Blumberg

PARTICIPANTS

Stephan C. Bischoff
Division of Clinical Nutrition/Prevention and Immunology
University of Hohenheim
Stuttgart
Germany

Richard S. Blumberg
Gastroenterology Division and
Laboratory of Mucosal Immunology
Brigham and Women's Hospital
Harvard, Massachusetts
USA

Per Brandtzaeg
Laboratory for Immunohistochemistry and Immunopathology (LIIPAT)
Institute of Pathology
University of Oslo
Oslo
Norway

Jonathan Braun
UCLA Hospital Center for Health Sciences
Los Angeles, California
USA

Hege S. Carlsen
Laboratory for Immunohistochemistry and Immunopathology (LIIPAT)
Institute of Pathology
University of Oslo
Oslo
Norway

Judy H. Cho
University of Chicago
Chicago, Illinois
USA

Yingzi Cong
Division of Gastroenterology and Hepatology
University of Alabama at Birmingham
Birmingham, Alabama
USA

Silvio Danese
Division of Gastroenterology
University Hospitals of Cleveland
Case Western Reserve University School of Medicine
Cleveland, Ohio
USA

Rainer Duchmann
Medizinische Klinik I
Charité-Universitätsmedizin Berlin
Campus Benjamin Franklin
Berlin
Germany

Charles O. Elson
Division of Gastroenterology and Hepatology
University of Alabama at Birmingham
Birmingham, Alabama
USA

Claudio Fiocchi
Division of Gastroenterology
University Hospitals of Cleveland
Case Western Reserve University School of Medicine
Cleveland, Ohio
USA

Ivan J. Fuss
Mucosal Immunity Section
Laboratory of Host Defense NIAID
National Institutes of Health
Bethesda, Maryland
USA

Thomas Gebhardt
Department of Gastroenterology, Hepatology and Endocrinology
Medical School of Hannover
Hannover
Germany

Trond S. Halstensen
Institute of Oral Biology
University of Oslo
Rikshospitalet, Oslo
Norway

Martin H. Holtmann
1st Department of Medicine
Johannes-Gutenberg-University
Mainz
Germany

Kenji Ina
Division of Gastroenterology
University Hospitals of Cleveland
Case Western Reserve University School of Medicine
Cleveland, Ohio
USA

Brian L. Kelsall
Laboratory of Molecular Immunology
National Institutes of Health
Bethesda, Maryland
USA

André Z.A. Leite
Division of Gastroenterology
University Hospitals of Cleveland
Case Western Reserve University School of Medicine
Cleveland, Ohio
USA

Francisco Leon
Laboratory of Molecular Immunology NIAID
National Institutes of Health
Bethesda, Maryland
USA

Thomas T. MacDonald
University of Southampton School of Medicine
Southhampton
UK

Giovanni Monteleone
Department of Gastroenterology
University Tor Vergata
Rome
Italy

Cathryn Nagler-Anderson
Department of Pediatrics (Immunology)
Massachusetts General Hospital
Harvard Medical School
Charlestown, Massachusetts
USA

Markus F. Neurath
1st Department of Medicine
Johannes-Gutenberg-University
Mainz
Germany

Edward E.S. Nieuwenhuis
Laboratory of Pediatrics
Department of Pediatric Gastroenterology
Erasmus MC, Rotterdam
The Netherlands

Reinhard Pabst
Center of Anatomy
Medical School of Hannover
Hannover
Germany

Hermann J. Rothkötter
Institute of Anatomy
University of Magdeburg
Magdeburg
Germany

R. Balfour Sartor
Division of Gastroenterology
and Hepatology
University of North Carolina
School of Medicine
Chapel Hill, North Carolina
USA

Phillip D. Smith
Division of Gastroenterology
and Hepatology
University of Alabama at Birmingham
Veterans Administration Medical Center
Birmingham, Alabama
USA

Lesley E. Smythies
Division of Gastroenterology
and Hepatology
Department of Medicine
University of Alabama at Birmingham
Birmingham, Alabama
USA

Warren Strober
Mucosal Immunity Section
Laboratory of Host Defense NIAID
National Institutes of Health
Bethesda, Maryland
USA

Stephan R. Targan
Cedars Sinai Medical Center
Los Angeles, California
USA

Jon D. Vogel
Division of Gastroenterology
University Hospitals of Cleveland
Case Western Reserve University
School of Medicine
Cleveland, Ohio
USA

Casey T. Weaver
Department of Pathology
University of Alabama at Birmingham
Birmingham, Alabama
USA

CONTENTS

1. STRUCTURE AND FUNCTION OF THE GUT MUCOSAL IMMUNE SYSTEM 1

Reinhard Pabst and Hermann J. Rothkötter

Introduction 1
Intraepithelial Lymphocytes (IEL) 1
Lamina Propria Lymphocytes (LPL) 3
Lymphocyte-Filled Villi (LFV) 4
Cryptopatches (CP) 5
Isolated Lymphoid Follicles (ILF) 5
Peyer's Patches (PP) 6
Lymphoid Structures in the Large Intestine 7
Mesenteric Lymph Nodes, Part of the Gut Immune System 7
Gut Lymphatics 8
Liver—Also a Part of the Gut Immune System? 8
Dynamics in the Gut Immune System 9
Conclusions 9
Note Added in Proof 9

2. FUNCTIONAL ASPECTS OF THE MUCOSAL IMMUNE SYSTEM 15

Cathryn Nagler-Anderson

Phenotypically and Functionally Unique Lymphocyte Populations in the Gut Associated Lymphoid Tissue 15
Mechanisms Governing Non-Responsiveness to Food Proteins and Commensal Bacteria 18

3. RECENT PROGRESS IN INFLAMMATORY BOWEL DISEASE GENETICS 24

Judy H. Cho

Introduction 24
IBD Epidemiology 24
The Human Genome Project and the Challenge of Complex Genetic Disorders 25
Linkage Studies in IBD 26
Association Studies in IBD 27
Future Directions 30

4. MICROBIAL AND DIETARY FACTORS IN THE PATHOGENESIS OF CHRONIC, IMMUNE-MEDIATED INTESTINAL INFLAMMATION 35

R. Balfour Sartor

Introduction 35
Evidence of Bacterial Induction and Maintenance of Intestinal Inflammation 38
Induction of Mucosal Immune Responses by Bacterial and Dietary Antigens and Adjuvants 42
Conclusion 48

5. EXPERIMENTAL MODELS OF MUCOSAL INFLAMMATION 55

Warren Strober and Ivan J. Fuss

Introduction 55
General Features of Models of Mucosal Inflammation 56
Representative Models of Mucosal Inflammation 68
Tabular Summary of Models of Mucosal Inflammation 80
Conclusion 80

6. OVERVIEW OF ROLE OF THE IMMUNE SYSTEM IN THE PATHOGENESIS OF INFLAMMATORY BOWEL DISEASE 98

Thomas T. MacDonald and Giovanni Monteleone

Abstract 98
Introduction 98
Crohn's Disease, but Not Ulcerative Colitis Bears the Stigmata of a Th1 Mediated Immune Response 99
The Role of T Cells in Ulcerative Colitis 100
The Role of Antibody in Ulcerative Colitis 100
Nonspecific Effector Mechanisms Down-Stream of the Immune System Cause Tissue Injury in Crohn's Disease and Ulcerative Colitis 102
The Immune System Fails to Negatively Regulate in CD and UC 103

7. THE ROLE OF THE EPITHELIAL BARRIER IN INFLAMMATORY BOWEL DISEASE 108

Edward E.S. Nieuwenhuis and Richard S. Blumberg

Introduction 108
Bacteria and the Epithelium 108
Bacteria and IBD 109
The Epithelial Cell as Effector Cell in IBD 110
Luminal Antigens, Fate and Processing 112
IEC and Mucosal Barrier Function/Permeability 113
Conclusion 113

8. INVOLVEMENT OF DENDRITIC CELLS IN THE PATHOGENESIS OF INFLAMMATORY BOWEL DISEASE **117**

Francisco Leon, Lesley E. Smythies, Phillip D. Smith and Brian L. Kelsall

Introduction **117**
Current Views on Dendritic Cell Biology **117**
Intestinal Dendritic Cells **119**
Dendritic Cells in Inflammatory Bowel Disease **122**
Intestinal Macrophages **126**
Intestinal Macrophages in IBD **126**
Summary **127**

9. ALTERATIONS OF T LYMPHOCYTES IN INFLAMMATORY BOWEL DISEASES **133**

Charles O. Elson, Yingzi Cong and Casey T. Weaver

$CD4^+$ T Cell Subsets in the Intestine **134**
Intestinal T Effector Cells **136**
Th2 Effectors in IBD **136**
Intestinal T Regulatory Cells **138**
$CD4^+CD45RB^{lo}$ T Cells **138**
T-Regulatory-1 Cells (Tr1) **139**
T-Helper-3 Cells (Th3) **139**
$\alpha_E\beta_7^+$ T Cells **140**
What Drives Effector and Regulatory T Cells in IBD? **141**
Conclusion **143**

10. THE B-CELL SYSTEM IN INFLAMMATORY BOWEL DISEASE **149**

Per Brandtzaeg, Hege S. Carlsen and Trond S. Halstensen

Abstract **149**
Introduction **150**
B Cell-Mediated Mucosal Homeostasis in the Normal Gut **150**
Humoral Immunity and Immune Exclusion in IBD Lesions **152**
Local Production of Antibodies to the Indigenous Gut Flora in IBD **153**
Local Production of Anti-Epithelial IgG1 Autoantibodies in Ulcerative Colitis **154**
Local Production and Diagnostic Aid of Other (Auto)Antibodies in IBD **156**
Mechanisms for Induction of Local Proinflammatory B-Cell Responses **157**
IBD Apparently Reflects Break of Oral Tolerance against Indigenous Bacteria **161**

11. ALTERATIONS OF MESENCHYMAL AND ENDOTHELIAL CELLS IN INFLAMMATORY BOWEL DISEASES 168

Claudio Fiocchi, Kenji Ina, Silvio Danese, André Z.A. Leite and Jon D. Vogel

Abstract 168
Introduction 168
Mesenchymal Cells 169
Smooth Muscle Cells 169
Fibroblasts and Myofibroblasts 169
Endothelial Cells 170
Human Intestinal Microvascular Endothelial Cells 171
Integration of Intestinal Immune-Nonimmune Cell Interactions by the CD40/CD40 Ligand System 172

12. ROLE OF MAST CELLS AND EOSINOPHILS IN NEUROIMMUNE INTERACTIONS REGULATING MUCOSAL INFLAMMATION IN INFLAMMATORY BOWEL DISEASE 177

Stephan C. Bischoff and Thomas Gebhardt

Introduction 177
Intestinal Mast Cells and Eosinophils 178
Role of Mast Cells and Eosinophils in Intestinal Inflammation 182
Structure and Function of The Enteric Nervous System 184
Mechansims of Mast Cell / Eosinophil-Neuroimmune-Crosstalk in the Gut 186
Neurophysiological Effects of Mast Cell and Eosinophil Activation 191
Alterations of Neuroimmune-Crosstalk in IBD 194
Conclusions 197

13. MULTIPARAMETER ANALYSIS OF IMMUNOGENETIC MECHANISMS IN CLINICAL DIAGNOSIS AND MANAGEMENT OF INFLAMMATORY BOWEL DISEASE 209

Jonathan Braun and Stephan R. Targan

Abstract 209
NOD2 and Innate Anti-Microbial Processes in IBD Diagnosis 209
Anti-Microbial Antibodies and Adaptive Immune Components in IBD Diagnosis 211
Autoantibodies in IBD Diagnosis 213
Multiparameter Analysis of Immunogenetic Traits in IBD Diagnosis 213

14. THE ROLE OF PROBIOTICS AND ANTIBIOTICS IN REGULATING MUCOSAL INFLAMMATION 219

Rainer Duchmann

Abstract 219
Probiotics: Animal Studies and Clinical Trials 219
Probiotics: Mechanisms 220
Antibiotics: Animal Studies and Clinical Trials 221
Antibiotics: Mechanisms 221
General Problems 222
Perspectives 222

15. FROM IMMUNOGENIC MECHANISMS TO NOVEL THERAPEUTIC APPROACHES IN INFLAMMATORY BOWEL DISEASE 227

Martin H. Holtmann and Markus F. Neurath

Abstract 227
Introduction 227
Genetic Factors 228
Physiological "Hyporeactivity" of the Intestinal Immune System 229
Activation of the Mucosal Immune System in IBD 229
Novel Therapeutic Approaches 234
Concluding Remarks 238

INDEX 243

Chapter 1

Structure and Function of the Gut Mucosal Immune System

Reinhard Pabst and Hermann J. Rothkötter

Introduction

For many decades immune reactions have been classified as humoral and cellular, innate and acquired, and the essential cells for protective and tolerogenic reactions were at first subdivided into TH1 and TH2. Later TH3 and TH0 cells were added and the antigen-presenting cell family is continuously growing. Many molecular pathways initiated by cytokine–receptor interactions have been clarified in vitro. However, in vivo all the players in this interacting orchestra of cells have to be at the right location to create the appropriate microenvironment as a basic requirement for meaningful immune reactions.[1] Therefore, the different compartments of the gut immune system will be described, focusing on the effects of age, species differences and influence of the nutritional and microbial content of the gut.

The gut immune system is integrated in the mucosal immune system in general. However, there is growing evidence for preferential routes from one organ to the other and immunization at one site does not result in protective effects in all other mucosal organs. Thus, the term "common mucosal immune system" should be replaced by "integrated mucosal immune system".[2] Examples will be mentioned to stress the heterogeneity of structures with at first sight similar appearance. The functional anatomy of the gut immune system is the basis for understanding the detailed description of the location of individual cells, which will be dealt with in the following chapters.

Intraepithelial Lymphocytes (IEL)

The first line of cellular defense is the epithelium (Fig. 1). In human adults there are about 20 IEL per 100 enterocytes in the healthy jejunum. In the ileum the numbers are 13 and in the colon 5 per 100 epithelial cells (for details see ref. 3). Based on the total number of enterocytes in the whole intestine the resulting total numbers of IEL are enormous. The proportion of γδ IEL seems to vary greatly between species: 10% in humans, 50% in cows and variations in the mouse of 20-80% (for review see refs. 4-5). The functional relevance of these species differences is not known. It has to be considered that there are also regional differences, e.g., obvious functional differences between IEL from crypts and those from villi in respect to IFNγ, and IL-5 production by IEL from crypts was reduced in comparison to those from villi.[6] There is no doubt that cytokines released from gut epithelial cells influence the IEL. IL-15 promotes the growth of γδ IEL in mice.[7] The loss of IL-7 results in a dramatic reduction of γδ IEL, and in IL-7R$^{-/-}$ mice a complete loss of γδ IEL.[8] The survival of activated γδ IEL is regulated by IL-2 and IL-15.[9] Obviously cell-cell interactions between enterocytes and γδ and αβ T cells via surface molecules are essential.[10] Further cell types stimulating IEL are fibroblasts and myofibroblasts, so far only shown in cell culture.[11] Thus, there is a network of interacting factors regulating IEL numbers. Human IEL are activated in situ but are tightly regulated by

Immune Mechanisms in Inflammatory Bowel Disease, edited by Richard S. Blumberg and Markus F. Neurath.

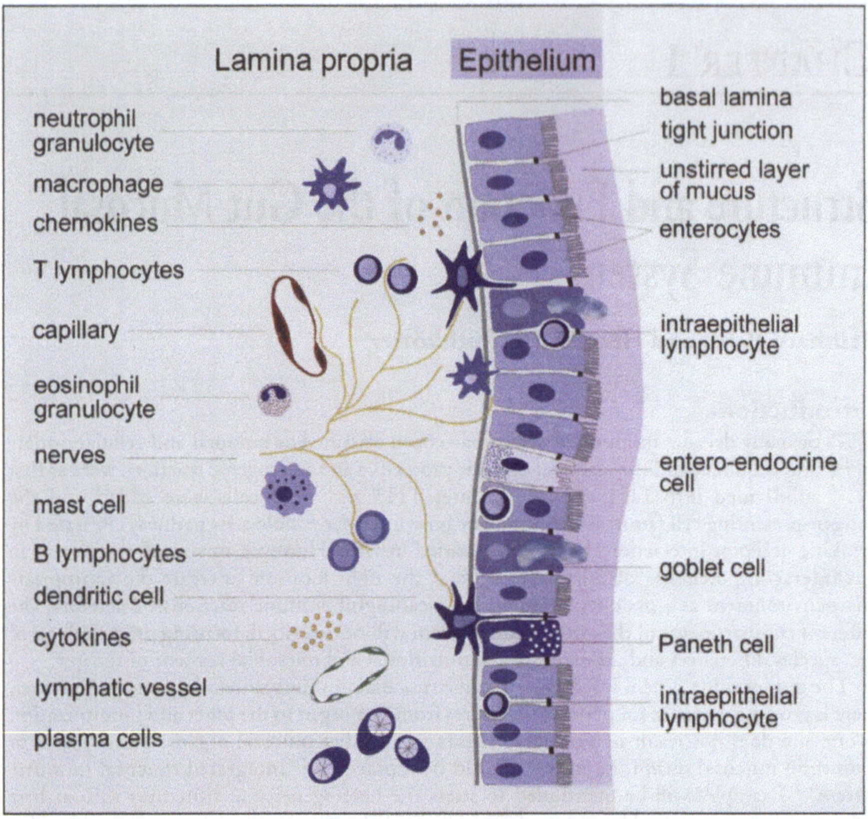

Figure 1. Schematic drawing of the gut epithelium and lamina propria with the different subsets of leukocytes and structural elements.

constitutional expression of Fas and FasL.[12] The number of IEL depends on the microbial colonization, as shown in mice and pigs which were germfree and then colonized as the number of proliferating $\alpha\beta$ and $\gamma\delta$ IEL increased.[13,14] The regional differences, on the other hand, did not depend on the microbial colonization in the rat.[15] An often neglected aspect is the great variability, e.g., in the proliferative rate and life span of $\alpha\beta$ and $\gamma\delta$ IEL.[14,16] IEL are an example for differences between in vivo and in vitro functions, as it was shown with normal human markers like granzyme, FasL, TNFα and IFNγ that IEL do not contribute to the apoptotic cell death of epithelial cells.[17]

Despite numerous studies on IEL many aspects of the functions of these cells are still not known. Hayday et al[18] recently suggested that there are at least two types: one primed to conventional antigen and the other with "ill-defined reactivities and origins" and placed them midway between adaptive and innate immunity. Thus, many more experiments in different species have to be performed, as even very basic aspects such as the route of transmigrating the basal lamina and that of returning to the lamina propria, dying in situ or release into the gut lumen are unknown. An interesting new aspect is that dendritic cells can also penetrate gut epithelial monolayers and express tight junction proteins and therefore probably reach the epithelium by their dendrites.[19]

Lamina Propria Lymphocytes (LPL)

In the intestinal lamina propria different leukocyte subsets are found: T (mostly CD4) cells, B cells and plasma cells (which mainly produce IgA), but also dendritic cells and macrophages as typical antigen-presenting cells, as well as eosinophils, neutrophils and mast cells representing the unspecific immune system (Fig. 1). These cells are not equally distributed throughout the lamina propria of the villi and crypt area: e.g., Ig-producing cells are much more frequent in the crypt area than in the lamina propria of the villi.[20] This is of major relevance when cell separation techniques for in vitro studies are performed and preferentially the villus lamina propria is digested.[21] We would therefore recommend preparing histological sections after different isolation steps of the remaining tissue. It has often been stated that LPL represent the largest lymphocyte pool in humans. This is not the case. Only when the number of plasma cells and in particular IgA-producing cells are counted, does the lamina propria achieve such an outstanding position for plasma cells (for review see ref. 22). In contrast to IEL which are partly descendent of locally produced T cells, LPL exchange rapidly as recently shown in parabiotic mice with a 50:50 mixture of both partners of LPL but only a low proportion among the CD8αα, CD8αβ and γδ T cells in the epithelium.[23] The differential emigration and homing of lymphocyte subsets have been documented in the pig.[24] LPL of the human gut express a specific pattern of adhesion molecules, and in particular blast cells bound several fold more effectively to high endothelial venules (HEV) in Peyer's patches than those of peripheral lymph nodes.[25] When the gene rearrangement of Ig genes was studied at different sites of the human gut, it was shown that intestinal IgV_H genes were highly mutated in plasma cells. Duodenal and colonic plasma cells were both highly mutated despite a great difference in the microbial flora of these two bowel segments.[26] The intestinal TCRδ showed increasing restriction with age in humans.[27]

Recent studies document that human LPL express certain chemokine receptors, e.g., CXCR3 and CCR5, but not others, e.g., CCR7, CXCR1 and 2.[28] Gut epithelial cells highly and selectively express the chemokine TECK (CCL25) and this seems to be a selective attractant for IgA antibody-secreting cells, thus possibly recruiting these cells to the gut LP.[29] A very recent study is of particular interest as it documents topographical heterogeneity in the lamina propria. In in vivo videomicroscopy LPL were seen to accumulate in microvessels at the tip of the villi and not in the crypt region, and this accumulation could be almost completely inhibited by anti-β7 integrin and to a lesser degree reduced by anti-MAdCAM-1 and anti-α4 integrin. This was different from the situation in HEV of Peyer's patches.[30] The lamina propria is a typical example for documenting nerve endings next to the immune system cells (for review see ref. 31), which is also indicated in Figure 1. Thus, the gut wall is an excellent example for the crosstalk between the immune system and the local constituents of the nervous system.[32-33]

There is no doubt that IgA is the prevailing antibody produced by plasma cells in the LP of the intestine. Some IgA-producing cells are long living (> 1 year) as has been shown recently.[34] In the human small intestine about 80% of all plasma cells are IgA+ (for review see ref. 35) but these show a surprising heterogeneity.[36] The precursors of these plasma cells can be divided into two types: the $CD5^+$ B1 cells are found in the peritoneal cavity[37] and these are precursors of cells producing a more unspecific IgA,[38] while the B2 cells are the precursors of the classical antigen-specific IgA-producing plasma cells which are dependent on IL-5, in contrast to B1 cells induced by IL-15 secreted from enterocytes.[39] Most data on B1 and B2 lymphocytes were obtained in the mouse and rat. However, human B1 cells have recently been characterized showing a similar phenotype but probably being less relevant for the immune system.[40] In addition to the regulatory effects of cytokines and antigen the intestinal motility also influences the IgA secretion in humans.[41] The IgA secretion occurs cyclically linked by motility-activated chloride secretion from intestinal crypts. Despite all the advances in IgA research a number of open or controversial aspects have to be clarified in future as stressed in two recent commentaries.[42,43]

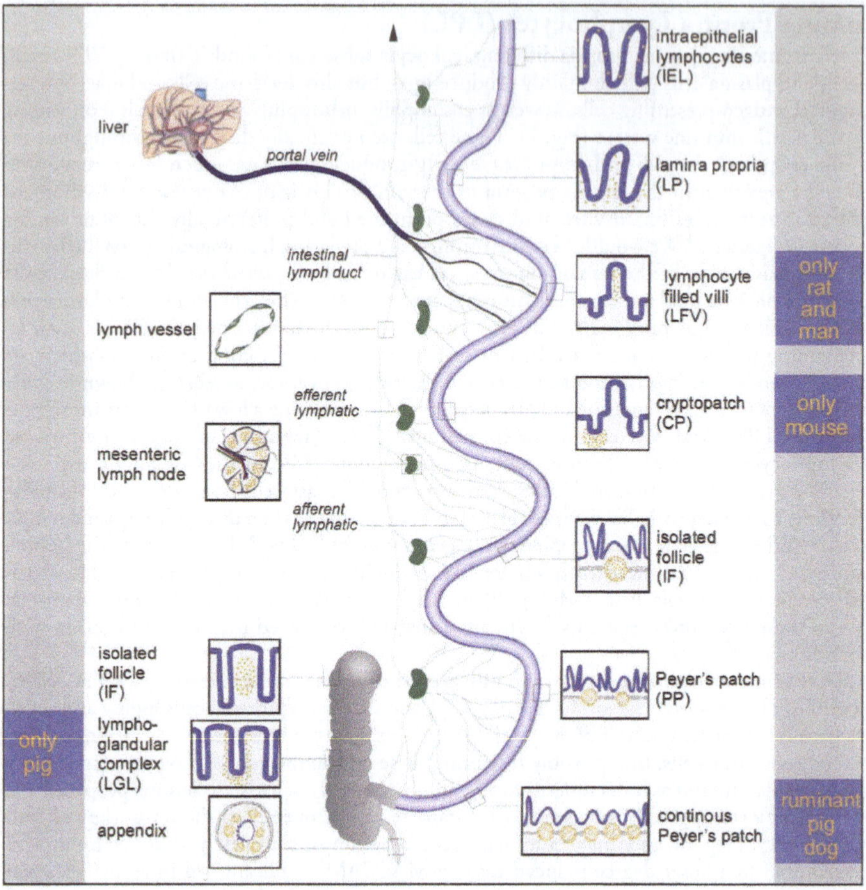

Figure 2. Overview of the different compartments of lymphocytes which are of relevance for immune reactions initiated or perpetuated in the gut immune system.

Lymphocyte-Filled Villi (LFV)

Only recently has a further solitary lymphoid structure been described in the human small intestine of six adult patients (mean age 45 years).[44] This structure was clearly different from isolated follicles which have a well-developed germinal center (Fig. 2). The LFV look as if the lamina propria of a villus is stuffed with lymphocytes (mainly memory T cells, variable numbers of B and dendritic cells, but no immature lymphocytes expressing c-kit or CD1a). There was approximately 1 LFV per 200 villi. Taking into account the enormous number of villi in the human small intestine a considerable total number of LFV will result. HEV, the specialized entry sites for lymphocytes in other organized lymphoid structures of the gut, were never observed. Structures similar to LFV have been described in abstract form for the rat.[45] We have examined total cross sections of samples from the small intestine of six children (2.5-14 months of age) who had died of sudden infant death or acute aspiration, but no LFV could be identified (Pabst and Debertin, unpublished). Sections from 30 small intestines of germfree piglets, piglets bred under SPF or conventional conditions (1 to 52 months of age) have also been screened on paraffin embedded material. No evidence of LFV was found here either (Pabst and Rothkötter, unpublished). More studies have to be performed in different species and at different ages to define the development and function of these structures in health and disease. Microbacteria might play a role as the LFV look similar to the "immune villi" described in the gut of newborn calves in infections with enterotoxin-forming E. coli.[46]

Cryptopatches (CP)

It was a great surprise after scientists have cut and evaluated the mouse gut histologically for generations, when in 1996 a lymphoid structure was described for the first time: the cryptopatches.[47] Over the following years this group further characterized their cell composition and functional role.[48-50] The CP consist of accumulations of lymphoid cells around crypts of the small intestine and colon, randomly distributed over the cross section and not opposite the mesentery like isolated follicles and Peyer's patches (Fig. 2). A cryptopatch consists of about 1000 lymphoid cells. Most of these are negative for B, CD3, CD4 and CD8 but they depend on IL-7 and the stem cell factor c-kit. The CP are also found in nude and SCID mice and they obviously play a critical role in the extrathymic T cell development of IEL. In a very recent study on T cell differentiation including TCR-β, -γ and α locus rearrangements, it was concluded that only approx. 3% of CP cells were clearly involved in T cell differentiation and thus these interesting structures may have an important additional role in gut immunity.[51] So far CP have only been documented in the mouse. We carefully evaluated histological sections of piglets of different breeding conditions (germfree, SPF, conventional) of a few weeks of age and could not find any CP (Pabst and Rothkötter, unpublished). In a further attempt we cut sections of paraffin embedded material from the human small intestine of six children (2.5 – 14 months of age), who had died of sudden infant death or acute aspiration. No evidence of CP could be found. Future studies have to clarify whether the CP are interesting but exotic structures, unique to the mouse. The CP are an excellent example of how careful one has to be when generalizing findings from one species to another.

Isolated Lymphoid Follicles (ILF)

Isolated aggregations of lymphoid cells with or without a typical germinal center in the lamina propria and a cap-like structure of lymphocytes to the lumen (like the dome in Peyer's patches) have been described in many species. They are usually taken as the simplest component of Peyer's patches. Therefore, the anatomical nomenclature talks of solitary lymphoid nodules for ILF and aggregated lymphoid nodules for Peyer's patches. Surprisingly, only very recently have the ILF been characterized in detail in the mouse.[52] In adult mice 100-200 ILF were identified on the anti-mesenteric wall of the small intestine. The follicles consisted of typical germinal centers and the covering epithelium contained M cells. Some isolated follicles reach deeper layers of the gut wall. Moghaddami et al[44] argue that these are two components of the same structure which can be verified by serial sections. These ILF develop after birth and are also found in germfree nude, RAG2,$^{-/-}$ TCR-β$^{-/-}$ and Igμ chain$^{-/-}$ mice. In lymphotoxin$^{-/-}$ and aly/aly mice neither ILF nor PP were seen. When the prenatal development of PP was suppressed by antibodies the ILF nevertheless developed. ILF are also found in the large intestine, about 3 per cm^2 in humans with an increase from the oral to the aboral parts of the colon.[53] It remains to be shown whether the lymphoid aggregates in ulcerative colitis are derived from preexisting ILF or are the result of chronic inflammatory stimuli.[54]

Taking into account the enormous number of ILF covered with an epithelium including M cells, the total number of these M cells might outnumber those in dome epithelium of PP. In studies on antigen uptake not only PP but also ILF should always be considered to avoid misinterpretations of experiments.

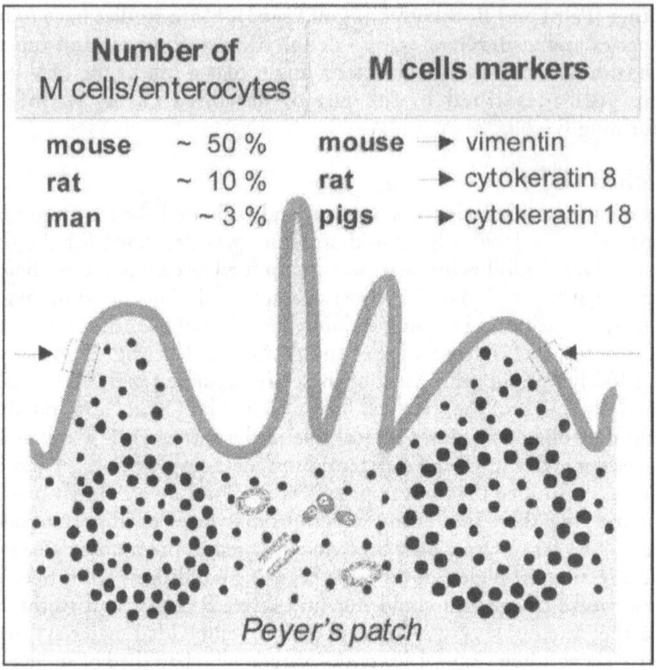

Figure 3. Basic structure of Peyer's patches including some species differences.[67]

Peyer's Patches (PP)

General Structure

The basic structure of PP—the aggregated lymphoid follicles—is well known (Fig. 3). There are different compartments: the germinal center surrounded by a corona or mantle zone, populated mainly by follicular dendritic cells, B and CD4 lymphocytes, the interfollicular area with the HEV being mainly a T cell area, and the dome bulging into the gut lumen with a mixture of different lymphocyte subsets, dendritic cells and macrophages. The diameter of the dome is usually much smaller than that of the follicle and therefore on many sections the follicle and interfollicular area only are seen. Serial sections, however, would always show a dome sitting like a small cap on a big head. This concept is also true for the human PP as elegantly shown by multiple color fluorescence.[55] There are enormous species differences in respect to the ontogeny of PP, the total number and also some structural composition as summarized by Griebel and Hein.[56] Not only in ruminants but also in carnivora, e.g., the dog, and omnivora, e.g., the pig, are there two main types of PP: in the upper small intestine a type similar to that of humans and rodents—the jejunal type PP—and in the terminal ileum a continuous, up to more than 1.5m long PP—the so-called ileal PP—which obviously has an important role in B lymphocyte production and has typical features of a primary lymphoid organ (reviewed in ref. 56). Therefore, data obtained in one species should not be generalized for all other species. More data are needed on the regulation of the total number and activity of PP. The localization and size of the individual PP did not change with age when they were mapped in individual pigs at three and nine months of age.[57] The adhesion molecule MAdCAM and its counterpart $\alpha4\beta7$ play a critical role in lymphocyte homing to PP as shown in several experiments, e.g., using intravital videomicroscopy.[58] The same group used a very unique approach combining nutritional aspects with lymphocyte migration to PP.[59] In rats fed on olive oil an increased transendothelial migration of T lymphocytes was observed, while octanoid oil stimulated lymphocyte rolling only and had no effect on adherence. Thus, dietary approaches might be used to regulate lymphocyte migration to PP.

Chemokines play a dual role in PP development and function, e.g., lymphotoxin $(LT)^{-/-}$ mice do not develop PP,[60] and chemokines are essential in the regulation of lymphocyte traffic into PP.[61] The lymphoid tissue chemokine (SLC) and its receptor CCR7 are critical for T lymphocyte rolling and adhesion, that is not necessary for B cells. A further interesting finding of that study was that HEV in the paracortical area differed from HEV near to follicles. Thus, there seem to be two different entrances for B and T lymphocytes into PP regulated by chemokines.[61] In recent years an increasing number of studies have been performed on human PP. Only a few interesting studies can be mentioned here. Marginal zone B cells from PP were studied in respect to the rearrangement of IgV_H genes and the data suggest that these are clearly memory B cells.[62] After exposure to a common dietary antigen β lactoglobulin human PP T cells did not produce IL-4, IL-5 or IL-10 but IL-12 in contrast to animal experiments.[63] These data furthermore stress the problems in extrapolating from mice to humans. This is of particular relevance in concepts of tolerance development, because in rodents the immunosuppressive TGFβ is produced, but in humans the proinflammatory IL-12.[64]

M Cells

The membraneous (M) epithelial cell in the epithelium covering the dome area attracted much interest after it had been shown to be an important entry site for particulates, bacteria, viruses etc. (for review see refs. 65-68). Nevertheless, there are still many questions such as why there are so many differences between species, e.g., in markers, or the number of M cells per epithelial cell (Fig. 3). Over the last few years the in vitro culture system has helped to understand the role of lymphocytes in the differentiation of M cells.[69-71] Sophisticated combinations of EM and immunohistochemistry with confocal microscopy have documented the development of M cells from distinct crypts.[72] A better knowledge of the regulatory processes to increase the number and/or activity of M cells in the gut before an oral vaccine might be of great clinical relevance. A further area of research will be the interaction of antigens taken up via M cells with different lymphocyte subsets in the M cell pockets[55] and dendritic cells as well as their migration to other compartments of PP, as recently outlined by Neutra et al.[73] Nevertheless it is still true that there are many "unsolved mysteries of intestinal M cells".[74]

Lymphoid Structures in the Large Intestine

It is well known that the appendix (not present in many species!) consists of tightly packed lymphoid follicles with a specialized epithelium covering the dome area. The appendix has often been compared to a large Peyer's patch. One extreme is the appendix in rabbits, which, however, has a different function in these coprophagic animals.

Diffusely distributed isolated follicles can be found. At the beginning of the last century Dukes and Bussey[53] published a careful study on the number of isolated follicles in the human large intestine: mean number $3/cm^2$ with more follicles in children (<10 years) $8/cm^2$. In pigs[75] and calves[76] lymphoglandular complexes have been described which embrace larger crypts. In 5-13 week old pigs more than 1200 lymphoglandular complexes have been counted resulting in a large total mass of lymphoid tissue.[75] It is of great clinical relevance that very high numbers of isolated follicles covered with an epithelium with M cells are found in the rectum enabling the uptake of HIV or feline immunodeficiency virus.[77-78]

Mesenteric Lymph Nodes, Part of the Gut Immune System

There is no difference in the basic structure of mesenteric lymph nodes compared to other lymph nodes, but the lymph arriving from the gut wall contains different molecules and cells. The route of the lymph and its components through the different compartments of the lymph nodes has often been neglected. Gretz et al[79-80] have described a conduit system of cords, channels and corridors directing this flow which also reaches the surrounding area of HEV, thus regulating one of the entry sites into the lymph node. Dendritic cells arriving at mesenteric lymph nodes carry apoptotic intestinal epithelial cells to T cell areas of mesenteric lymph nodes which might be a critical part of inducing self-tolerance.[81] There is growing evidence that the microenvironment within mesenteric lymph nodes differs from that in peripheral lymph nodes. This microenvironment governs the survival (proliferation and apoptosis) of lymphocytes arriving at the lymph nodes via HEV.[82-83] Much less is known about how the exit from lymph nodes into the efferent lymphatics is regulated. It might be similar to the mechanisms shown in peripheral lymph nodes in sheep, where TNFα reduced the output up to 90% within 6-10 h following stimulation.[84] In respect to species differences, it has to be mentioned that in pig mesenteric lymph nodes lymphocytes do not only enter the parenchyma via HEV but also exit via HEV and hardly via efferent lymphatics (for review see ref. 85). A central role of mesenteric lymph nodes in the expansion of lymphocytes stimulated in the gut wall (T cells and precursors of IgA-producing cells) has often been proposed but the effects on immune reactions after resection of mesenteric lymph nodes will have to be studied in detail.

Gut Lymphatics

Not only the lamina propria but also Peyer's patches are drained by lymphatics which run to mesenteric lymph nodes. In addition to the fluid and absorbed nutrients such as fat, large numbers of lymphoid and nonlymphoid cells, e.g., dendritic cells, use this route to leave the gut wall. Nothing is known about which molecules are essential for regulating this exit. The lymphatic microvessels of PP[86] and the composition of lymphocyte subsets in the lumen (as an indicator of which cells emigrate via this route) have been studied not only in the rat[87] but also in humans.[88] Many memory cells and most B cell blasts in these lymphatics showed an adhesion cell pattern of $\alpha4\beta7^{high}$ L-selectin.^{low} When mesenteric lymph nodes are excised the afferent and efferent gut lymphatics regenerate in pigs, the intestinal lymph trunk can be cannulated for up to two weeks and the lymph coming from the gut wall collected without the filter effect of the mesenteric lymph nodes.[89] The total number of lymphocytes, their subsets and proportion of newly formed cells could be quantified in pigs.[90] In absolute numbers more newly formed T lymphocytes (2×10^6/h) than IgA$^+$ (0.4×10^6/h) and IgM$^+$ (0.5×10^6/h) left the gut wall. This approach has the additional advantage that dendritic cells leaving the gut wall can also be collected.[91] Thus, this technique for studying the gut lymphatics enables a quantification of the exit from the gut wall.

Liver—Also a Part of the Gut Immune System?

Very little is known about whether immune cells enter blood vessels within the gut mucosa and the organized lymphoid structures of the gut wall. It is likely that small molecules like cytokines and chemokines and their soluble receptors will diffuse into capillaries and venules, thus reaching the liver via the portal vein. Lymphocyte subsets in the human liver have recently been characterized and it was shown that these differ phenotypically from lymphocytes in the blood.[92] In rats it has been documented that activated lymphoid cells accumulated in the periportal area, while mature "memory" and naïve T cells rapidly entered and later left the liver after i.v. injection.[93] Thus, there is not only the enormous phagocytic capacity of the Kupfer cells in the liver but also the lymphoid cells. They might play a role as one of the final barriers of the protective function of the gut not only in health but probably after damage to the gut epithelium, which is much more relevant. In future studies the role of the liver in gut immune functions should no longer be ignored.

Dynamics in the Gut Immune System

It is well accepted that enormous numbers of lymphocytes migrate through the gut lamina propria and organized lymphoid tissues, and this is taken as the basis for an effective protection of the gut wall. Only selected data on human material published over the last few years will be mentioned; for general aspects we refer to several reviews.[94-96] The expression of β7 integrins in the human gut wall[97-98] has been characterized by interesting combinations of multiple fluorescent labeled antibodies. As cell transfer studies with sequential biopsies cannot be performed in humans, experiments on cultured human microvascular endothelial cells[99] or immunohistology have to be used.[98,100] Samples of the peripheral blood can also be taken as an indicator of immune reactions in the normal gut wall[101] and in situations with mucosal inflammation.[102-103] After arriving in different compartments of the gut, cytokines produced for example by epithelial cells like IL-15[104] will either stimulate the lymphoid cells, induce proliferation or apoptosis (for review see refs. 105-107). The knockout technology has enabled the role of individual cytokines to be studied, e.g., IFNγ$R^{-/-}$ mice had an impaired gut immune response but normal oral tolerance.[108] The general precautions in extrapolating from data of knockout mice to other species, however, should not be neglected. The microenvironment within the compartments of the gut wall is largely dependent on direct cell-cell contacts, e.g., dendritic cells with lymphoid cells,[109-111] thus initiating either tolerance or protective immunity.[112-114] A further family of molecules is of growing relevance for gut immunology: the chemokines and their receptors,[115-116] not only in organogenesis[117-118] but also in the migration and localization of dendritic cells[119] and lymphocytes.[120-121] Thus, the number of molecules (only schematically shown in Fig. 2) known to influence the local microenvironment is steadily increasing and their role in animal models and patients with chronic inflammatory bowel disease has to be defined in future.

Conclusions

The different components of the gut immune system are much more heterogenous than previously believed. There are significant species differences and the effects of the bacterial and antigenic content of the gut lumen are of major relevance. Little interest has been shown so far in age-related effects from immaturity around birth to ageing.[14,21-22,122-124] The microenvironment differs between compartments and not only depends on the spatial organization of structural and immune cells but also the epithelium, and all this is regulated by a multitude of cytokines and chemokines as well as their receptors. Data obtained from isolated cells are important but their clinical relevance can only be demonstrated in the physiological microenvironment with its rapid, dynamic modifications of the network of regulatory factors. For new therapeutic approaches this fascinating interplay of molecules and cells must be considered.

Note Added in Proof

In a recent review the different compartments of the gut immune system with the afferent and efferent loop, and the enormous consequences of not using precise nomenclature in respect to mucosal immunology have been stressed.[125] The heterogeneity and/or overlapping functions of structures such as ILF and cryptopatches have been outlined.[126] The role of the cryptopatches as the main origin of intraepithelial lymphocytes has been further studied using other techniques and discussed in detail.[127,128] A real challenge of the concept of the presence of M cells only in the epithelium covering structural lymphoid aggregations was the documentation of some M cells scattered in the normal gut epithelium of the mouse.[129] The potential functional role has been critically discussed.[130] The functional role of the soluble and cellular content of the afferent lymph directed via the conduit system to the space around the HEV has been documented in much more detail and the important role for immune reactions discussed.[131] Dendritic cells in the gut wall, their migratory route and regulation by chemokines have been further characterized,[132,133] and their heterogeneity stressed.

References

1. Brandtzaeg P, Baekkevold ES, Farstad IN et al. Regional specialization in the mucosal immune system: What happens in the microcompartments? Immunol Today 1999; 20:141-151.
2. Brandtzaeg P, Farstad IN, Haraldsen G. Regional specialization in the mucosal immune system: Primed cells do not always home along the same track. Immunol Today 1999; 20:267-277.
3. Pabst R. The anatomical basis for the immune function of the gut. Anat Embryol 1987; 176:135-144.
4. Mowat AM, Viney JL. The anatomical basis of intestinal immunity. Immunol Rev 1997; 156:145-166.
5. Lefrancois L, Puddington L. Basic aspects of intraepithelial lymphocyte immunobiology. In: Ogra PL, Mestecky J, Lamm ME et al, eds. Mucosal immunology. 2nd ed. San Diego: Academic Press, 1999:413-428.
6. Kawabata S, Boyaka PN, Coste M et al. Intraepithelial lymphocyte from villus tip and crypt portions of the murine small intestine show distinct characteristics. Gastroenterology 1998; 115:866-873.
7. Inagaki-Ohara K, Nishimura H, Mitani A et al. Interleukin-15 preferentially promotes the growth of intestinal intraepithelial lymphocytes bearing γδ T cell receptor in mice. Eur J Immunol 1997; 27:2885-2891.
8. Fujihashi K, McGhee JR, Yamamoto M et al. An interleukin-7 internet for intestinal intraepichelial T cell development: Knockout of ligand or receptor reveal differences in the immunodeficient state. Eur J Immunol 1997; 27:2133-2138.
9. Chu CL, Chen SS, Wu TS et al. Differential effects of IL-2 and IL-15 on the death and survival of activated $TCR\gamma\delta^+$ intestinal intraepithelial lymphocytes. J Immunol 1999; 162:1896-1903.
10. Yamamoto M, Fujihashi K, Kawabata K et al. A mucosal intranet: Intestinal epithelial cells down-regulate intraepithelial, but not peripheral, T lymphocytes. J Immunol 1998; 160:2188-2196.
11. Roberts AI, Nadler SC, Ebert EC. Mesenchymal cells stimulate human intestinal intraepithelial lymphocytes. Gastroenterology 1997; 113:144-150.
12. Morimoto Y, Hizuta A, Ding EX et al. Functional expression of Fas and Fas ligand on human intestinal intraepithelial lymphocytes. Clin Exp Immunol 1999; 116:84-89.
13. Imaoka A, Matsumoto S, Setoyama H et al. Proliferative recruitment of intestinal intraepithelial lymphocytes after microbial colonization of germ-free mice. Eur J Immunol 1996; 26:945-948.
14. Rothkötter HJ, Möllhoff S, Pabst R. The influence of age and breeding conditions on the number and proliferation of intraepithelial lymphocytes in pigs. Scand J Immunol 1999; 50:31-38.
15. Helgeland L, Vaage JT, Rolstad B et al. Regional phenotypic specialization of intraepithelial lymphocytes in the rat intestine does not depend on microbial colonization. Scand J Immunol 1997; 46:349-357.
16. Penney L, Kilshaw PJ, MacDonald TT. Regional variation in the proliferative rate and lifespan of $\alpha\beta$ TCR^+ and $\gamma\delta$ TCR^+ intraepithelial lymphocytes in the murine small intestine. Immunology 1995; 86:212-218.
17. Chott A, Gerdes D, Spooner A et al. Intraepithelial lymphocytes in normal human intestine do not express proteins associated with cytolytic function. Am J Pathol 1997; 151:435-442.
18. Hayday A, Theodoridis E, Ramsburg E et al. Intraepithelial lymphocytes: Exploring the Third Way in immunology. Nature Immunol 2001; 2:997-1003.
19. Rescigno M, Urbano M, Valzasina B et al. Dendritic cells express tight junction proteins and penetrate gut epithelial monolayers to sample bacteria. Nat Immunol 2001; 2:361-367.
20. Rothkötter HJ, Ulbrich H, Pabst R. The postnatal development of gut lamina propria lymphocytes: Number, proliferation and T and B cell subsets in conventional and germ-free pigs. Pediatr Res 1991; 29:237-242.
21. Rothkötter HJ, Kirchhoff T, Pabst R. Lymphoid and nonlymphoid cells in the epithelium and lamina propria of intestinal mucosa of pigs. Gut 1994; 35:1582-1589.
22. Brandtzaeg P. Development of the mucosal immune system in humans. In: Bindels JG, Goedhart AC, Visser HKA, eds. Recent Developments in Infant Nutrition. London: Kluwer Academic, 1996:349-376.
23. Sugahara S, Shimizu T, Yoshida Y et al. Extrathymic derivation of gut lymphocytes in parabiotic mice. Immunology 1999; 96:57-65.
24. Rothkötter HJ, Hriesik C, Barman NN et al. B and also T lymphocytes migrate via gut lymph to all lymphoid organs and the gut wall, but only IgA cells accumulate in the lamina propria of the intestinal mucosa. Eur J Immunol 1999; 29:327-333.
25. Jalkanen S, Nash GS, De los Toyos J et al. Human lamina propria lymphocytes bear homing receptors and bind selectively to mucosal lymphoid high endothelium. Eur J Immunol 1989; 19:63-68.

26. Dunn-Walters DK, Boursier L, Spencer J. Hypermutation, diversity and dissemination of human intestinal lamina propria plasma cells. Eur J Immunol 1997; 27:2959-2964.
27. Holtmeier W, Witthöft T, Hennemann A et al. The TCR-γ repertoire in human intestine undergoes characteristic changes during fetal to adult development. J Immunol 1997; 158:5632-5641.
28. Agace WW, Roberts AI, Wu L et al. Human intestinal lamina propria and intraepithelial lymphocytes express receptors specific for chemokines induced by inflammation. Eur J Immunol 2000; 30:819-826.
29. Bowman EP, Kuklin NA, Youngman KR et al. The intestinal chemokine thymus-expressed chemokine (CCL25) attracts IgA antibody-secreting cells. J Exp Med 2002; 195:269-275.
30. Fujimori H, Miura S, Koseki S et al. Intravital observation of adhesion of lamina propria lymphocytes to microvessels of small intestine in mice. Gastroenterology 2002; 122:734-744.
31. Downing JEG, Miyan JA. Neural immunoregulation: Emerging roles for nerves in immune homeostasis and disease. Immunol Today 2000; 21:281-289.
32. Frieling T, Weber E, Schemann M. Inflammatory mediators influencing submucosal secretory reflexes. Ann NY Acad Sci 2000; 915:98-101.
33. Sharkey KA, Kroese AB. Consequences of intestinal inflammation on the enteric nervous system: Neuronal activation induced by inflammatory mediators. Anat Rec 2001; 262:79-90.
34. Slifka MK, Antia R, Whitmire JK et al. Humoral immunity due to long-lived plasma cells. Immunity 1998; 8:363-372.
35. Brandtzaeg P, Farstad IN. The human mucosal B cell system. In: Ogra PL, Mestecky J, Lamm ME et al, eds. Mucosal immunology. 2nd ed. San Diego: Academic Press, 1999:439-468.
36. Farstad IN, Carlsen H, Morton HC et al. Immunoglobulin a cell distribution in the human small intestine: Phenotypic and functional characteristics. Immunology 2000; 101:354-363.
37. Martin F, Kearney JF. B1 cells: Similarities and differences with other B cell subsets. Curr Opinion Immunol 2001; 13:195-201.
38. Su I, Tarakhovsky A. B1 cells: Orthodox or conformist? Curr Opinion Immunol 2000; 12:191-194.
39. Hiroi T, Kiyono H. Role of interleukin 15 and IL-15R for common mucosal immune system - independent IgA B1 cells. Mucosal Immunol Update 2000; 8:14-15.
40. Donze HH, Lue C, Julian BA et al. Human peritoneal B-1 cells and the influence of continuous ambulatory peritoneal dialysis on peritoneal and peripheral blood mononuclear cell (PBMC) composition and immunoglobulin levels. Clin Exp Immunol 1997; 109:356-361.
41. Mellander A, Mattsson A, Svennerholm AM et al. Relationship between interdigestive motility and secretion of immunoglobulin A in human proximal small intestine. Dig Dis Sci 1997; 42:554-567.
42. Mestecky J, Russell MW, Elson CO. Intestinal IgA: Novel views on its function in the defence of the largest mucosal surface. Gut 1999; 44:2-5.
43. Lamm ME, Phillips-Quagliata JM. Origin and homing of intestinal IgA antibody-secreting cells. J Exp Med 2002; 195:F5-F8.
44. Moghaddami M, Cummins A, Mayrhofer G. Lymphocyte-filled villi: Comparison with other lymphoid aggregations in the mucosa of the human small intestine. Gastroenterology 1998; 115:1414-1425.
45. Mayrhofer G, Brooks A. Lymphopoiesis in lymphocyte-filled villi in the small intestine of the rat. Clin Immunol Immunopathol 1995; 76:S55.
46. Günther H, Schulze F, Heilmann P. Zum pathogenetischen Verhalten verschiedener enterotoxinbildender Escherichia-coli-Stämme im Darmkanal neugeborener Kälber. Monatsheft Vet Med 1983; 38:96-103.
47. Kanamori Y, Ishimaru K, Nanno M et al. Identification of novel lymphoid tissues in murine intestinal mucosa where clusters of c-kit$^+$ IL-7R$^+$ Thy1$^+$ lympho-hemopoietic progenitors develop. J Exp Med 1996; 184:1449-1459.
48. Saito H, Kanamori Y, Takemori T et al. Generation of intestinal T cells from progenitors residing in gut cryptopatches. Science 1998; 280:275-278.
49. Oida T, Suzuki K, Nanno M et al. Role of gut cryptopatches in early extrathymic maturation of intestinal intraepithelial T cells. J Immunol 2000; 164:3616-3626.
50. Suzuki K, Oida T, Hamada H et al. Gut cryptopatches: Direct evidence of extrathymic anatomical sites for intestinal T lymphopoiesis. Immunity 2000; 13:691-702.
51. Lambolez F, Azogui O, Joret AM et al. Characterization of T cell differentiation in the murine gut. J Exp Med 2002; 195:437-449.
52. Hamada H, Hiroi T, Nishiyama Y et al. Identification of multiple isolated lymphoid follicles on the antimesenteric wall of the mouse small intestine. J Immunol 2002; 168:57-64.
53. Dukes C, Bussey HJR. The number of lymphoid follicles of the human large intestine. Pathol Bacteriol 1926; 29:111-116.
54. Yeung MMW, Melgar S, Baranov V et al. Characterisation of mucosal lymphoid aggregates in ulcerative colitis: Immune cell phenotype and TcR-γδ expression. Gut 2000; 47:215-227.

55. Farstad IN, Halstensen TS, Fausa O et al. Heterogeneity of M-cell-associated B and T cells in human Peyer's patches. Immunology 1994; 83:457-464.
56. Griebel PJ, Hein WR. Expanding the role of Peyer's patches in B-cell ontogeny. Immunol Today 1996; 17:30-39.
57. Rothkötter HJ, Pabst R. Lymphocyte subsets in jejunal and ileal Peyer's patches of normal and gnotobiotic minipigs. Immunology 1989; 67:103-108.
58. Tsuzuki Y, Miura S, Suematsu M et al. α4-integrin plays a critical role in early stages of T lymphocyte migration in Peyer's patches of rats. Int Immunol 1996; 8:287-295.
59. Tsuzuki Y, Miura S, Kurose I et al. Enhanced lymphocyte interaction in postcapillary venules of Peyer's patches during fat absorption in rats. Gastroenterology 1997; 112:813-825.
60. Spahn TW, Fontana A, Faria AMC et al. Induction of oral tolerance to cellular immune responses in the absence of Peyer's patches. Eur J Immunol 2001; 31:1278-1287.
61. Warnock RA, Campbell JJ, Dorf ME et al. The role of chemokines in the microenvironmental control of T versus B cell arrest in Peyer's patch high endothelial venules. J Exp Med 2000; 191:77-88.
62. Dunn-Walters DK, Isaacson PG, Spencer J. Sequence analysis of rearranged IgV_H genes from microdissected human Peyer's patch marginal zone B cells. Immunology 1996; 88:618-624.
63. Nagata S, McKenzie C, Pender SLF et al. Human Peyer's patch T cells are sensitized to dietary antigen and display a Th cell type 1 cytokine profile. J Immunol 2000; 165:5315-5321.
64. MacDonald TT, Monteleone G. IL-12 and Th1 immune responses in human Peyer's patches. Trends Immunol 2001; 22:244-247.
65. Neutra MR, Frey A, Kraehenbuhl. Epithelial M cells: Gateways for mucosal infection and immunization. Cell 1996; 86:345-348.
66. Gebert A, Rothkötter HJ, Pabst R. M cells in Peyer's patches of the intestine. Int Rev Cytol 1996; 167:91-159.
67. Gebert A. The role of M cells in the protection of mucosal membranes. Histochem Cell Biol 1997; 108:455-470.
68. Davis IC, Owen RL. The immunopathology of M cells. Springer Sem Immunopathol 1997; 18:421-448.
69. Kernéis S, Pringault E. Plasticity of the gastrointestinal epithelium: The M cell paradigm and opportunism of pathogenic microorganisms. Semin Immunol 1999; 11:205-215.
70. Debard N, Sierro F, Kraehenbuhl JP. Development of Peyer's patches, follicle-associated epithelium and M cell: Lessons from immunodeficient and knockout mice. Semin Immunol 1999; 11:183-191.
71. Schulte R, Kerneis S, Klinke S et al. Translocation of Yersinia enterocolitica across reconstituted intestinal epithelial monolayers is triggered by Yersinia invasin binding to β1 integrins apically expressed on M-like cells. Cellular Microbiol 2000; 2:173-185.
72. Gebert A, Fassbender S, Werner K et al. The development of M cells in Peyer's patches is restricted to specialized dome-associated crypts. Am J Pathol 1999; 154:1573-1582.
73. Neutra MR, Mantis NJ, Kraehenbuhl JP. Collaboration of epithelial cells with organized mucosal lymphoid tissues. Nature Immunol 2001; 2:1004-1009.
74. Nicoletti C. Unsolved mysteries of intestinal M cells. Gut 2000; 47:735-739.
75. Morfitt DG, Pohlenz JFL. Porcine colonic lymphoglandular complex: Distribution, structure, and epithelium. Am J Anat 1989; 184:41-51.
76. Liebler EM, Paar M, Pohlenz JF. M cells in the rectum of calves. Res Vet Sci 1991; 51:107-114.
77. Neutra MR. Interactions of viruses and microparticles with apical plasma membranes of M cells: Implications for human immunodeficiency virus transmission. J Infect Dis 1999; 179(Suppl 3):441-443.
78. Bishop SA, Stokes CR, Gruffydd-Jones TJ et al. Vaginal and rectal infection of cats with feline immunodeficiency virus. Vet Microbiol 1996; 51:217-227.
79. Gretz JE, Anderson AO, Shaw S. Cords, channels, corridors and conduits: Critical architectural elements facilitating cell interactions in the lymph nodes cortex. Immunol Rev 1997; 156:11-24.
80. Gretz JE, Norbury CC, Anderson AO et al. Lymph-borne chemokines and other low molecular weight molecules reach high endothelial venules via specialized conduits while a functional barrier limits access to the lymphocyte microenvironments in lymph node cortex. J Exp Med 2000; 192:1425-1439.
81. Huang FP, Platt N, Wykes M et al. A discrete subpopulation of dendritic cells transports apoptotic intestinal epithelial cells to T cell areas of mesenteric lymph nodes. J Exp Med 2000; 191:435-443.
82. Bode U, Duda C, Weidner F et al. Activated T cells enter rat lymph nodes and Peyer's patches via high endothelial venules: Survival by tissue-specific proliferation and preferential exit of CD8+ T cell progeny. Eur J Immunol 1999; 29:1487-1495.

83. Bode U, Sparmann G, Westermann J. Gut-derived effector T cells circulating in the blood of the rat: Preferential redistribution by TGFβ-1 and IL-4 maintained proliferation. Eur J Immunol 2001; 31:2116-2125.
84. Young AJ, Seabrook TJ, Marston WL et al. A role for lymphatic endothelium in the sequestration of recirculating $\gamma\delta$ T cells in TNF-α-stimulated lymph nodes. Eur J Immunol 2000; 30:327-334.
85. Binns RM, Pabst R. Lymphoid tissue structure and lymphocyte trafficking in the pig. Vet Immunol Immunopathol 1994; 43:79-87.
86. Regoli M, Borghesi C, Bertelli E et al. Arrangement of the small intestine lymphatic network in the Peyer's patches of the mouse. A light and transmission electron microscopic study. Ann Anat 1995; 177:229-235.
87. Nagata H, Miyairi M, Sekizuka E et al. In vivo visualization of lymphatic microvessels and lymphocyte migration through rat Peyer's patches. Gastroenterology 1994; 106:1548-1553.
88. Farstad IN, Norstein J, Brandtzaeg P. Phenotypes of B and T cells in human intestinal and mesenteric lymph. Gastroenterology 1997; 112:163-173.
89. Thielke KH, Pabst R, Rothkötter HJ. Quantification of proliferating lymphocyte subsets appearing in the intestinal lymph and the blood. Clin Exp Immunol 1999; 117:277-284.
90. Rothkötter HJ, Hriesik C, Pabst R. More newly formed T than B lymphocytes leave the intestinal mucosa via lymphatics. Eur J Immunol 1995; 25:866-869.
91. MacPherson GG, Liu LM. Dendritic cells and Langerhans cells in the uptake of mucosal antigens. Curr Top Microbiol Immunol 1999; 236:33-53.
92. Norris S, Collins C, Doherty DG et al. Resident human hepatic lymphocytes are phenotypically different from circulating lymphocytes. J Hepatology 1998; 28:84-90.
93. Luettig B, Pape L, Bode U et al. Naive and memory T lymphocytes migrate in comparable numbers through normal rat liver: Activated T cells accumulate in the periportal field. J Immunol 1999; 163:4300-4307.
94. Butcher EC, Williams M, Youngman K et al. Lymphocyte trafficking and regional immunity. Adv Immunol 1999; 72:209-253.
95. Westermann J, Engelhardt B, Hoffmann JC. Migration of T cells in vivo: Molecular mechanisms and clinical implications. Ann Intern Med 2001; 135:279-295.
96. Kunkel EJ, Ramos CL, Steeber DA et al. The roles of L-selectin, β_7 integrins, and P-selectin in leukocyte rolling and adhesion in high endothelial venules of Peyer's patches. J Immunol 1998; 161:2449-2456.
97. Farstad IN, Halstensen TS, Lien B et al. Distribution of β7 integrins in human intestinal mucosa and organized gut-associated lymphoid tissue. Immunology 1996; 89:227-237.
98. Farstad IN, Halstensen TS, Kvale D et al. Topographic distribution of homing receptors on B and T cells in human gut-associated lymphoid tissue. Am J Pathol 1997; 150:187-199.
99. Haraldsen G, Kvale D, Lien B et al. Cytokine-regulated expression of E-selectin, intercellular adhesion molecule-1 (ICAM-1), and vascular cell adhesion molecule-1 (VCAM-1) in human intestinal microvascular endothelial cells. J Immunol 1996; 156:2558-2565.
100. Briskin M, Winsor-Hines D, Shyjan A et al. Human mucosal addressin cell adhesion molecule-1 is preferentially expressed in intestinal tract and associated lymphoid tissue. Am J Pathol 1997; 151:97-110.
101. Kantele JM, Kantele A, Arvilommi H. Circulating immunoglobulin-secreting cells are heterogeneous in their expression of maturation markers and homing receptors. Clin Exp Immunol 1996; 104:525-530.
102. Meenan J, Spaans J, Grool TA et al. Altered expression of $\alpha_4\beta_7$, a gut homing integrin, by circulating and mucosal T cells in colonic mucosal inflammation. Gut 1997; 40:241-246.
103. Kantele JM, Arvilommi H, Kontianinen S et al. Mucosally activated circulating human B cells in diarrhea express homing receptors directing them back to the gut. Gastroenterology 1996; 110:1061-1067.
104. Ebert EC. Interleukin 15 is a potent stimulant of intraepithelial lymphocytes. Gastroenterology 1998; 115:1439-1445.
105. Hedges SR, Agace WW, Svanborg C. Epithelial cytokine responses and mucosal cytokine networks. Trends Microbiol 1995; 3:266-270.
106. Slifka MK, Whitton JL. Antigen-specific regulation of T cell-mediated cytokine production. Immunity 2000; 12:451-457.
107. Fujihashi K, Kweon MN, Kiyono H et al. A T cell/B cell/epithelial cell internet for mucosal inflammation and immunity. Springer Semin Immunopathol 1997; 18:477-494.
108. Kjerrulf M, Grdic D, Ekman L et al. Interferon-γ receptor-deficient mice exhibit impaired gut mucosal immune responses but intact oral tolerance. Immunology 1997; 92:60-68.

109. Iwasaki A, Kelsall BL. Localization of distinct Peyer's patch dendritic cell subsets and their recruitment by chemokines macrophage inflammatory protein (MIP)-3α, MIP-3β, and secondary lymphoid organ chemokine. J Exp Med 2001; 191:1381-1393.
110. Blanas E, Davey GM, Carbone FR et al. A bone marrow-derived APC in the gut-associated lymphoid tissue captures oral antigens and presents them to both CD4+ and CD8+ T cells. J Immunol 2000; 164:2890-2896.
111. Ardavin C, Martinez del Hoyo G, Martin P et al. Origin and differentiation of dendritic cells. Trends Immunol 2001; 22:691-700.
112. Weiner HL. The mucosal milieu creates tolerogenic dendritic cells and T_R1 and T_H3 regulatory cells. Nature Immunol 2001; 2:671-672.
113. Shortman K, Heath WR. Immunity of tolerance? That is the question for dendritic cells. Nature Immunol 2001; 2:988-989.
114. Fu YX, Chaplin DD. Development and maturation of secondary lymphoid tissues. Annu Rev Immunol 1999; 17:399-433.
115. Yoshie O, Imai T, Nomiyama H. Chemokines in immunity. Adv Immunol 2001; 78:57-110.
116. Jung S, Littman DR. Chemokine receptors in lymphoid organ homeostasis. Curr Opin Immunol 1999; 11:319-325.
117. Körner H, Cook M, Riminton DS et al. Distinct roles for lymphotoxin-α and tumor necrosis factor in organogenesis and spatial organization of lymphoid tissue. Eur J Immunol 1997; 27:2600-2609.
118. Ansel KM, Cyster JG. Chemokines in lymphopoiesis and lymphoid organ development. Curr Opinion Immunol 2001; 13:172-179.
119. Cook DN, Prosser DM, Forster R et al. CCR6 medicates dendritic cell localization, lymphocyte homeostasis, and immune responses in mucosal tissue. Immunity 2000; 12:495-503.
120. Campbell JJ, Butcher EC. Chemokines in tissue-specific and microenvironment-specific lymphocyte homing. Curr Opin Immunol 2000; 12:336-341.
121. Ansel KM, Harris RBS, Cyster JG. CXCL13 is required for B1 cell homing, natural antibody production, and body cavity immunity. Immunity 2002; 16:67-76.
122. Thoreux K, Owen RL, Schmucker DL. Intestinal lymphocyte number, migration and antibody secretion in young and old rats. Immunology 2000; 101:161-167.
123. Koga T, McGhee JR, Kato H et al. Evidence for early aging in the mucosal immune system. J Immunol 2000; 165:5352-5359.
124. Schmucker DL, Heyworth MF, Owen RL et al. Impact of aging on gastrointestinal mucosal immunity. Dig Dis Sci 1996; 41:1183-1193.
125. Brandtzaeg P, Pabst R. Let's go mucosal: communication on slippery ground. Trends Immunol 2004; 25:570-577.
126. Pabst O, Herbrand H, Worbs T et al. Cryptopatches and isolated lymphoid follicles: dynamic lymphoid tissues dispensable for the generation of intraepithelial lymphocytes. Eur J Immunol 2005; 35:98-107.
127. Eberl G, Littman DR. Thymic origin of intestinal αβT cells revealed by fate mapping of RORγt+ cells. Science 2004; 305:248-251.
128. Guy-Grand D, Azogui O, Celli S et al. Extrathymic T cell lymphopoiesis: ontogeny and contribution to gut intraepithelial lymphocytes in athymic and euthymic mice. J Exp Med 2003; 197:333-341.
129. Jang MH, Kweon MN, Iwatani K et al. Intestinal villous M cells: an antigen entry site in the mucosal epithelium. PNAS 2004; 101:6110-6115.
130. Guy-Grand D, Vassalli P. Tracing an orphan's genealogy. Science 2004; 305:185-187.
131. Sixt M, Kanazawa N, Seig M et al. The conduit system transports soluble antigens from the afferent lymph to resident dendritic cells in the T cell area of the lymph node. Immunity 2005; 22:19-29.
132. Kobayashi H, Miura S, Nagata H et al. In situ demonstration of dendritic cell migration from rat intestine to mesenteric lymph nodes: relationships to maturation and role of chemokines. J Leukoc Biol 2004; 75:434-442.
133. Bilsborough J, Viney JL. Gastrointestinal dendritic cells play a role in immunity, tolerance, and disease. Gastroenterol 2004; 127:300-309.

Chapter 2

Functional Aspects of the Mucosal Immune System

Cathryn Nagler-Anderson

The mucosal immune system is faced with a daunting challenge. It must quickly and efficiently protect the epithelial barrier from invasion by microbes while avoiding a response to antigenic stimuli from the commensal bacteria or food proteins that constantly bombard it.[1] It meets this challenge through tightly controlled overlapping regulatory mechanisms brought about by specialized populations of B and T cells. When these control mechanisms go awry, luminal bacteria drive the chronic intestinal inflammatory response that characterizes inflammatory bowel disease. This review summarizes the unique features of mucosal lymphocytes that help prevent the inappropriate generation of an inflammatory response.

Phenotypically and Functionally Unique Lymphocyte Populations in the Gut Associated Lymphoid Tissue

Peyer's patches (PP), aggregations of lymphoid follicles in the distal ileum of the small intestine, are the most readily discernible lymphoid structure in the gut associated lymphoid tissue (GALT). Peyer's patches have B cell follicles and germinal centers surrounded by areas that contain predominantly T cells, similar to those found in lymph nodes. 100-200 small, well-developed intestinal lymphoid follicles (ILF), complete with germinal centers, are present along the anti-mesenteric wall of the murine small intestine.[2] Like Peyer's patches, ILF appear to be a source of intestinal IgA to luminal antigens. Throughout the small and large intestine, clusters of T cell progenitors are found in the crypt lamina propria.[3] These cryptopatches are indispensable for the thymus independent development of intraepithelial lymphocytes (IEL), a subpopulation of T cells which resides between epithelial cells, above the basement membrane. The enterocytes themselves produce the IL-7 that is both important and sufficient for the development of these IEL.[4] Loosely organized lymphoid cells are also abundant within the lamina propria of the intestinal villi. Remarkably, each of these unusual mucosal lymphoid structures seems to have its own set of developmental requirements, location and effector function (Table 1). Both the Peyer's patch and the ILF are covered by a distinctive follicle associated epithelium containing M cells, specialized enterocytes that lack the apical brush border glycocalyx that characterizes villus enterocytes. The M cell's microfolds are more accessible to luminal antigens, facilitating their adherence and transport to antigen presenting cells in the underlying GALT.[5]

The mucosal immune system contains subpopulations of T cells not found at other sites. In both mice[6,7] and humans[8,9] T cells are attracted to the small intestinal epithelium by its abundant expression of thymocyte expressed chemokine (TECK, CCL25), the same chemokine which attracts developing thymocytes to the thymic epithelium, via their expression of its ligand CCR9. In mice, the preferential expression of TECK (CCL25) in the villus crypts also suggests a role in the extrathymic maturation of IELs.[6] Interestingly, many IEL

Immune Mechanisms in Inflammatory Bowel Disease, edited by Richard S. Blumberg and Markus F. Neurath.

Table 1. Lymphocyte populations and structures in GALT

Peyer's patches (PP)	Aggregations of lymphoid follicles in the distal ileum of small intestine
Intestinal lymphoid follicle (ILF)	Small follicles found on the anti-mesenteric wall of the murine small intestine
Cryptopatches	Clusters of T cell progenitors in the crypt lamina propria in both large and small intestine
Intraepithelial lymphocytes (IEL)	Atypical subpopulations of T cells that reside between enterocytes, above the basement membrane
Intestinal lamina propria	Connective tissue between the epithelium and the muscularis mucosa that forms the villus core. This area is abundantly populated with loosely organized immune effector cells
Mesenteric lymph node (MLN)	Lymph node(s) that drain GALT, located in the mesentery of the small intestine

are also constitutively cytolytic ex vivo, an observation that has long puzzled mucosal immunologists. What keeps these activated effector cells from lysing the surrounding epithelium? Leishman et al have shed new light on this conundrum while also elucidating a role for the unusual form of CD8 expressed by these cells.[10] In mice, the CD8αα chain interacts with high affinity with a nonclassical MHC Class I molecule, the thymus leukemia antigen (TL), the expression of which appears to be restricted to intestinal epithelial cells. This antigen independent interaction inhibits IEL cytotoxicity and proliferation but enhances cytokine release. Interestingly, $CD4^+$ T cells often acquire CD8αα expression (and presumably functional activity) after migration to the intestine. Taken together these observations have led to the hypothesis that CD8αα/TL binding in the intestinal epithelium helps to regulate both IEL survival and the homeostasis of this unusual microenvironment.[10,11] Although they did not examine the role of TL directly, the work of Poussier et al also supports this idea.[12] Using a murine model of colitis induced by the transfer of $TCR\alpha\beta^+CD4^+CD45RB^{hi}$ T cells into severe combined immunodeficient (SCID) mice these authors showed that $CD8\alpha^+\beta^-TCR\ \alpha\beta^+$ IEL can safeguard the mucosa from the chronic inflammation this transfer typically induces. Reconstitution of SCID mice with this (but not other) subsets of IEL prior to transfer of splenic $TCR\alpha\beta^+CD4^+CD45RB^{hi}$ T cells protected against the subsequent development of colitis. Protection was dependent upon IL-10 and was associated with the presence of reduced numbers of $CD4^+$ T cells in the colonic lamina propria.

IgA is produced in enormous quantity at the mucosal surface but little is found in the circulation. Because of this highly localized production it is often not appreciated that IgA (not IgG) is the major isotype of Ig synthesized by the body. Binding to the poly-Ig receptor (pIgR) expressed constitutively on the basolateral surface of enterocytes results in the transport of IgA to the apical surface. At the apical surface, the portion of the pIgR attached to the Fc region of IgA is enzymatically cleaved and remains bound to the dimeric IgA molecule as the secretory component (SC). Once secreted into the lumen, secretory IgA (sIgA) protects the epithelium against microbial colonization and invasion (immune exclusion). SC had been thought to function primarily to prevent proteolytic damage to secretory IgA in the harsh luminal environment. New work suggests a direct role for SC in IgA mediated immune exclusion of bacteria. Phalipon et al have shown that the carbohydrate residues on SC function in its tissue localization by anchoring sIgA in the mucus that lines the epithelial surface.[13] IgA can also use the pIgR to actively transport antigens that do penetrate the barrier out of the lamina propria to the enterocyte's apical surface (and back into the lumen).[14]

IgA secreting plasma cells are committed to the secretion of this isotype long before they reach the intestinal lamina propria. Moreover, trafficking to the lamina propria (LP) is antigen independent.[15] How, then, do IgA secreting plasma cells become localized to mucosal surfaces? Epithelial cell expression of the chemokine TECK (CCL25) selectively attracts IgA secreting B cells from the spleen, mesenteric lymph node and Peyer's patch and thus acts as a chemoattractant for their localization to this site.[16] Switching to IgA and antigen specific clonal expansion occurs in the germinal centers of these organized lymphoid tissues prior to trafficking into the lamina propria. However, this paradigm may only apply to "conventional" bone marrow-derived B2 B cells. Another recent report has demonstrated that switching and differentiation into IgA secreting cells can take place in situ in the intestinal lamina propria itself.[17] The precursors of these locally derived IgA$^+$ cells are B220$^+$IgM$^+$ lymphocytes. Experimentally, LPS, IL-5 and TGFβ are sufficient to induce switching from IgM to IgA by this subset of B cells. In vivo, LP stromal cell derived cytokines appear to control the switch from IgM to IgA. Unlike what is typically observed in the Peyer's patch, class switching in this environment is T cell independent and does not involve germinal center formation or a requirement for signaling via CD40. These B cells are likely to be derived from the B1 subset, a subpopulation of self-renewing B cells in the peritoneal and pleural cavities with an unusual specificity for self and microbial antigens. B1 B cells typically do not enter lymphoid follicles. Class switching by B1 B cells in the lamina propria may occur in response to antigen presented by DC which have sampled the luminal contents by extending their dendrites through epithelial cell tight junctions.[17-19] Although the natural antibodies made by B1 B cells are typically encoded by unmutated germline V genes, IgA secreting B1 B cells specific for commensal bacteria exhibit evidence of somatic mutation.[20] Others have described a mechanism for the induction of a mucosal IgA response to commensal bacteria that requires neither T cell help nor organized follicular lymphoid tissue.[21] Plasma cells secreting IgA specific for antigens in the cell walls of commensal bacteria are found diffusely distributed throughout the intestinal lamina propria. The anti-commensal IgA response requires the presence of the intestinal flora and correlates with bacterial load. It is now apparent that this T independent IgA pathway is present even in μMT mice which lack the IgM or IgD expression that had previously been thought to be essential for class switching.[22] The IgA secreting B cells present in μMT mice have some (but not all) of the features of B1 B cells. It will be interesting to examine the relationship of this T cell independent pathway for IgA secretion to that described by Fagarasan et al[17] and determine whether switching to anti-commensal IgA in the μMT mutant mice also occurs in the intestinal lamina propria. Other evidence that the intestinal lamina propria can induce IgA production in a manner independent of Peyer's patches or mesenteric lymph nodes comes from the work of Kang et al.[23] They have shown that the presence, in the lamina propria, of lymphotoxin (LT), a cytokine in the TNF family, is sufficient for IgA production. LT plays a critical role in the development of organized lymphoid structures; LT-/- mice are IgA deficient and lack PP and MLN. However, transplantation of bone marrow from wild type mice, or a segment of intestine from RAG 1-/- mice, is sufficient to reverse the IgA deficiency of LT mutant mice. Large numbers of IgA secreting cells are detectable in the transplanted intestinal segment, but not in the adjacent recipient intestine.

As indicated above, chemokines are known to govern the trafficking of lymphocytes into the lamina propria. Binding of the mucosal addressin cell adhesion molecule (MadCAM-1) to lymphocytes expressing the intestinal homing receptor α4β7 also plays a central role in migration of lymphocytes to this site.[24] Presentation of antigen to naïve T cells in the intestinal MLN induces the upregulation of α4β7 (and TECK responsiveness) prior to their exit into the circulation, ensuring that potential effector cells are targeted to the intestinal mucosa.[25] This targeting is likely to facilitate a vigorous anamnestic response upon antigen reexposure.[26,27] Recent work has shown that denderitic cells (DC) are the antigen presenting cell (APC) population in the GALT required to 'imprint' gut homing specificity.[28,29] Interestingly, the expression of

MadCAM-1, as well as two chemokines involved in B cell migration (B cell chemoattractant (BLC/CXCL13) and secondary lymphoid tissue chemokine (SLC)) is reduced in the lamina propria of LT-/- mice.[23] Adoptive transfer studies suggest that, in the absence of LT, the reduced expression of these chemokines/adhesion molecules impairs the migration of IgA secreting B cells from the peritoneal cavity to this site. Other work has shown that although LT is required for the secretion of CXCL13 by stromal cells in lymphoid follicles, the expression of CXCL13 by peritoneal macrophages is independent of B cells and of LTα, LTβ and TNF.[30] Homing of B1 B cells to the peritoneal cavity is therefore CXCL13 dependent but LT independent. Additional work will be needed to determine if the subsequent migration of IgA secreting B1 B cells to the lamina propria involves direct trafficking from the peritoneal cavity and requires the LT dependent secretion of CXCL13 by gut stromal cells. LT-LTβR interactions during embryogenesis are required for formation of the spleen and lymph node. In the lamina propria, however, post-gestational signaling via this ligand-receptor pair directs lymphocyte subsets to this site.[31] The lamina propria of LT-/- mice lacks B cells but has a normal composition of T cells. LT sufficient B cells and an LT-βR^+ stromal cell are required for B cells (primarily B-2 cells) to reside in the lamina propria.[31] However, fecal IgA can be produced in the virtual absence of B cells in the LP, perhaps by B-1 B cells arising from another site. B1 and B2 B cells appear to migrate into the intestine in response to different signals and/or chemokines.

Mechanisms Governing Non-Responsiveness to Food Proteins and Commensal Bacteria

In addition to bacteria (particularly commensals) the other major source of chronic antigenic stimulation in the GALT comes from food proteins. Experimentally, oral administration of soluble antigens induces systemic nonresponsiveness to peripheral antigen challenge.[32] Typically known as "oral tolerance", the induction of both local and systemic nonresponsiveness to nonpathogenic dietary or commensal antigens plays a vital physiological role in preventing hyperreactive immune responses to the luminal contents (reviewed in ref. 33). Antigen presentation in the presence of low levels of costimulation and the active suppression of immune responses by the secretion of immunoregulatory cytokines are two of the major pathways by which tolerance is induced. However they need not be viewed as mutually exclusive. Indeed, it is likely that they function synergistically (or at least in tandem) (Table 2).

Productive, antigen specific, immune responses are induced when antigenic peptide are presented in the context of MHC proteins and a "costimulatory" signal generated by the innate immune system. In the best characterized pathway, inflammatory stimuli upregulate the expression of CD80/86 costimulatory molecules and, concomitantly, induce the maturation and migration of dendritic cells to the T cell areas of lymph nodes. Antigen presentation to $CD28^+$ T cells initiates the clonal expansion required for the induction of an adaptive immune response. By contrast, soluble antigens, which do not trigger the innate immune system, are presented in the presence of low levels of costimulatory signals and preferentially bind the high affinity CD80/86 ligand CTLA-4. The initial response to both tolerogenic and immunogenic forms of orally administered antigen is the same; both induce transient T cell activation and proliferation that is observed primarily in the GALT.[34-36] However, the proliferative capacity of antigen specific cells from antigen fed mice is reduced at each subsequent antigen exposure, when compared to mice that have received antigen in an immunogenic form.[36] CTLA-4 signaling contributes to the induction of nonresponsiveness to soluble antigens administered both peripherally[37,38] and orally[39] by regulating the progression of T cells through the cell cycle during their initial response to antigen.

Dendritic cells (DC) are the APC responsible for the presentation of antigen to naïve T cells. In their immature form DC are highly phagocytic cells optimized for antigen uptake. As mentioned above, exposure to inflammatory stimuli induces the "maturation" of DC to a form that functions primarily to present antigen to T cells. The migration of DC has been linked to

Table 2. Mechanisms contributing to the induction of non-responsiveness at mucosal surfaces

Antigen presentation
- a. in the presence of low levels of costimulation (CTLA-4 signaling)
- b. by "tolerogenic" DC populations, with novel phenotypic markers, that secrete IL-10 (and not IL-12)

Regulatory T cells
- a. $CD4^+CD25^+$ "natural" Tregs bearing cell-surface TGF-β
- b. (luminal?) antigen driven Tr1 that secrete IL-10

this maturation process. In the gut, however, immature mucosal DC continually pick up remnants of apoptotic enterocytes, shed from the villus tips as the epithelium regenerates, and transport them to the T cell areas of the mesenteric lymph node.[40] Sentinel DC also constitutively transport antigen from the mucosal epithelium of the lungs in the absence of overt inflammatory stimulation.[41,42] The ability of DC to extend their dendrites between enterocytes allows them to sample luminal antigens without inducing their own maturation.[18] Immature, gut-derived DC may then constitutively migrate from the mucosal epithelium to the MLN (or other peripheral sites), and induce tolerance by presenting antigen themselves or by transferring antigen to a "tolerance inducing" subset of DC resident in T cell areas of the LN.[43] Although, in the mouse, at least five phenotypic subsets of DC have been defined, DC have been broadly divided into two major phenotypic and functional subsets.[44-46] In the GALT, myeloid ($CD11c^+CD11b^+$) DC are prominently localized in the sub-epithelial dome of the Peyer's patch while lymphoid ($CD11c^+$, $CD8\alpha^+$) DC are found in T cell areas.[47] Initially, presentation of antigen by myeloid DC was associated with a Th2 response and lymphoid DC were implicated in Th1 responses. Newer studies have emphasized the functional plasticity of these DC subsets and suggest that T cell priming and polarization is also strongly influenced by antigen dose, immunoregulatory mediators and tissue microenvironment.[48] In the Peyer's patch, novel subsets of DC that predominantly produce IL-10 have been identified.[45,47] It has been postulated that mucosal DC which fail to mature and secrete IL-12 may be central to the generation of regulatory T cells, particularly in the GALT.[49] Groux et al have recently described a novel population of $CD11c^{low}CD45RB^{hi}$ DC that secrete IL-10 when stimulated with Toll-like receptor ligands such as LPS or CpG.[50] These $CD11c^{low}$ $CD45RB^{hi}$ DC have a stable immature phenotype and prime for regulatory T cells that secrete IL-10. However, other work has shown that, in a microenvironment bathed in immunoregulatory cytokines, DC expressing high levels of costimulatory molecules can also induce nonresponsiveness.[51] In the respiratory mucosa, phenotypically mature IL-10 producing DC are required for both the induction of tolerance and the generation of IL-10 secreting regulatory T cells.[51] Secretion of IL-10 by other APC populations may also contribute to immunoregulation in the GALT. Chronic intestinal inflammation induces a subset of regulatory B cells that secretes IL-10.[52] The enhanced production of IL-10 is associated with the upregulation of CD1d. These regulatory B cells appear after intestinal inflammation has been established and participate in slowing its progression.

Functionally nonresponsive "anergic" T cells retain the ability to secrete immunoregulatory mediators; IL-10 and TGF-β are among the most important of these in the GALT. IL-10's central role in the maintenance of immune homeostasis is emphasized by the development of colitis in IL-10 -/- mice.[53] IL-10 secreting T regulatory 1 cells (Tr1) were initially identified as a subset of $CD4^+$ T cells effective at preventing chronic intestinal inflammation.[54] Subsequent work has established the role of Tr1 cells in preventing inappropriate responses to the luminal antigens (including the bacterial flora).[55-57] TGF-β has long been implicated in maintenance of oral tolerance.[33] More recently, regulatory T cells dependent on both

TGF-β and signaling via CTLA-4 have been described.[58,59] The release of TGF-β by apoptotic T cells suggests one mechanism by which TGF-β may contribute to peripheral nonresponsiveness.[60] Cell surface TGF-β is also important in cell contact dependent immunosuppression.[61] New work suggests that soluble and cell surface TGF-β may have specialized immunoregulatory functions. Soluble TGF-β inhibits the Tec kinase Itk to limit the differentiation of naïve T cells, while cell surface TGF-β inhibits antigen-experienced T effector cells.[62] More than their shared localization in the GALT links IL-10 and TGF-β. Strober's laboratory has used a doxycycline-regulatable system for TGF-β production (in mice) to show that TGF-β induces SMAD-4 mediated production of IL-10.[63] TGF-β and IL-10 then act coordinately to regulate immune responsiveness. Data has also rapidly accumulated in support of a major subset of naturally occurring $CD4^+CD25^+$ regulatory T cells with a central role in the prevention of autoimmunity.[64] The association of this subset with the transcription factor Foxp3 has defined it as a distinct T cell lineage.[65] The ability of $CD4^+CD25^+$ $CD45RB^{lo}$ regulatory T cells to control the colitis induced by the transfer of $CD45RB^{hi}$ T cells to immunodeficient recipients is well documented (reviewed in ref. 66). The suppression of innate immune pathology by $CD4^+CD25^+$ Tregs is dependent on both IL-10 and TGF-β.[67] Whether or not these Tregs are related to the $Foxp3^+$ lineage is not yet clear. It is likely that, in the GALT, multiple, overlapping layers of immunoregulatory cells and mediators function in the suppression of immune hyperreactivity.[68]

Basal anti-inflammatory cyclooxygenase-2 (COX-2) dependent prostaglandin (PGE_2) production by non bone marrow-derived stromal cells in the lamina propria is another major contributor to the anti-inflammatory tone of the GALT.[69,70] Interestingly, the spontaneous and constitutive production of the enzyme COX-2 (which is required to metabolize PGE_2) is not dependent upon the presence of the luminal flora or on inflammatory mediators such as LPS, IL1-β, IFN-γ, TNF or IL-12. PGE_2 suppresses innate immune responses to luminal antigen by signaling via the prostaglandin E receptor EP4.[71] In a murine model of colitis induced by the administration of DSS, treatment with an EP4 antagonist exacerbated the course of disease. Inhibition of EP4 signaling resulted in the sustained activation of $CD4^+$ T cells and the concomitant production of inflammatory cytokines. The immunosuppressive tone of the GALT may also help prevent inappropriate responses to innocuous antigen by raising the threshold for T cell activation. Kellerman et al have reported that the PP microenvironment induces hyporesponsiveness to chemokine-stimulated T cell chemotaxis.[72] Other evidence suggests that lamina propria T cells exhibit slower cell cycling kinetics than peripheral blood T cells due to their endogenous upregulation of the inhibitory protein p53.[73] Clonal expansion is delayed but not prevented, allowing for a vigorous memory response upon antigen challenge. Interestingly, this tightly controlled microenvironment also appears to serve as a repository for memory effector cells.[26,27]

The responsiveness of the T and B cells present in the GALT is determined by both the unique functional properties of these mucosal lymphocytes and stringently controlled, redundant immunoregulatory mechanisms. Inappropriate responses to innocuous antigens are typically held tightly in check, but memory T cells stand ready to respond quickly and effectively to an infectious challenge. Dysregulation of these control mechanisms at any of a number of different levels can result in a chronic inflammatory response. A clearer understanding of mucosal immunoregulation will be vital for the development of new therapeutic modalities for the treatment of IBD.

Acknowledgements

This work was supported by RO-1 DK 55678 and by the Center for the Study of Inflammatory Bowel Disease at Mass. General Hospital and the Clinical Nutrition Research Center at Harvard. The author thanks Bobby Cherayil, Peter Andersen, Hai Ning Shi, Donald Smith and Guenolee Prioult for critical review of the manuscript.

References

1. Nagler-Anderson C. Man the barrier! Strategic defenses in the intestinal mucosa. Nat Rev Immunol 2001; 1:59-67.
2. Hamada H et al. Identification of multiple isolated lymphoid follicles on the antimesenteric wall of the mouse small intestine. J Immunol 2002; 168:57-64.
3. Suzuki K et al. Gut cryptopatches: Direct evidence for extrathymic anatomical sites for intestinal T lymphopoiesis. Immunity 2000; 13:691-702.
4. Laky K et al. Enterocyte expression of interleukin 7 induces development of γδ T cells and Peyer's patches. J Exp Med 2000; 191:1569-1580.
5. Kraehenbuhl JP, Neutra MR. Epithelial M cells: Differentiation and function. Ann Rev Cell Dev Biol 2000; 16:301-332.
6. Wurbel MA et al. The chemokine TECK is expressed by thymic and intestinal epithelial cells and attracts double and single-positive thymocytes expressing the TECK receptor CCR9. Eur J Immunol 2000; 30:262-271.
7. Kunkel EJ et al. Lymphocyte CC chemokine receptor 9 and epithelial thymus-derived chemokine (TECK) expression distinguish the small intestinal immune compartment: Epithelial expression of tissue specific chemokines as an organizing principle in regional immunity. J Exp Med 2000; 192:761-767.
8. Zabel BA et al. Human G protein-coupled receptor GPR-9-6/CC chemokine receptor 9 is selectively expressed on intestinal homing T lymphocytes, mucosal lymphocytes, and thymocytes and is required for thymus-expressed chemokine-mediated chemotaxis. J Exp Med 1999; 190:1241-1256.
9. Papadakis KA et al. The role of thymus-expressed chemokine and its receptor CCR9 on lymphocytes in the regional specialization of the mucosal immune system. J Immunol 2000; 165:5069-5076.
10. Leishman AJ et al. T cell responses modulated through interaction between CD8αα and the nonclassical MHC Class I molecule TL. Science 2001; 294:1936-1939.
11. Lambolez F, Rocha B. A molecular gut reaction. Science 2001; 294:1848-1849.
12. Poussier P, Ning T, Banerjee D et al. A unique subset of self-specific intraintestinal T cells maintains gut integrity. J Exp Med 2002; 195:1491-1497.
13. Phalipon A et al. Secretory component: A new role in secretory IgA-mediated immune exclusion in vivo. Immunity 2002; 17:107-115.
14. Robinson JK, Blanchard TG, Levine AD et al. A mucosal IgA-mediated excretory immune system in vivo. J Immunol 2001; 166:3688-3692.
15. Lamm ME, Phillips-Quagliata JM. Origin and homing of intestinal IgA antibody-secreting cells. J Exp Med 2002; 195:F5-F8.
16. Bowman EP et al. The intestinal chemokine thymus-expressed chemokine (CCL25) attracts IgA antibody-secreting cells. J Exp Med 2002; 195:269-275.
17. Fagarasan S, Kinoshita K, Muramatsu M et al. In situ class switching and differentiation to IgA-producing cells in the gut lamina propria. Nature 2001; 413:639-643.
18. Rescigno M et al. Dendritic cells express tight junction proteins and penetrate gut epithelial monolayers to sample bacteria. Nature Immunology 2001; 2:361-367.
19. Brandtzaeg P, Baekkevold ES, Morton HC. From B to A the mucosal way. Nature Immunol 2001; 2:1093-1094.
20. Fagarasan S, Honjo T. T-independent immune response: New aspects of B cell biology. Science 2000; 290:89-92.
21. Macpherson AJ et al. A primitive T cell-independent mechanism of intestinal mucosal IgA responses to commensal bacteria. Science 2000; 288:2222-2226.
22. MacPherson AJS et al. IgA expression without μ or δ chain expression in developing B cells. Nature Immunol 2001; 2:625-631.
23. Kang HS et al. Signaling via LTβR on the lamina propria stromal cells of the gut is required for IgA production. Nature Immunol 2002; 3:576-582.
24. Rott LS, Briskin MJ, Andrew DP et al. A fundamental subdivision of circulating lymphocytes defined by adhesion to mucosal addressin cell adhesion molecule-1. Comparison with vascular cell adhesion molecule-1 and correlation with beta 7 integrins and memory differentiation. J Immunol 1996; 156:3727-3736.
25. Campbell DJ, Butcher EC. Rapid acquisition of tissue-specific homing phenotypes by $CD4^+$ T cells activated in cutaneous or mucosal lymphoid tissues. J Exp Med 2002; 195:135-141.
26. Masopust D, Vezys V, Marzo AL et al. Preferential localization of effector memory cells in nonlymphoid tissue. Science 2001; 291:2413-2417.
27. Reinhardt RL, Khoruts A, Merica R et al. Visualizing the generation of memory CD4 T cells in the whole body. Nature 2001; 410:101-105.

28. Mora JR et al. Selective imprinting of gut-homing T cells by Peyer's patch dendritic cells. Nature 2003; 424:88-93.
29. Johansson-Lindbom B et al. Selective generation of gut tropic T cells in gut-associated lymphoid tissue (GALT): Requirement for GALT dendritic cells and adjuvant. J Exp Med 2003; 198:963-969.
30. Ansel KM, Harris RBS, Cyster JG. CXCL13 is required for B1 cell homing, natural antibody production and body cavity immunity. Immunity 2002; 16:67-76.
31. Newberry RD, McDonough JS, McDonald KG et al. Postgestational lymphotoxin/lymphotoxin β receptor interactions are essential for the presence of intestinal B lymphocytes. J Immunol 2002; 168:4988-4997.
32. Nagler-Anderson C, Shi HN. Peripheral nonresponsiveness to orally administered soluble protein antigens. Crit Rev Immunol 2001; 21:121-131.
33. Mowat AM, Weiner HL. Oral tolerance. In: Ogra PL, Mestecky J, Lamm ME et al, eds. Physiological basis and clinical applications. Mucosal Immunology 1999; 587-618.
34. Sun J, Dirden-Kramer B, Ito K et al. Antigen-specific T cell activation and proliferation during oral tolerance induction. J Immunol 1999; 162:5868-5875.
35. Blanas E, Davey GM, Carbone FR et al. A bone marrow-derived APC in the gut-associated lymphoid tissue captures oral antigens and presents them to both $CD4^+$ and $CD8^+$ T cells. J Immunol 2000; 164:2890-2896.
36. Shi HN, Liu HY, Nagler-Anderson C. Enteric infection acts as an adjuvant for the response to a model food antigen. J Immunol 2000; 165:6174-6182.
37. Greenwald RJ, Boussiotis VA, Lorsbach RB et al. CTLA-4 regulates induction of anergy in vivo. Immunity 2001; 14:145-155.
38. Greenwald RJ, Oosterwegel MA, van der Woude D et al. CTLA-4 regulates cell cycle progression during a primary immune response. Eur J Immunol 2002; 32:366-373.
39. Fowler S, Powrie F. CTLA-4 expression on antigen-specific cells but not IL-10 secretion is required for oral tolerance. Eur J Immunol 2002; 32:2997-3006.
40. Huang FP et al. A discrete subpopulation of dendritic cells transports apoptotic intestinal epithelial cells to T cell areas of mesenteric lymph nodes. J Exp Med 2000; 191:435-443.
41. Vermaelen KY, Carro-Muino I, Lambrecht BN et al. Specific migratory dendritic cells rapidly transport antigen from the airways to the thoracic lymph nodes. J Exp Med 2001; 193:51-60.
42. Stumbles PA et al. Regulation of dendritic cell recruitment into resting and inflamed airway epithelium: Use of alternative chemokine receptors as a function of inducing stimulus. J Immunol 2001; 167:228-234.
43. Sallusto F, Lanzavecchia. A Mobilizing dendritic cells for tolerance, priming and chronic inflammation. J Exp Med 1999; 189:611-614.
44. Henri S et al. The dendritic cell populations of mouse lymph nodes. J Immunol 2001; 167:741-748.
45. Iwasaki A, Kelsall BL. Unique functions of $CD11b^+$, $CD8\alpha^+$, and double-negative Peyer's patch dendritic cells. J Immunol 2001; 166:4884-4890.
46. Pulendran B, Maraskovsky E, Banchereau J et al. Modulating the immune response with dendritic cells and their growth factors. Trends Immunol 2001; 22:41-47.
47. Iwasaki A, Kelsall BL. Freshly isolated Peyer's patch, but not spleen, dendritic cells produce interleukin 10 and induce the differentiation of T helper type 2 cells. J Exp Med 1999; 190:229-239.
48. Kelsall BL, Biron CA, Sharma O et al. Dendritic cells at the host-pathogen interface. Nature Immunol 2002; 3:699-702.
49. Roncarolo MG, Levings MK, Traversari C. Differentiation of T regulatory cells by immature dendritic cells. J Exp Med 2001; 193:F5-F9.
50. Wakkach A, Fournier N, Brun V et al. Characterization of dendritic cells that induce tolerance and T regulatory 1 cell differentiation in vivo. Immunity 2003; 18:605-617.
51. Akbari O, DeKruyff RH, Umetsu DT. Pulmonary dendritic cells producing IL-10 mediate tolerance induced by respiratory exposure to antigen. Nature Immunol 2001; 2:725-731.
52. Mizoguchi A, Mizoguchi E, Takedatsu H et al. Chronic intestinal inflammatory condition generates IL-10-producing regulatory B cell subset characterized by CD1d upregulation. Immunity 2002; 16:219-230.
53. Kuhn R, Lohler J, Rennick D et al. Interleukin-10 deficient mice develop chronic enterocolitis. Cell 1993; 75:263-274.
54. Groux H et al. A $CD4^+$ T cell subset inhibits antigen-specific T cell responses and prevents colitis. Nature 1997; 389:737-742.
55. Cottrez F, Hurst SD, Coffman RL et al. T regulatory cells 1 inhibit a Th2-specific response in vivo. J Immunol 2000; 165:4848-4853.

56. Foussat A, Cottrez F, Brun V et al. A comparative study between T regulatory type 1 and CD4+CD25+ T cells in the control of inflammation. J Immunol 2003; 171:5018-5026.
57. Cong Y, Weaver CT, Lazenby A et al. Bacterial-reactive T regulatory cells inhibit pathogenic immune responses to the enteric flora. J Immunol 2002; 169:6112-6119.
58. Chen W, Jin W, Wahl SM. Engagement of cytotoxic T lymphocyte-associated antigen 4 (CTLA-4) induces transforming growth factor β (TGF-β) production by murine CD4+ T cells. J Exp Med 1998; 188:1849-1857.
59. Read S, Malmstrom V, Powrie F. Cytotoxic T lymphocyte-associated antigen 4 plays an essential role in the function of CD25+CD4+ regulatory cells that control intestinal inflammation. J Exp Med 2000; 192:295-302.
60. Chen WJ, Frank ME, Jin W et al. TGF-β released by apoptotic T cells contributes to an immunosuppresive milieu. Immunity 2001; 14:715-725.
61. Nakamura K, Kitani A, Strober W et al. Cell contact dependent immunosuppression by CD4+CD25+ regulatory T cells is mediated by cell surface bound transforming growth factor β. J Exp Med 2001; 194:629-644.
62. Chen CH et al. Tranforming growth factor β blocks Tec kinase phosphorylation, Ca^{++} influx and NFATc translocation causing inhibition of T cell differentiation. J Exp Med 2003; 197:1689-1699.
63. Kitani A, Fuss I, Nakamura K et al. Transforming growth factor (TGF)-β1-producing regulatory T cells induce Smad-mediated interleukin-10 secretion that facilitates coordinated immunoregulatory activity and amelioration of TGF-β1-mediated fibrosis. J Exp Med 2003; 198:1179-1188.
64. Shevach EM. Certified professionals: CD4+CD25+ suppressor T cells. J Exp Med 2001; 193:F41-F45.
65. Ramsdell F. Foxp3 and natural regulatory T cells: Key to a cell lineage? Immunity 2003; 19:165-168.
66. Annacker O, Powrie F. Homeostasis of intestinal immune regulation. Microbes Infect 2002; 4:567-574.
67. Maloy KJ, Salaun L, Cahill R et al. CD4+CD25+ T_R cells suppress innate immune pathology through cytokine-dependent mechanisms. J Exp Med 2003; 197:111-119.
68. Nagler-Anderson C, Bhan AK, Podolsky DK et al. Control freaks: Immune regulatory cells. Nat Immunol 2004; 5:119-122.
69. Newberry RD, Stenson WF, Lorenz RG. Cyclooxygenase-2-dependent arachidonic acid metabolites are essential modulators of the intestinal immune response to dietary antigen. Nat Med 1999; 5:900-906.
70. Newberry RD, McDonough JS, Stenson WF et al. Spontaneous and continuous cyclooxygenase-2-dependent prostaglandin E_2 production by stromal cells in the murine small intestine lamina propria: Directing the tone of the intestinal immune reponse. J Immunol 2001; 166:4465-4472.
71. Kabashima K et al. The prostaglandin receptor EP4 suppresses colitis, mucosal damage and CD4 cell activation in the gut. J Clin Invest 2002; 109:883-893.
72. Kellermann SA, McEvoy LM. The Peyer's patch microenvironment suppresses T cell responses to chemokines and other stimuli. J Immunol 2001; 167:682-690.
73. Sturm A, Itoh J, Jacobberger JW et al. p53 negatively regulates intestinal immunity by delaying mucosal T cell cycling. J Clin Invest 2002; 109:1481-1492.

CHAPTER 3

Recent Progress in Inflammatory Bowel Disease Genetics

Judy H. Cho

Introduction

One approach toward understanding the pathophysiology of Crohn's disease (CD) and ulcerative colitis (UC) is through genetic linkage and association studies. These approaches have provided support for general genomic regions (linkage studies) potentially containing inflammatory bowel disease (IBD) genes, as well as specific genetic variants observed at higher allelic frequencies in IBD patients compared to case-matched controls (association studies). Both approaches have been utilized to characterize multiple likely disease associations in IBD. Of particular importance is the association of NOD2/CARD15 variants with CD.[1,2] That NOD2/CARD15 likely functions as an intracellular sensing mechanism for a component of bacterial cell wall peptidoglycan[3,4] is fully consistent with existing murine models of IBD, where the role of intraluminal bacteria is critical in disease pathogenesis.[5] The underlying engine for these advances has been the large amount of sequence information now available on inter-individual differences in genomic DNA structure. In this chapter, we review the epidemiology of IBD, in particular with respect to the insight these data provide on genetic pathogenesis. Following a brief review of the Human Genome Project, we discuss the present state of understanding of IBD pathogenesis as defined by linkage and association studies. Finally, future challenges of defining IBD pathogenesis through genetic approaches are discussed.

IBD Epidemiology

The incidence estimates in North American European cohorts have ranged between 2.2-14.3/100,000/year for UC and 3.1-14.6/100,000/year for CD. The corresponding prevalence estimates range between 37.5-229/100,000 for UC and 26.0-198.5/100,000 for CD.[6] Rates of IBD have generally been estimated to be lower in nonwhite populations. In a health maintenance organization-based mail survey, the CD prevalences per 100,000 were 43.6 among Caucasians, 29.8 among African Americans, 4.1 among Hispanics, and 5.6 among Asians.[7] However, a recent study among African-Americans in Georgia estimated a relatively high CD incidence of 7-12/100,000, more comparable to estimates for Caucasian cohorts.[8] While these estimates place IBD as an important public health priority, like other common, chronic inflammatory disorders, its prevalence is relatively low compared to other intensively studied genetic disorders such as diabetes, obesity, and asthma.

Clustering of IBD cases within families reflects shared genetic, developmental and environmental contributions. In population-based cohorts, it is estimated that 5-10% of all IBD cases have a positive family history.[9-12] The relative risk of both UC and CD among first degree relatives of developing the same disease is 10-15.[13,14] The risk to first degree relatives of a CD proband of developing UC (or conversely) is also increased, but significantly lower than that for the same disease.[13] These data are consistent with present association findings in IBD,

Immune Mechanisms in Inflammatory Bowel Disease, edited by Richard S. Blumberg and Markus F. Neurath.

namely, that some risk alleles are specific to either UC or CD (e.g., NOD2/CARD15), whereas others are common to both disorders (e.g., IBD5 on chromosome 5q) [15,16]

Twin studies provide a means of defining the significance of genetic factors by comparing the concordance rates between monozygotic (MZ) and dizygotic (DZ) twins. That the MZ concordance rates generally and in comparison to DZ concordance is higher for CD than for UC supports the general concept that genetic factors play a more significant role in CD pathogenesis. MZ twin concordance estimates also provide an upper or cumulative limit of disease penetrance for genetic variation. By definition, given a genetic background (defined either by a particular risk allele, or by the organism as a whole), disease penetrance refers to that fraction of individuals who manifest disease. For CD and UC, the MZ concordance has been estimated between 42-58% and 6-17%, respectively.[17-19] That these values are significantly below 100% highlights the importance of environmental and developmental factors in disease pathogenesis. By comparison, the CD penetrance among whites of double-dose carriers of NOD2/CARD15 risk alleles is estimated below 10%. The difference between the penetrance estimates of NOD2/CARD15 risk alleles alone and for CD MZ twins generally reflects the contributions of other, as yet unidentified, risk alleles. The very low penetrance of the NOD2/CARD15 risk alleles also would predict the absence of spontaneous CD in murine NOD2/CARD15 knockout/knockin models.

The Human Genome Project and the Challenge of Complex Genetic Disorders

As opposed to monogenic disorders, where disease gene identification follows a basic paradigm of precise localization of genetic linkage signals, genuine progress in multigenic complex disorders will rely heavily on the large amount of largely sequence information on genetic variation currently being generated. Cross-species sequence comparison has provided a central means of hypothesis generation and model development, as many fundamental cellular processes are highly conserved between species. Most relevant in this regard are the pathogen recognition proteins of the innate immune system, which are highly conserved in hosts from humans to plants. The well-characterized plant R proteins mediate plant resistance to various microbial pathogens and contain both the nucleotide binding domain and leucine-rich repeat domains found within the NOD2/CARD15 gene.[20] This sequence similarity provided an important means by which the essential function of NOD2/CARD15 is currently hypothesized.

Resequencing to Characterize Human Variation

Following the initial sequencing of the human genome has been resequencing efforts to comprehensively characterize human genomic variation. Single nucleotide polymorphisms (SNPs) are single base substitutions present abundantly throughout the human genome.[21] Most of these SNPs are di-allelic variants and are believed to represent single mutational events which have survived and expanded within a given population. As a very general rule, the higher the allelic frequency of a given SNP (e.g., a SNP may approach a 50% allelic frequency, where both alleles are equally frequent in a given population), the more evolutionarily ancient it is. In contrast, low frequency SNPs (such as the NOD2/CARD15 mutations) very generally have occurred more recently in evolutionary history and thus may be found uniquely in specific populations.[22] Clearly, important population differences exist which may provide special insight into functional variants/polymorphisms. For this reason, large-scale resequencing efforts are systematically including multiple population groups for comparative purposes.

Structural and Functional Characterization of Genetic Variation

While the complete cataloguing of all relatively common human genetic variation represents a critical foundation, organization of this mass of information is required. For disease gene identification projects, it is not practicable for all common variants to be typed in all tested individuals. Helpful in this regard is the fact that many genetic variants "travel together," or, in genetic parlance, are in linkage disequilibrium. It is typical for a series of genetic variants located within a short distance of each other to be genetically equivalent, so that, in order to capture the genetic differences at this series of variants between individuals, only one of these variants would need to be typed. Haplotype maps make use of the observation that meiotic recombination events do not occur randomly as a function of sequence distance, but rather are focused on recombinational hotspots. Between these hotspots are haplotype blocks, each of which are characterized by only a few different, relatively common haplotypes (contiguous stretches of DNA), the characterization of which would largely define the genetic variation contained within a given block.[23] The identification and characterization of the local haplotype block structure was instrumental in identifying that a region within chromosome 5q31 (IBD5) is associated with IBD.[16]

While structural organization of genetic variation will assist in identifying disease associations, proof of disease causality requires identifying which of multiple, often genetically equivalent, variants directly affect cellular and organismal function. It has been proposed that SNPs which alter amino acid composition (cSNPs) are more likely to have functional consequences[22] and the disease-associated NOD2/CARD15 variants are consistent with this model. As opposed to immune receptors of the acquired immune system where receptor variation is generated through somatic recombination events and is reinvented each generation,[24] it may be that heritable, germ-line amino acid variants of innate immune receptors represent an important means by which inter-individual differences in response to environmental (e.g., microbial) factors occur. While cSNPs may be of particular importance in innate immune receptors, it is anticipated that many of the most important risk alleles for complex multigenic disorders will involve noncoding region variation subtly affecting gene regulation and function. The characterization of which of the mass of genetic variation represents functional variants is a major, critical function for subsequent genetic initiatives. The paradigm from single gene disorders of proceeding from genetic linkage to refinement of localization and disease gene identification has not been inevitably extended to complex multigenic disorders. Therefore, the systematic, broad-scale characterization of functional human polymorphisms may well provide the key toward more systematic progress in the genetic definition of complex disorders.

Linkage Studies in IBD

In genetic linkage, families with more than one member affected by IBD are typed at genetic markers throughout the genome to identify general genomic regions shared in excess of statistical expectation between relatives harboring the same disease.[25] Compared to monogenic disorders, the magnitude of the linkage peaks for multigenic diseases is relatively modest. Accordingly, linkage findings from studies are not universally replicated from one study to another. An additional challenge with linkage studies in multigenic diseases is that the resolution of the linkage signals is relatively poor, with linkage signals being observed over broad genomic regions containing hundreds of genes. Despite these general caveats, linkage studies in IBD have been uniquely successful in identifying significant and replicated results which have been instrumental in ultimately identifying specific disease gene associations.

Table 1 lists the some of the more established genomic regions of linkage in IBD (http://www.ncbi.nlm.nih.gov:80/entrez/dispomim.cgi?id=266600). For some of the genomic regions, evidence for linkage is most significant in pure CD-CD affected relative pairs, notably the *IBD1* locus in the pericentromeric region of chromosome 16.[26] The linkage signal at *IBD1* is due at least partly to coding region polymorphisms within the NOD2/CARD15 gene.[1,2] In an Australian cohort demonstrating highly significant linkage to IBD1,[27] the three major NOD2/CARD15 risk alleles account for only a fraction of the linkage signal,[28] indicating that other noncoding region variation within the NOD2/CARD15 gene itself, or other genes in the general region may account for the linkage signal at *IBD1*. It may be that a general paradigm for complex disorders is that multiple genes/risk alleles in a general genomic region are required to produce significant and replicated linkage signals.

Table 1. Linkage and association regions in IBD

Locus	Location	Linkage	Association	Reference
IBD1	16cen	CD	NOD2/CARD15	26
IBD2	12q	UC>CD	-	31
IBD3	6p	CD and UC	HLA, TNFα	29
IBD4	14q	CD	-	75, 76
IBD5	5q	CD	cytokine cluster, CD & UC	77
IBD6	19p	CD and UC	-	77
IBD7	1p	CD and UC	-	78
IBD8	16p	CD and UC	three locus haplotype	79

For other genomic regions, such as *IBD3*[29,30] on chromosome 6p encompassing the MHC (major histocompatibility complex) locus, evidence for linkage is observed in both CD and UC. Given the enormous genetic and immunologic complexity in this broad region, it seems inevitable that multiple distinct genes and risk alleles account for the linkage signal at *IBD3*. For other regions, notably the *IBD2* locus, while the evidence for linkage has been observed in both CD and UC,[31] there is some evidence to suggest that the linkage is most significant in UC-UC affected relative pairs.[32] These linkage findings are consistent with the epidemiologic finding that cases of CD and UC are often observed within the same family, indicating that at least some IBD susceptibility genes are common to CD and UC.

The following sections review the most established disease associations for IBD that were initially implicated through genetic linkage studies, notably the NOD/CARD15 gene and *IBD1*, the cytokine 5q31 cluster (*IBD5*), and the HLA and TNFα studies for the *IBD3* locus.

Association Studies in IBD

IBD1: NOD2/CARD15 CD Associations

The NOD2/CARD15 gene is a cytosolic protein comprised of an N-terminus caspase activation recruitment domain (CARD), a central nucleotide-binding domain, and a C-terminus leucine rich repeat (LRR) domain.[33,34] Its expression was initially observed in peripheral blood monocytes, however more recent studies demonstrate expression in macrophages, dendritic cells, and intestinal epithelial cells.[35] The NOD2/CARD15 gene is highly homologous to the family of plant R gene family, which mediates host resistance to microbial pathogens.[20] In this family of proteins, the LRR domain represents the critical domain mediating response to microbial components. Similarly, the plasma membrane toll-like receptors represent a family of membrane proteins containing extracellular LRR domains which mediate responsiveness to a broad array of pathogen associated molecular patterns (PAMP) such as bacterial lipopolysaccaride, viral dsRNA, and bacterial Cpg DNA.[36] In this sense, NOD2/CARD15 represents a pattern recognition receptor of the innate immune system. It is now know that the specific microbial product which mediates activation of NOD2/CARD15 is a peptidoglycan breakdown product, muramyl dipeptide (MDP).[3,4] Treatment of NOD2/CARD15-containing cells with MDP results in oligomerization, followed by recruitment of the CARD domain-containing protein, RICK/RIP2,[37-39] which in turn activates NF-κB via IKKγ (NEMO).[33,34,40]

The NOD2/CARD15 gene is located in the center of the *IBD1* region, and coding region polymorphisms clustered near the LRR domain of NOD2/CARD15 have been reported. Three major polymorphisms, Leu1007fsinsC, Gly908Arg, Arg702Trp, are associated with CD, but not UC. Table 2 list the range of reported allele frequencies of various Caucasian cohorts. The NOD2/CARD15 CD association has been observed in a variety of European white cohorts,

Table 2. Allele frequencies among CD cohorts for the three major NOD2/CARD15 risk alleles

Population	Arg702Trp	Gly908Arg	L1007fsinsC	Reference
Europe	11%	6%	12%	2
Europe	9%	3%	7%	56
United Kingdom	12.5%	3.3%	9.4%	60
Australia	11%	2%	7%	28
WNJ North America	8.5%	5.9%	8.2%	41
Jewish, North America	2.4%	10.2%	7.1%	41
WNJ North America	13.8%	5.4%	7.8%	42
Jewish, North America	7.1%	12.5%	10.7%	42
Quebec, sporadic	12.3%	5.3%	11.7%	80
Finland, sporadic	3.3%	0.5%	3.5%	81
Ireland, sporadic	7%	3%	4%	82

WNJ, white non-Jewish

however significant allele frequency differences have been observed in specific populations. Compared to nonJewish whites, Jewish CD patients have a higher allele frequency for the Gly908Arg variant, and a lower carriage of the Arg702Trp variant.[41-43] In addition, a significantly lower allele frequency for all three variants have been observed in Finnish and Irish populations. At this time, there is no consistent trend to suggest that NOD2/CARD15 risk allele carriage is significantly higher in familial compared to sporadic CD cases, however, important population-specific distinctions may exist.

Heterozygous carriage of any of the three major risk alleles confers a two to four fold increased risk of developing CD, whereas homozygous carriage or compound heterozygous carriage (double-dose carriers) confers a twenty to forty fold increased CD risk. Therefore, these variants function in an additive, to largely recessive manner. In addition to the three major NOD2/CARD15 risk alleles, a large number of very rare amino acid polymorphisms within or near the LRR domain have been reported, which in composite represent private mutations likely associated with CD.

The low allele frequencies for the three major CD associated NOD2/CARD15 variants would suggest that these mutations might represent recent evolutionary events. Accordingly, it is not unexpected that the three major variants appear to be specific to European white populations. Specifically, the three major NOD2/CARD15 mutations have not been observed in either Japanese[43-45] or Korean.[46] CD. Similarly, the three variants are present in African-American CD,[47] but at significantly lower rates that are consistent with the 11-20% European admixture estimated for African-Americans. Whether this indicates that altered functional activity of NOD2/CARD15 does not contribute to CD in these populations, or whether different mutations within the NOD2/CARD15 gene (likely within noncoding intronic or promoter regions) may exist in these other populations has yet to be established. Of interest in this regard is that the three major NOD2/CARD15 mutations account for only half of the evidence for linkage at *IBD1* in an Australian CD cohort.[28] This would indicate either that noncoding region variants within the NOD2/CARD15 gene, or that the other risk alleles in separate genes in the general region account for the observed evidence for linkage in this general area.

Functional and Phenotypic Associations of the NOD2/CARD15 CD Variants

Because the CD-associated mutations cluster within the LRR domain required for MDP responsiveness,[1,2,48] it is not surprising all three major mutations are associated with decreased

MDP-induced NF-κB activation.[3,41,49] How decreased NF-κB activation of this small component of the intestinal immune response results in increased CD susceptibility remains to be fully defined. In contrast to the studies in humans, of note is the murine knock-in mutation of the frameshift mutation which was associated as a positive regulator of NF-κB activation.[49a] Of interest in this regard is a recent report that intestinal epithelial cells transfected with the frameshift variant of NOD2/CARD15 have a decreased capacity to kill invasive *S. typhimurium*.[50] These findings are consistent with a model where an inadequate innate immune response in turn results in a chronic inflammation characteristic of IBD.

Immunolocalization studies have demonstrated the presence of NOD2/CARD15 within Paneth cells located near the base of small intestinal crypts.[51,52] Treatment of intestinal crypt preparations with MDP, as well as LPS, lipoteichoic acid, lipid A and gram positive and negative bacteria,[53] results in the release of antimicrobial peptides which, teleologically, have been hypothesized to protect the replicative zones of small intestinal crypts from bacterial injury. That NOD2/CARD15 is highly expressed in Paneth cells of the small intestine (with metaplastic Paneth cell expression within colon observed) could account for the co-segregation of NOD2/CARD15 carriage with ileal CD (see below). Furthermore, homozygous carriers of NOD2/CARD15 normally express the protein within Paneth cells,[52] suggesting that altered function, and not defects in cellular processing, account for disease susceptibility. Interestingly, the NOD2/CARD15 knockout is associated with decreased cryptdin expression, further implicating its function in Paneth cells as central to disease pathogenesis.[52a] In addition, NOD2/CARD15 expression is increased with TNFα treatment,[35,54] raising the possibility that functional alterations from the CD mutations are most marked under inflammatory conditions. The expression of NOD2/CARD15 in monocytes and macrophages is consistent with the finding of granulomas exclusively in CD, and not UC.

In most studies, NOD2/CARD15 risk allele carriage co-segregates with ileal location and younger age of onset.[48,55-57] The identification of definitive risk alleles provides an opportunity to reclassify disease on a more molecular basis. Some studies have demonstrated the highest risk allele frequencies in ileal only disease, intermediate values in ileocolonic disease, with the lowest frequencies in colon-only CD. Disease location associations are most marked for double-dose carriers. Similarly, the trend toward younger age of onset (between two to three years) is most marked for double-dose carriers. By multivariate analysis, tobacco use, the other major risk factor for ileal disease, increases ileal risk independently from NOD2/CARD15 carriage.[57]

IBD5: Association of CD and UC with the Chromosome 5q31 Cytokine Cluster

Multiple cohorts have demonstrated an association of a 250 kB haplotype containing multiple immune-associated genes on chromosome 5q31.[15,16,54,58] Initially reported to be associated with CD, the association has now been extended to UC as well.[15] The associated haplotype is common, being observed at a haplotype frequency of 37 percent in European white populations.[16] Recently, two functional polymorphisms within the adjacent, closely related organic cation transporter genes, OCTN1/SLC22A4 and OCTN2/SLC22A5 have been identified that alter transporter activity and transcriptional activation, respectively. It cannot be established that the IBD association in this regions occurs **solely** through these functional polymorphisms. However that functional polymorphisms in these organic cation transporter genes on this risk haplotype contribute to IBD susceptibility is strongly supported by these studies.[58a] While the associated haplotype is relatively common (thus conferring a high population attributable risk) differences in haplotype frequencies between cases and controls are relatively modest. In one large European cohort, significant association for the 5q cytokine cluster was only observed in NOD2/CARD15 positive CD patients,[58] but the putative association between IBD5 and NOD2/CARD15 has not been observed in all cohorts.[15,16] Whether a genuine mechanistic association exists between NOD2/CARD15 and the risk allele(s) at IBD5, or whether the reported stratified associations merely represent a modest enrichment of the IBD5 association through elimination of phenocopies will require further studies in very large cohorts.

IBD3: MHC and Other Associations at Chromosome 6p

After the *IBD1* locus on chromosome 16, the linkage at *IBD3* [29] represents the most well-replicated linkage region. The evidence for linkage has been observed in both CD and UC affected relative pairs. Contained within this broad area of linkage is the MHC region containing a number of immune-associated genes. A number of case-control association studies of the TNFα gene have been reported with modest association reported for CD for variants in the TNFα promoter. However, given the high frequencies of the putatively associated promoter variants, these alleles cannot account for the observed evidence for linkage, indicating the presence of additional risk alleles. A number of case control studies for MHC class II variants have been reported for CD and UC. A meta-analysis of the large number of individual case control studies have demonstrated highly significant association of UC to DR2, DR9 and DRB1*0103, whereas a negative association was found for DR4. For CD, a positive association was found for DR7, DRB3*0301 and DQ4 and a negative association with DR2 and DR3.[59] Of great interest is that among NOD2/CARD15 negative CD patients there is a positive association of the DRB1*0701 allele with developing ileal CD.[60] As additional risk alleles are identified for IBD, a precise stratification of these risk alleles will be achieved, perhaps allowing for the more precise assessment of disease risk, disease course, and response to therapy. The different genetic associations observed between colon only CD and ileal CD for both NOD2/CARD15 and the HLA alleles would indicate that they are pathophysiologically distinct entities.

Future Directions

Using Genetic Information to Guide Patient Management

It is not established at this time that genotyping of the NOD2/CARD15 or any other risk alleles by themselves is of clinical value. Because NOD2/CARD15 risk alleles are not associated with colon only CD disease, their use would not be of value in classifying those cases of indeterminate colitis. Rather, serologic tests (pANCA, ASCA) have been utilized to characterize cases of inflammation confined to the colon. Some, but not all studies, have observed an association of NOD2/CARD15 risk allele carriage with fibrostenosing complications. However, the certainty of these correlations is somewhat limited by the retrospective studies reported thus far. In particular, fibrostenotic and fistulizing complications of CD accumulate through time, and the fraction of CD complications must be corrected for disease duration.[61] Furthermore, different criteria for defining fibrostenosing and fistulizing complications likely account for the wide range of estimates reported for these complications. Of note is that one study of NOD2/CARD15 risk allele carriers did not observe an increased rate of reoperation following the initial resectional surgery compared to CD NOD2/CARD15 noncarriers.[60] A more definitive assessment of the clinical utility of the NOD2/CARD15 variants awaits further prospective studies. If it can be established that particular genetic variants are associated with an increased or earlier development of CD complications, the earlier institution of more aggressive therapies would then need to be studied.

Apart from utilizing genetic testing to broadly characterize disease prognosis and severity, a separate future goal of genetic testing in diagnosed IBD patients would be to predict response to specific therapies. It is possible that the genetic heterogeneity of IBD (e.g., not all CD patients harbor known NOD2/CARD15 risk alleles) reflects pathogenetic heterogeneity potentially responding differently to various therapies. However, the majority of CD patients responding to anti-TNF therapies do not have a significantly different rate of NOD2/CARD15 carriage compared to patients not responding to these agents.[62,63] However, given the highly pleiotropic functions of TNFα, combined with the sufficiency of excess TNFα alone to produce ileitis in murine models,[64] it is quite possible that a genetic basis for response/nonresponse to anti-TNF therapies does not exist. Finally, genetic variation in genes affecting drug pharmacokinetics is increasingly being defined (pharmacogenetics). For example, the

MDR1 (multidrug resistance) gene codes for a membrane protein mediating transport of lipophilic agents, in some cases resulting in chemotherapeutic resistance.[65] Genetic variants within MDR1 have been reported which alter either functional activity[66] or expression levels,[67] and which may therefore alter therapeutic response to drugs transported by this protein. Of particular relevance with respect to IBD is that mdr1-deficient mice spontaneously develop colitis,[68] and therefore human genetic variants within MDR1 may both contribute to IBD susceptibility as well as affect response to specific therapeutic agents, such as corticosteroids[69-71]

Utilizing Existing Disease Associations to Identify Additional Disease Genes, Establish Additional IBD Models and Develop New Therapies

A classic approach in genetics to identify additional disease genes once initial associations are established is to examine close homologs and other members of affected cellular signaling pathways. The closest homolog of NOD2/CARD15 is NOD1/CARD4, which is expressed in epithelial cells.[33,34] While no coding region variants within NOD1/CARD4 have been associated with IBD,[72] it is still possible that noncoding variants may contribute to disease. While NOD2/CARD15 responds to MDP treatment, NOD1/CARD4 signals in response to a diaminopimelate-containing muramyl tripeptide (MTP), a component of gram-negative bacterial cell wall peptidoglycan.[73,74] Why NOD2/CARD15 is associated with CD and NOD1/CARD4 is not, may reflect differences in tissue expression, important functional distinctions between MDP and MTP, or may merely reflect the stochastic nature of evolutionary mutational events. That NOD2/CARD15 rather specifically and uniquely responds to MDP would suggest that other genes affecting cellular and organismal response to MDP and/or peptidoglycan may include additional CD-susceptibility genes. Given the generally accepted importance of gene-environment interactions in complex, multigenic disorders, we speculate that the most proximate members of functional pathways are most likely to be associated with disease risk, as opposed to the more critical, downstream members common to numerous signaling paths.

While murine models of IBD have been indispensable thus far in defining mechanisms of disease pathogenesis and suggesting candidate genes for human disease, the importance of such in vivo models will likely increase as the complex mosaic of human disease associations is increasingly defined. Given the insufficiency of any single risk allele to produce disease, the identification of gene-gene associations which synergistically increase disease risk represents a major, additional goal of IBD genetics. If defined, such multigene associations can be replicated in murine crosses to more precisely define pathogenesis and test potential new therapies.

References

1. Ogura Y, Bonen DK, Inohara N et al. A frameshift mutation in NOD2 associated with susceptibility to Crohn's disease. Nature 2001; 411:603-6.
2. Hugot JP, Chamaillard M, Zouali H et al. Association of NOD2 leucine-rich repeat variants with susceptibility to Crohn's disease. Nature 2001; 411:599-603.
3. Inohara N, Ogura Y, Fontalba A et al. Host recognition of bacterial muramyl dipeptide mediated through NOD2. Implications for Crohn's disease. J Biol Chem 2003; 278:5509-12.
4. Girardin SE, Boneca IG, Viala J et al. Nod2 is a general sensor of peptidoglycan through muramyl dipeptide (MDP) detection. J Biol Chem 2003.
5. Strober W, Fuss IJ, Blumberg RS. The immunology of mucosal models of inflammation. Annu Rev Immunol 2002; 20:495-549.
6. Loftus Jr EV, Sandborn WJ. Epidemiology of inflammatory bowel disease. Gastroenterol Clin North Am 2002; 31:1-20.
7. Kurata JH, Kantor-Fish S, Frankl H et al. Crohn's disease among ethnic groups in a large health maintenance organization. Gastroenterology 1992; 102:1940-8.
8. Ogunbi SO, Ransom JA, Sullivan K et al. Inflammatory bowel disease in African-American children living in Georgia. J Pediatr 1998; 133:103-7.
9. Farmer RG, Michener WM, Mortimer EA. Studies of family history among patients with inflammatory bowel disease. Clin Gastroenterol 1980; 9:271-7.
10. Monsen U, Bernell O, Johansson C et al. Prevalence of inflammatory bowel disease among relatives of patients with Crohn's disease. Scand J Gastroenterol 1991; 26:302-6.

11. Monsen U, Brostrom O, Nordenvall B et al. Prevalence of inflammatory bowel disease among relatives of patients with ulcerative colitis. Scand J Gastroenterol 1987; 22:214-8.
12. Russel MG, Pastoor CJ, Janssen KM et al. Familial aggregation of inflammatory bowel disease: A population-based study in South Limburg, The Netherlands. The South Limburg IBD Study Group. Scand J Gastroenterol Suppl 1997; 223:88-91.
13. Orholm M, Munkholm P, Langholz E et al. Familial occurrence of inflammatory bowel disease. N Engl J Med 1991; 324:84-8.
14. Peeters M, Nevens H, Baert F et al. Familial aggregation in Crohn's disease: Increased age-adjusted risk and concordance in clinical characteristics. Gastroenterology 1996; 111:597-603.
15. Giallourakis C, Stoll M, Miller K et al. IBD5 is a general risk factor for inflammatory bowel disease: Replication of association with Crohn disease and identification of a novel association with ulcerative colitis. Am J Hum Genet 2003; 73:205-11.
16. Rioux JD, Daly MJ, Silverberg MS et al. Genetic variation in the 5q31 cytokine gene cluster confers susceptibility to Crohn disease. Nat Genet 2001; 29:223-8.
17. Subhani JMS, Ounder RE, Wakefield AJ. Concordance rates oftwins and siblings in inflammatory bowel disease. Gut 1998; 42(Suppl 1):A40.
18. Thompson NP, Driscoll R, Pounder RE et al. Genetics versus environment in inflammatory bowel disease: Results of a British twin study. Bmj 1996; 312:95-6.
19. Tysk C, Lindberg E, Jarnerot G et al. Ulcerative colitis and Crohn's disease in an unselected population of monozygotic and dizygotic twins. A study of heritability and the influence of smoking. Gut 1988; 29:990-6.
20. Inohara N, Nunez G. NODs: Intracellular proteins involved in inflammation and apoptosis. Nat Rev Immunol 2003; 3:371-82.
21. Sachidanandam R, Weissman D, Schmidt SC et al. A map of human genome sequence variation containing 1.42 million single nucleotide polymorphisms. Nature 2001; 409:928-33.
22. Chakravarti A. Population genetics—making sense out of sequence. Nat Genet 1999; 21:56-60.
23. Daly MJ, Rioux JD, Schaffner SF et al. High-resolution haplotype structure in the human genome. Nat Genet 2001; 29:229-32.
24. Medzhitov R, Janeway Jr C. Innate immunity. N Engl J Med 2000; 343:338-44.
25. Lander ES, Schork NJ. Genetic dissection of complex traits. Science 1994; 265:2037-48.
26. Hugot JP, Laurent-Puig P, Gower-Rousseau C et al. Mapping of a susceptibility locus for Crohn's disease on chromosome 16. Nature 1996; 379:821-3.
27. Cavanaugh JA, Callen DF, Wilson SR et al. Analysis of Australian Crohn's disease pedigrees refines the localization for susceptibility to inflammatory bowel disease on chromosome 16. Ann Hum Genet 1998; 62:291-8.
28. Cavanaugh JA, Adams KE, Quak EJ et al. CARD15/NOD2 risk alleles in the development of Crohn's disease in the Australian population. Ann Hum Genet 2003; 67:35-41.
29. Hampe J, Schreiber S, Shaw SH et al. A genomewide analysis provides evidence for novel linkages in inflammatory bowel disease in a large European cohort. Am J Hum Genet 1999; 64:808-16.
30. Hampe J, Shaw SH, Saiz R et al. Linkage of inflammatory bowel disease to human chromosome 6p. Am J Hum Genet 1999; 65:1647-55.
31. Satsangi J, Parkes M, Louis E et al. Two stage genome-wide search in inflammatory bowel disease provides evidence for susceptibility loci on chromosomes 3, 7 and 12. Nat Genet 1996; 14:199-202.
32. Parkes M, Barmada MM, Satsangi J et al. The IBD2 locus shows linkage heterogeneity between ulcerative colitis and Crohn disease. Am J Hum Genet 2000; 67:1605-10.
33. Ogura Y, Inohara N, Benito A et al. Nod2, a Nod1/Apaf-1 family member that is restricted to monocytes and activates NF-kappaB. J Biol Chem 2001; 276:4812-8.
34. Inohara N, Koseki T, del Peso L et al. Nod1, an Apaf-1-like activator of caspase-9 and nuclear factor-kappaB. J Biol Chem 1999; 274:14560-7.
35. Gutierrez O, Pipaon C, Inohara N et al. Induction of Nod2 in myelomonocytic and intestinal epithelial cells via nuclear factor-kappa B activation. J Biol Chem 2002; 277:41701-5.
36. Janeway Jr CA, Medzhitov R. Innate immune recognition. Annu Rev Immunol 2002; 20:197-216.
37. Thome M, Hofmann K, Burns K et al. Identification of CARDIAK, a RIP-like kinase that associates with caspase-1. Curr Biol 1998; 8:885-8.
38. Inohara N, del Peso L, Koseki T et al. RICK, a novel protein kinase containing a caspase recruitment domain, interacts with CLARP and regulates CD95-mediated apoptosis. J Biol Chem 1998; 273:12296-300.
39. McCarthy JV, Ni J, Dixit VM. RIP2 is a novel NF-kappaB-activating and cell death-inducing kinase. J Biol Chem 1998; 273:16968-75.

40. Inohara N, Koseki T, Lin J et al. An induced proximity model for NF-kappa B activation in the Nod1/RICK and RIP signaling pathways. J Biol Chem 2000; 275:27823-31.
41. Bonen DK, Ogura Y, Nicolae DL et al. Crohn's disease-associated NOD2 variants share a signaling defect in response to lipopolysaccharide and peptidoglycan. Gastroenterology 2003; 124:140-6.
42. Sugimura K, Taylor KD, Lin YC et al. A novel NOD2/CARD15 haplotype conferring risk for crohn disease in Ashkenazi Jews. Am J Hum Genet 2003; 72:509-18.
43. Zhou Z, Lin XY, Akolkar PN et al. Variation at NOD2/CARD15 in familial and sporadic cases of Crohn's disease in the Ashkenazi Jewish population. Am J Gastroenterol 2002; 97:3095-101.
44. Yamazaki K, Takazoe M, Tanaka T et al. Absence of mutation in the NOD2/CARD15 gene among 483 Japanese patients with Crohn's disease. J Hum Genet 2002; 47:469-72.
45. Inoue N, Tamura K, Kinouchi Y et al. Lack of common NOD2 variants in Japanese patients with Crohn's disease. Gastroenterology 2002; 123:86-91.
46. Croucher PJ, Mascheretti S, Hampe J et al. Haplotype structure and association to Crohn's disease of CARD15 mutations in two ethnically divergent populations. Eur J Hum Genet 2003; 11:6-16.
47. Bonen DK, Nicolae DL, Moran T et al. Racial differences in NOD2 variation: Characterization of NOD2 in african-americans with Crohn's disease [Abstract 248]. Gastroenterology 2002; 122(Suppl):A-29.
48. Lesage S, Zouali H, Cezard JP et al. CARD15/NOD2 mutational analysis and genotype-phenotype correlation in 612 patients with inflammatory bowel disease. Am J Hum Genet 2002; 70:845-57.
49. Chamaillard M, Philpott D, Girardin SE et al. Gene-environment interaction modulated by allelic heterogeneity in inflammatory diseases. Proc Natl Acad Sci USA 2003; 100:3455-60.
49a. Maeda S, Hsu LC, Liu H et al. Nod2 mutation in Crohn's disease potentiates NF-kappaB activity and IL-1beta processing. Science 2005; 4:734-8.
50. Hisamatsu T, Suzuki M, Reinecker HC et al. CARD15/NOD2 functions as an anti-bacterial factor in human intestinal epithelial cells. Gastroenterology 2003.
51. Lala S, Ogura Y, Osborne C et al. Crohn's disease and the NOD2 gene: A role for paneth cells. Gastroenterology 2003; 125:47-57.
52. Ogura Y, Xin W, Smith E et al. Expression of NOD2 in paneth cells: A possible link to Crohn's Ileitis. Gut, In press.
52a. Kobayashi KS, Chamaillard M, Ogura Y et al. Nod2-dependent regulation of innate and adaptive immunity in the intestinal tract. Science 2005; 4:731-4.
53. Ayabe T, Satchell DP, Wilson CL et al. Secretion of microbicidal alpha-defensins by intestinal Paneth cells in response to bacteria. Nat Immunol 2000; 1:113-8.
54. Rosenstiel P, Fantini M, Brautigam K et al. TNF-alpha and IFN-gamma regulate the expression of the NOD2 (CARD15) gene in human intestinal epithelial cells. Gastroenterology 2003; 124:1001-9.
55. Ahmad T, Satsangi J, McGovern D et al. Review article: The genetics of inflammatory bowel disease. Aliment Pharmacol Ther 2001; 15:731-48.
56. Cuthbert AP, Fisher SA, Mirza MM et al. The contribution of NOD2 gene mutations to the risk and site of disease in inflammatory bowel disease. Gastroenterology 2002; 122:867-74.
57. Brant SR, Picco MF, Achkar JP et al. Crohn's disease: Role of NOD2/CARD15 gene mutations in clinical heterogeneity. Am J Hum Genet 2002; (Abstract) In presss.
58. Mirza MM, Fisher SA, King K et al. Genetic evidence for interaction of the 5q31 cytokine locus and the CARD15 gene in Crohn disease. Am J Hum Genet 2003; 72:1018-22.
58a. Peltekova VD, Wintle RF, Rubin LA et al. Functional variants of OCTN cation transporter genes are associated with Crohn disease. Nat Genet 2004; 36:471-5.
59. Stokkers PC, Reitsma PH, Tytgat GN et al. HLA-DR and -DQ phenotypes in inflammatory bowel disease: A meta- analysis. Gut 1999; 45:395-401.
60. Ahmad T, Armuzzi A, Bunce M et al. The molecular classification of the clinical manifestations of Crohn's disease. Gastroenterology 2002; 122:854-66.
61. Louis E, Michel V, Hugot JP et al. Early development of stricturing or penetrating pattern in Crohn's disease is influenced by disease location, number of flares, and smoking but not by NOD2/CARD15 genotype. Gut 2003; 52:552-7.
62. Vermeire S, Louis E, Rutgeerts P et al. NOD2/CARD15 does not influence response to infliximab in Crohn's disease. Gastroenterology 2002; 123:106-11.
63. Mascheretti S, Hampe J, Croucher PJ et al. Response to infliximab treatment in Crohn's disease is not associated with mutations in the CARD15 (NOD2) gene: An analysis in 534 patients from two multicenter, prospective GCP-level trials. Pharmacogenetics 2002; 12:509-15.

64. Kontoyiannis D, Pasparakis M, Pizarro TT et al. Impaired on/off regulation of TNF biosynthesis in mice lacking TNF AU- rich elements: Implications for joint and gut-associated immunopathologies. Immunity 1999; 10:387-98.
65. Choi KH, Chen CJ, Kriegler M et al. An altered pattern of cross-resistance in multidrug-resistant human cells results from spontaneous mutations in the mdr1 (P-glycoprotein) gene. Cell 1988; 53:519-29.
66. Kim RB, Leake BF, Choo EF et al. Identification of functionally variant MDR1 alleles among European Americans and African Americans. Clin Pharmacol Ther 2001; 70:189-99.
67. Hoffmeyer S, Burk O, von Richter O et al. Functional polymorphisms of the human multidrug-resistance gene: Multiple sequence variations and correlation of one allele with P-glycoprotein expression and activity in vivo. Proc Natl Acad Sci USA 2000; 97:3473-8.
68. Panwala CM, Jones JC, Viney JL. A novel model of inflammatory bowel disease: Mice deficient for the multiple drug resistance gene, mdr1a, spontaneously develop colitis. J Immunol 1998; 161:5733-44.
69. Schinkel AH, Wagenaar E, van Deemter L et al. Absence of the mdr1a P-Glycoprotein in mice affects tissue distribution and pharmacokinetics of dexamethasone, digoxin, and cyclosporin A. J Clin Invest 1995; 96:1698-705.
70. Lown KS, Mayo RR, Leichtman AB et al. Role of intestinal P-glycoprotein (mdr1) in interpatient variation in the oral bioavailability of cyclosporine. Clin Pharmacol Ther 1997; 62:248-60.
71. Farrell RJ, Murphy A, Long A et al. High multidrug resistance (P-glycoprotein 170) expression in inflammatory bowel disease patients who fail medical therapy. Gastroenterology 2000; 118:279-88.
72. Zouali H, Lesage S, Merlin F et al. CARD4/NOD1 is not involved in inflammatory bowel disease. Gut 2003; 52:71-4.
73. Girardin SE, Boneca IG, Carneiro LA et al. Nod1 detects a unique muropeptide from gram-negative bacterial peptidoglycan. Science 2003; 300:1584-7.
74. Chamaillard M, Hashimoto M, Horie Y et al. An essential role for NOD1 in host recognition of bacterial peptidoglycan containing diaminopimelic acid. Nat Immunol 2003; 4:702-7.
75. Duerr RH, Barmada MM, Zhang L et al. High-density genome scan in Crohn disease shows confirmed linkage to chromosome 14q11-12. Am J Hum Genet 2000; 66:1857-62.
76. Ma Y, Ohmen JD, Li Z et al. A genome-wide search identifies potential new susceptibility loci for Crohn's disease. Inflamm Bowel Dis 1999; 5:271-8.
77. Rioux JD, Silverberg MS, Daly MJ et al. Genomewide search in Canadian families with inflammatory bowel disease reveals two novel susceptibility loci. Am J Hum Genet 2000; 66:1863-70.
78. Cho JH, Nicolae DL, Gold LH et al. Identification of novel susceptibility loci for inflammatory bowel disease on chromosomes 1p, 3q, and 4q: Evidence for epistasis between 1p and IBD1. Proc Natl Acad Sci USA 1998; 95:7502-7.
79. Hampe J, Frenzel H, Mirza MM et al. Evidence for a NOD2-independent susceptibility locus for inflammatory bowel disease on chromosome 16p. Proc Natl Acad Sci USA 2002; 99:321-6.
80. Vermeire S, Wild G, Kocher K et al. CARD15 genetic variation in a Quebec population: Prevalence, genotype- phenotype relationship, and haplotype structure. Am J Hum Genet 2002; 71:74-83.
81. Helio T, Halme L, Lappalainen M et al. CARD15/NOD2 gene variants are associated with familially occurring and complicated forms of Crohn's disease. Gut 2003; 52:558-62.
82. Bairead E, Harmon DL, Curtis AM et al. Association of NOD2 with Crohn's disease in a homogenous Irish population. Eur J Hum Genet 2003; 11:237-44.

Chapter 4

Microbial and Dietary Factors in the Pathogenesis of Chronic, Immune-Mediated Intestinal Inflammation

R. Balfour Sartor

Introduction

Although genetic background is an important prerequisite for Crohn's disease, ulcerative colitis and experiment intestinal inflammation, yet to be identified environmental factors profoundly influence genetic susceptibility. The influence of the environment is documented by the relatively low concordance of disease in identical twins, rapid changes in the incidence of idiopathic Inflammatory bowel diseases (IBD) in a given population within 3-4 decades, alterations in disease frequency following migration of a population to a new environment, and striking differences in aggressiveness and phenotype of experimental ileocolitis when a susceptible rodent colony is moved to a new environment or when their housing conditions are changed.[1] Although environmental influences are complex, commensal enteric bacteria, episodic or persistent infections and the diet are the most obvious environmental factors capable of modulating genetic susceptibility to chronic immune-mediated intestinal inflammation.

The distal ileum and colon, which are the preferential sites of human IBD and experimental intestinal inflammation in susceptible rodents, are colonized with an incredibly complex mixture of predominantly anaerobic bacteria as well as commensal fungi[2] (Table 1). Moreover, we are all intermittently exposed to self-limited enteric and systemic infections that can transiently induce intestinal inflammation, break the mucosal barrier, alter the balance of pro and anti-inflammatory cytokines and activate effector innate and cognate immune cells. In certain settings, persistent enteric pathogens can influence chronic gastrointestinal inflammation, as documented by the influence of *Helicobacter pylori* in peptic ulcer disease. To further increase the complexity of the intestinal microenvironment, certain subsets of bacteria preferentially adhere to the intestinal epithelial cells, others integrate into the mucus layers, thereby constituting a biofilm, while yet others appear to solely colonize the lumen.[2-4] Moreover, certain bacterial species selectively colonize different intestinal regions. Thus, unfortunately, stool samples do not accurately reflect the mucosally associated organisms that most likely provide the adjuvants and antigens responsible for activating mucosal immune responses.

This chapter explores evidence supporting the hypothesis that luminal commensal bacteria, and perhaps dietary components, provide the dominant adjuvants and antigens that induce pathogenic mucosal immune responses culminating in IBD in genetically susceptible hosts, but stimulate protective innate and acquired immune responses in normal hosts (Fig. 1). Environmental triggers (infections, dietary toxins and nonsteroidal antiinflammatory drugs [NSAIDs]), initiate and reactivate this inflammatory response with variable outcomes in hosts with different genetically programmed immune responses. Considerable evidence derived from experimental models and human IBD supports this hypothesis (Table 2). We stress recent

Immune Mechanisms in Inflammatory Bowel Disease, edited by Richard S. Blumberg and Markus F. Neurath.

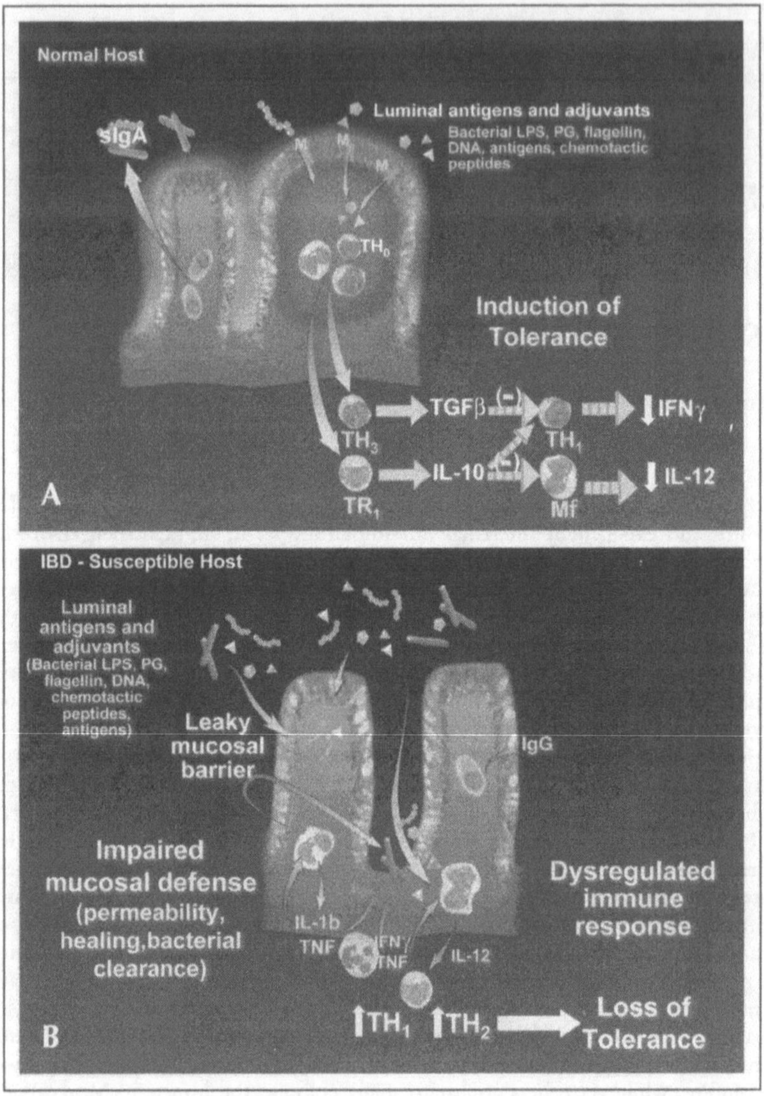

Figure 1. Induction of tolerogenic or pathologic immune responses by commensal enteric bacteria in genetically resistant or susceptible hosts. A) Initiation of protective immune responses in normal hosts. A complex mix of dominantly anaerobic bacteria populate the distal ileum and colon. Controlled uptake of luminal bacterial adjuvants and antigens across the specialized epithelial cells in organized lymphoid aggregates activate naïve antigen presenting cells (APC), especially dendritic cells (DC). APC process and present antigen to naïve T cells. Under the influence of secreted IL-10, these bacterially-activated regulatory cells differentiate, mature and expand. Regulatory T cells suppress pathogenic TH1 and TH2 responses to commensal bacteria secreting IL-10 and TGFβ, and possibly via membrane-bound TGFβ. B) Initiation of Inflammation. Enhanced mucosal uptake of luminal commensal bacteria adjuvants and antigens due to an acquired (transient infection or ingested toxin) or intrinsic (host genetic) defect in mucosal barrier function activates pathogenic resident innate immune responses in genetically susceptible hosts. Genetically programmed immunoregulatory defects or continued antigenic stimulation (lack of mucosal healing or clearance of invading bacteria) lead to pathogenic (TH1 or TH2) immunologic responses to commensal bacterial antigens (loss of tolerance). From AGA clinical teaching slide set 2003, with permission.

Table 1. Composition and concentration of commensal luminal bacteria in the gastrointestinal tract of normal humans

Region	Concentration	Composition (dominant organisms)
Stomach	0-10^2	*Lactobacillus, Candida, Streptococcus, Helicobacter pylori, Peptostreptococcus*
Duodenum	10^2	*Streptococcus, Lactobacillus*
Jejunum	10^2	*Streptococcus, Lactobacillus*
Proximal ileum	10^3	*Streptococcus, Lactobacillus*
Distal ileum	10^5-10^6	*Clostridium, Bacteroides sp.*, coliforms
Colon	10^{11}	*Bacteroides, Bifidobacterium, Clostridium coccoides, Clostridium lepium/Fusobacterium*

From reference 3, with copyright permission from Elsevier.

Table 2. Evidence that commensal enteric bacteria induce experimental and human chronic intestinal inflammation

	Experimental Enterocolitis	Crohn's Disease	UC	Pouchitis
Inflammation in areas of ↑ bacteria	Colon, TI	TI, colon	Colon	Ileal pouch
Disease in sterile environment	No	?	?	?
↑ mucosal uptake of bacterial products	Yes	Yes	Yes	Yes
Response to anaerobic antibiotics	Yes	Yes	No	Yes
Response to aerobic antibiotics	Yes	Yes	No	Yes
↓ inflammation with bypass	Yes	Yes	No	Yes
Protection with probiotics	Yes	?	Yes	Yes
Immune response to bacteria	Yes	Yes	Yes	?
Exacerbation by pathogens	Yes	Yes	Yes	?
Implicated commensal species	*Bacteroides vulgatus* *E. coli* *Enterococcus faecium* *Helicobacter hepaticus* *Klebsiella pneumoniae*	*E. coli* *Bacteroides* *Enterococci* *Eubacteria* *Peptostreptococcus* *Fusobacterium*	*E. coli*	Aerobic and anaerobic species

Modified from reference 3, with copyright permission from Elsevier.

observations in rodent models of chronic immune-mediated intestinal inflammation, where mechanisms have been more thoroughly evaluated and where individual components of the complex luminal microecology can be more precisely manipulated. The reader is referred to recent reviews with a more extensive list of older references.[3;5-10]

Evidence of Bacterial Induction and Maintenance of Intestinal Inflammation

Experimental Models

Requirement for Commensal Bacteria

Normal commensal bacteria are required in every model of chronic, T cell-mediated enterocolitis studied to date (Table 3). In these models, disease is either absent or markedly attenuated (IL-2-/- and Samp-1/Yit mice) in the germ-free (sterile) state.[11-17] Moreover, the degree of intestinal inflammation depends on the luminal concentration of commensal bacteria and opportunistic pathogens. For example, IL-10-/- mice develop aggressive, lethal enterocolitis when housed in conventional rodent facilities, nonlethal colitis in specific pathogen free (SPF) conditions, attenuated disease when treated with broad spectrum antibiotics, and no inflammation or immune activation in a sterile environment.[11,18,19] Similarly, HLA B27 transgenic rats have attenuated colitis with broad spectrum antibiotics[20] or when the cecum is bypassed, thereby decreasing total and anaerobic bacterial luminal concentrations, but exhibit more aggressive colitis when bacterial overgrowth is induced by a self-filling cecal blind loop.[21] In keeping with this concept, IL-10-/- mice on a resistant C57BL/6 background develop TH1-mediated colitis when colonized with *Helicobacter hepaticus*,[22] although this opportunistic pathogen does not influence disease in the susceptible 129 SvEv background.[23] Genetic background is a key feature in interpreting results in experimental models, since genetically determined susceptibility governs the immunologic response to bacterial stimuli. In all of these models, wild type mice and rats do not develop enterocolitis due to immunologic tolerance to the same enteric commensal organisms that induce disease in congenic strains with defective immunoregulation or epithelial repair mechanisms. In addition to inducing colitis, chronic luminal bacterial antigenic stimulation is necessary to perpetuate disease, since activated bacterial responsive T cells from mice with colitis fail to induce inflammation when adoptively transferred to germ-free T cell deficient recipients.[16] Similarly, *H. pylori*-reactive CD4+ T cells fail to induce colitis in Rag 2-/- recipients not colonized with this organism[24] and ovalbumin (OVA)-reactive T cells do not induce colitis unless recipient T cell-deficient mice are colonized with *E. coli* engineered to produce OVA.[25] Although specific T cell responses to luminal bacteria have been well documented (see below), luminal bacteria are also important in migration and retention of T cells in the colon.[26]

Similar results are found in induced models of more acute, innate immune-mediated intestinal inflammation (Table 3). In almost all models, disease is attenuated or absent in germ-free rodents, prevented by antibiotics and markedly decreased by bypass.[20,27-29] The exception to this rule is variable reports of colitis in germ-free mice treated with dextran sodium sulfate (DSS).[30,31] It is possible that bacterial metabolism of DSS diminishes its epithelial toxicity.

Selective Induction of Colitis by Defined Bacterial Species

Results from gnotobiotic rodents and from selective colonization of knockout mice with opportunistic pathogens demonstrate that enteric bacteria differ in their ability to cause intestinal inflammation. Some bacterial species can selectively induce colitis, other commensal bacteria are neutral and an additional subgroup can prevent inflammation (Table 4). Of considerable potential clinical importance, each genetically distinct host appears to have a unique group of bacterial species that induce disease, and even within a single host, different bacterial species can cause different phenotypes of colitis. For example, *Bacteroides vulgatus* selectively causes colitis (but not gastritis) in monoassociated HLA B27 transgenic rats and

Table 3. The normal luminal bacterial environment influences intestinal inflammation in animal models

Model	Species	SPF	Germ-Free	Antibiotics	Intestinal Bypass
A. Induced disease					
Indomethacin	Rat	Acute SB, colonic and gastric ulcers, chronic SB ulcers	Attenuated acute, absent chronic	↓ by metronidazole, tetracycline	ND
Carrageenan	Guinea pig	Cecal inflammation	No colitis	↓ by metronidazole, clindamycin	ND
DSS	Mouse	Colitis	↑ or ↓ colitis	↓ by cipro, imipenum/ vancomycin (acute phase)	ND
TNBS	Rat	Colitis	ND	↓ by amoxicillin/ clavulanic acid	ND
B. Genetically engineered					
HLA-B_{27} transgenic	Rat	Gastritis, colitis, arthritis	No GI or joint inflammation	↓ by metronidazole, cipro, vancomycin/ imipenum	↓ bypass, ↑ blind loop
CD45 RB^{hi} → SCID	Mouse	Colitis	No colitis*	↓ by streptomycin and bacitracin	ND
IL-2-/-	Mouse	Colitis, gastritis, hepatitis	Attenuated inflammation	ND	ND
IL-10-/-	Mouse	Colitis	No colitis	↓ by metronidazole, cipro, vancomycin/ imipenum	ND
TCRα-/-	Mouse	Colitis	No colitis	ND	↓ resection cecal tip
mdr1α-/-	Mouse	Colitis	ND	↓ by streptomycin neomycin, bacitracin, amphotericin	ND
BM → CD3ε transgenic	Mouse	Colitis	No colitis	ND	ND
C. Spontaneous mutations					
Samp 1/Yit	Mouse	Ileitis	Attenuated ileitis	↓ metronidazole/ cipro	ND
Cotton top Tamarin	Marmoset	Colitis	ND	ND	↓ Thiry-Vella loop

SB, small bowel; GI, gastrointestinal; ND, not done; ↓, attenuated inflammation; ↑, potentiated disease; BM, bone marrow transplant. * Simplified microflora, not germ-free (sterile). Modified from reference 3, with copyright permission from Elsevier.

Table 4. Commensal enteric bacteria selectively influence experimental colitis

Model	Bacterial Species Inducing Disease	No Response	Protection
A. Mouse			
IL-10-/- (129 SvEv)	*E. coli, Enterococcus faecalis, Klebsiella pneumoniae, Bifidobacterium animalis*	*B. vulgatus, C. sordelii, Streptococcus viridans, Candida albicans*	*Lactobacillus plantarum, L. reuteri,* VSL 3
IL-10-/- (C57 BL/6)	*Helicobacter hepaticus*	–	–
IL-2-/-	*E. coli*	–	*B. vulgatus*
Mdr-1-/-	*Helicobacter bilis*	–	*H. hepaticus*
TCRα-/-	*B. vulgatus*	–	–
CD45RB^{high}→SCID	*H. muridarum*	*Filamentous bacterium, Ochrobactrum anthropi, Morganella morganii*	
BM → CD3εTg	–	*E. coli, B. vulgatus, E. faecalis*	–
B. Rat			
HLA B27 Tg	*B. vulgatus, Bacteroides thetaiotamicron, Bacteroides fragilis*	*E. coli, E. faecalis*	*Lactobacillus GG*
C. Guinea pig			
Carageenan-fed	*B. vulgatus*	Other commensals	–

Tg, transgenic; BM, bone marrow transplant; VSL-3, combination of 8 probiotic species.

guinea pigs fed carrageenan,[27,32] but does not cause disease in IL-10-/- mice or CD3ε transgenic mice (S. Kim and R.B. Sartor, unpublished data) and prevents colitis in IL-2 knockout mice coassociated with *E. coli.*[33] *B. vulgatus* has also been implicated in colitis in TLRα-/- mice.[34] Similarly, *E. coli* can cause immune-mediated colitis in monoassociated IL-2-/- and IL-10-/- mice, but not in HLA B27 transgenic rats or CD3ε transgenic mice.[32,33] Importantly, these commensal bacterial species are not pathogens, since they do not cause inflammation in wild type hosts. However, enteric pathogens can cause TH1-mediated colitis in normal mice. Higgins et al[35] demonstrated that *Citrobacter freundii* could induce hyperplastic colitis in normal mice and that colitis was due to intimin. In subsequent studies, rectal administration of purified *C. freundii* intimin induced TH1-mediated colitis.[36] Bacterial selectivity extends even within a bacterial genus, since *B. vulgatus* and *Bacteroides thetaiotamicron* but not *Bacteroides distasonis* cause colitis in monoassociated HLA B27 transgenic rats.[32,37] *Helicobacter bilis* but not *H. hepaticus* accelerates colitis in mdr-1-/- mice[38] and murine *E. coli* but not *E. coli* Nissle produces colitis in IL-2-/- mice.[33] The ability of different bacteria to cause different phenotypes of disease in the same host is demonstrated by early onset cecal-dominated inflammation in IL-10-/- mice monoassociated with *E. coli* and slow onset distal colitis in *E. faecalis*-monoassociated IL-10-/- mice.[39] Of importance, colonization with a single bacterial species causes less aggressive disease than that induced in the same host by more complex bacterial populations, suggesting additive or synergistic roles for various commensal species. This concept is supported by more active colitis with a more rapid onset in IL-10-/- mice dually associated with *E. coli* and *E. faecalis.*[40]

These observations have important clinical implications from both diagnostic and therapeutic perspectives. Serologic or T cell responses to individual bacterial species or groups of commensal bacteria could identify clinically important subsets of IBD patients who selectively respond to various antibiotic, probiotic or prebiotic treatment or who have a predictable disease phenotype or natural history (see below). Therapeutic approaches to altering the balance of beneficial and aggressive commensal bacteria with antibiotics, probiotics and prebiotics show promise in animal models of enterocolitis. This topic is covered in detail in Chapter 14.

Alterations of Commensal Bacterial Composition with Colitis

One theory of IBD pathogenesis is that environmental or genetic influences alter the commensal bacterial ecology to suppress beneficial bacteria and foster growth of more aggressive species.[3] Using molecular techniques, Wilson et al demonstrated that *E. coli*, *Klebsiella* and *Clostridium ramosum* selectively expand, while 2 dominant Clostridial groups contract, in ex germ-free IL-10-/- mice colonized with SPF fecal bacteria, compared with wild type mice colonized with identical fecal material.[41] Colonization of gnotobiotic IL-10-/- mice with the 3 expanded bacterial species caused colitis, but the intensity of inflammation was not more than that in *E. coli*-monoassociated IL-10-/- mice (J. Tsang and R.B. Sartor, unpublished observations).

Human IBD

Commensal Bacteria

Considerable clinical evidence supports an etiologic role for commensal enteric bacteria in the pathogenesis of Crohn's disease and pouchitis, but a less obvious influence in ulcerative colitis (Table 2). Disease occurs in areas of highest luminal concentrations of coliform and anaerobic bacteria. Both pouchitis and colonic Crohn's disease respond to metronidazole, ciprofloxacin or broad spectrum antibiotics, although these agents are not effective in the majority of cases of ileal Crohn's disease and ulcerative colitis (summarized in Chapter 14).[42] However, probiotics appear to be more effective in preventing recurrence of quiescent ulcerative colitis and pouchitis than in Crohn's disease (Chapter 14).[43,44] Diversion of luminal contents by a proximal ileostomy improves active Crohn's colitis and prevents inflammation in the neoterminal ileum after surgical resection.[45,46] In both situations, restoring bowel continuity after reversal of the ostomy or infusion of ileostomy contents leads to rapid onset of mucosal inflammation. Harper et al implicated microbial products rather than dietary components by reinfusing ultrafiltrated ileostomy contents.[45] Similarly, pouchitis does not develop until luminal contents bathe the pouch, where relative stagnation of motility leads to bacterial overgrowth.[47] Finally, approximately 2/3 of patients with active ileal Crohn's disease, but not those with Crohn's colitis or ulcerative colitis, respond to bowel rest by total parenteral nutrition, elemental or polymeric diets.[48] As with bacterial diversion, reinstitution of a regular diet leads to rapid recurrence of disease activity. While these studies may indicate that dietary antigens induce pathogenic immune responses, an alternative explanation is that food components, especially poorly digestible polysaccharides, oligosaccharides and refined sugars, profoundly influence the composition of commensal enteric bacteria.

In active IBD enhanced mucosal permeability leads to increased uptake of luminal microbial (and possibly dietary) antigens and adjuvants that can stimulate mucosal immune responses, thereby potentiating the inflammatory response, causing tissue injury and leading to septic complications.[3] Enteric bacteria were cultured from the serosal surface or mesenteric lymph nodes (MLN) of 56% of resected tissues from Crohn's patients but from only 17% of control tissues[49] and *E. coli* or streptococcal antigen was found adjacent to fistulae and ulcers in 2/3 of Crohn's disease patients.[50] Sutton used molecular techniques to identify a novel transcriptional regulation gene, I2, of *Pseudomonas fluorescens*, subsequently identified to be a superantigen, in lamina propria mononuclear cells of Crohn's disease tissues.[51,52]

Whether uptake of these bacterial products represent a primary or secondary event remains controversial.[53] A population based study in Iceland suggested that occult mucosal inflammation in Crohn's disease is mediated at a genetic level.[54] However, several lines of evidence indicate that commensal bacteria adhere with higher frequency to the mucosa of IBD patients. Swidsinski et al showed 40-100 fold increases in mucosally adherent aerobe and anaerobic commensal organisms, particularly Enterobacteriaceae (*E. coli*, Klebsiella, etc) and *Bacteroides species*, in active Crohn's disease.[55] Lower concentrations of mucosally adherent bacteria were found in ulcerative colitis and infectious colitis. Bacterial concentrations correlated with disease activity, suggesting that these abnormalities were secondary to inflammation rather than primary events. In a separate study, mucosal bacterial invasion was documented in 83% of ulcerative colitis surgical resections, 56% of ileal Crohn's disease tissues and 25% of Crohn's colonic specimens, but 0% of controls.[56] The most common invading species were Proteobacteria, Enterobacteriaceae and Bacteroides/Prevotella. Possible mechanisms causing these abnormalities include loss of mucus, secreted antimicrobial products (defensins), or bacterial virulence factors enhancing mucosal adherence. Selective defects in defensin production have been described in Crohn's disease.[57] Defective secretion of these antimicrobial peptides by epithelial cells could lead to increased luminal bacterial concentrations, particularly within the normally sterile crypts. A molecular mechanism is suggested by selective expression of CARD 15/NOD 2 in Paneth cells, the site of α-defensin production[58] (see below).

An alternative mechanism for enhanced bacterial adherence and invasion is selective expression of bacterial virulence factors in IBD patients. *Bacteroides* and other enteric bacterial species secrete mucolytic enzymes that could degrade intestinal mucins. Adherent/invasive *E. coli* have been recovered from ileal tissues of Crohn's disease patients with early postoperative recurrence.[59] These *E. coli* strains express novel adhesions that mediate adherence to differentiated Caco2 cells and invade macrophages by an actin monofilament and microtubule-dependent mechanism, allowing them to replicate intracellularly.[60] Preliminary studies show *E. coli* within lamina propria macrophages in IBD patients.[61] A genetically determined host defect in clearance of invasive intracellular bacteria is proposed by Hisamatsu et al,[62] who demonstrated that the truncated CARD 15/NOD 2 mutation of Crohn's disease was associated with defective clearance of *Salmonella typhimurium* by colonic epithelial cells. This defect in microbial clearance/killing with Crohn's associated CARD 15/NOD 2 polymorphisms is consistent with the hypothesis of defective bacterial killing by phagocytic cells espoused by Korzenick and Dieckgraefe[63] and the observed beneficial effect of granulocyte-macrophage colony stimulating factor (GM-CSF) in Crohn's disease.[64]

Pathogens

Pathogenic organisms could impact Crohn's disease by several mechanisms: environmental triggers that initiate a cascade of events culminating in idiopathic IBD in genetically susceptible hosts or causing a flare of quiescent disease, or perhaps due to persistent pathogens inducing chronic inflammation. This topic is beyond the scope of this treatise, but the reader is referred to several recent reviews thoroughly discussing this topic.[3,5] However, it should be noted that several putative pathogens including *Mycobacterium paratuberculosis*, *Listeria monocytogenes*, and *Mycoplasma pneumoniae* are obligate intracellular pathogens with cell wall components that activate innate immune responses by binding to toll-like receptors (TLR).[50,65,66]

Induction of Mucosal Immune Responses by Bacterial and Dietary Antigens and Adjuvants

Innate Immune Responses

A variety of cells contribute to mucosal innate immune responses, including epithelial cells, macrophages, neutrophils, dendritic cells, mesenchymal cells and endothelial cells.[67-70] These cells provide the primary and secondary defense mechanisms to invading pathogens, contain

Table 5. Microbial components selectively bind to cellular pattern recognition receptors

Pattern Recognition Receptor	Microbial Source	Ligand
TLR 1 (dimerizes with TLR 2)	Gram pos bacteria, mycobacteria mycoplasma	lipopeptides, modulins
TLR 2 (can dimerize with TLR 1 or TLR 6)	Gram pos bacteria	peptidoglycan, lipoteichoic acid, modulins, mannuronic acid polymers
	Gram neg bacteria	peptidoglycan, fimbrae, porin
	Mycobacteria, mycoplasma	lipopeptides, lipoproteins, lipoarabinomannan
	Yeast	zymosan
	Spirochetes	glycolipids, LPS
	Measles	hemaglutinin
TLR 3	Virus	double stranded DNA
TLR 4	Gram neg bacteria	LPS, HSP 60, fimbrae
	Gram pos bacteria	lipoteichoic acid, mannuronic acid polymers
	Respiratory syncytial virus	F protein
TLR 5	Gram neg bacteria	flagellin
	Gram pos bacteria	flagellin
TLR 6 (dimerizes with TLR 2)	Gram pos bacteria	diacylated lipopeptides, modulins
	Mycoplasma	lipopeptides (diacylated)
TLR 7	?	? antiviral compounds
TLR 8	?	? antimicrobial compounds
TLR 9	Gram neg bacteria	unmethylated CpG DNA
	Gram pos bacteria	unmethylated CpG DNA
CARD 4/NOD 1	Bacteria	diaminopimelic acid
CARD 15/NOD 2	Bacteria	muramyl dipeptide

translocating commensal bacteria after the mucosal barrier is breached, and transmit inductive signals to regulatory T lymphocytes. Homologous pattern recognition receptors provide the primary mechanism by which innate immune cells recognize a wide variety of bacterial components (Table 5).[71,72] Ligation of these bacterial, mycoplasma and viral components is mediated in some situations by a combination of accessory binding proteins and homo and heterodimerization of TLR molecules. For example, lipopolysaccharide (LPS) binding to TLR-4 is dramatically augmented by soluble LPS binding protein and membrane-bound CD14. Similarly, lipopeptide binding to TLR-2 is facilitated by heterodimerization with TLR-1 or TLR-6.[65,66] In addition to surface ligation by various membrane-bound TLR molecules, intracellular bacterial products bind to CARD 4/NOD 1 or CARD 15/NOD 2 (Table 5). Muramyl dipeptide (MDP), the minimally active component of peptidoglycan, binds to the leucine-rich repeat (LRR) region of CARD 15/NOD[2,73] The 3 primary polymorphisms of CARD 15/NOD 2 associated with Crohn's disease occur in or near the LRR binding region. Inefficient ligation of MDP presumably results in the defective NOD 2 signaling that is observed in Crohn's mutations.[74] Similarly, CARD 4/NOD 1 binds the bacterial product diaminopimelic acid.[75] Because multiple other intracellular CARD homologues express LRR regions,[76] it is likely that a number of other intracellular receptors can mediate intracellular bacterial signaling. TLR and CARD molecules are variably expressed on mucosal epithelial, lamina propria and tissue innate immune cells.[67,68] Relatively low expression of TLR and CARD 15/NOD 2 molecules on intestinal epithelial cells determines relative nonresponsiveness to luminal bacterial products. However, proinflammatory molecules such as TNF and possibly IFNγ upregulate CARD 15/NOD 2 expression,[62] which could potentiate and perpetuate the inflammatory response and provide a mechanism for loss of tolerance to commensal bacteria. Similarly, induction of Class 2 MHC molecules by IFNγ could lead to more efficient bacterial antigen presentation by intestinal epithelial cells in Crohn's disease. Constitutive expression of CARD 15/NOD 2 in Paneth cells[58] may explain the predilection of Crohn's disease to the distal ileum and provide a potential mechanism for defective production of α defensins in Crohn's disease patients. In addition to upregulating pattern recognition receptors and MHC Class 2 molecules by proinflammatory cytokines, migration of monocytes into the inflammatory site potentiates responses to commensal bacterial adjuvants. Newly immigrated macrophages express higher concentrations of CD14 and secrete more proinflammatory cytokines than resident macrophages in IBD tissues.[77] In addition, LPS from *B. vulgatus* or *E. coli* activates NFκB through TLR-4 and induces expression of ICAM-1, COX-2 and IL-6 in intestinal epithelial cells[78] *B. vulgatus* colonization of gnotobiotic rats transiently induces colonic epithelial cell NFκB activation, which is subsequently downregulated by induction of TGFβ in lamina propria lymphocytes.[78,79]

Bacterial adjuvants, particularly LPS, peptidoglycan and CpG DNA motifs, activate naïve dendritic cells, and stimulate IL-12 p40 expression leading to secretion of IL-12 and IL-23 and expression of costimulatory molecules and MHC Class 2 molecules. Thus bacterial adjuvants are required for efficient induction of mucosal TH1 responses (see Chapter 8). The role of commensal bacteria in inducing intestinal dendritic cells was elegantly documented by Becker et al,[80] who reported that distal ileal lamina propria CD11c+ dendritic cells selectively express IL-12 p40, produce IL-23, and engulf luminal bacteria. The ileal dendritic cells of germ-free mice did not express p40, strongly supporting a bacterial stimulation of IL-23 expression. Intestinal dendritic cells are imminently poised to sample luminal bacteria through intraepithelial processes and a subepithelial location in organized lymphoid aggregates, in addition to phagocytosing translocated bacteria, bacterial adjuvants and antigens after epithelial injury.[69,70,81] Microbial peptides are presented by mucosal APC, as evidenced by identification of antigens bound to MHC Class 2 grooves from the intestines of normal and IBD patients.[82] Over 90% of bound peptides were exogenous, and many were from *E. coli, Saccharomyces cervicisiae* or *Caenorhabditis elegans.*

Upon ligation of TLR molecules by bacterial products, signaling converges on central pathways that are shared proximally by the scaffolding proteins associated with the IL-1 receptor and distally by the CARD/NOD pathways (Fig. 2). Signaling of most TLR molecules is dependent on MyD88, although MyD88-independent pathways have been documented. MyD88-dependent and independent TLR signaling and CARD/NOD signaling through RIP 2 converge on the IκB kinase (IKK) complex. IKK phosphorylates IκBα, releasing NFκB heterodimers to translocate to the nucleus. In parallel, TLR ligation activates the mitogen-activated protein kinase (MAPK) cascade, leading to activation of JNK, p38 and/or ERK 1/2. Individually or in concert these transcription factors transcriptionally regulate production of a large number of proinflammatory cytokines, chemokines, adhesion molecules, costimulatory molecules and MHC Class 2 molecules that mediate innate and acquired immune responses in intestinal inflammation.[67,68,83] Simultaneously, NFκB stimulates

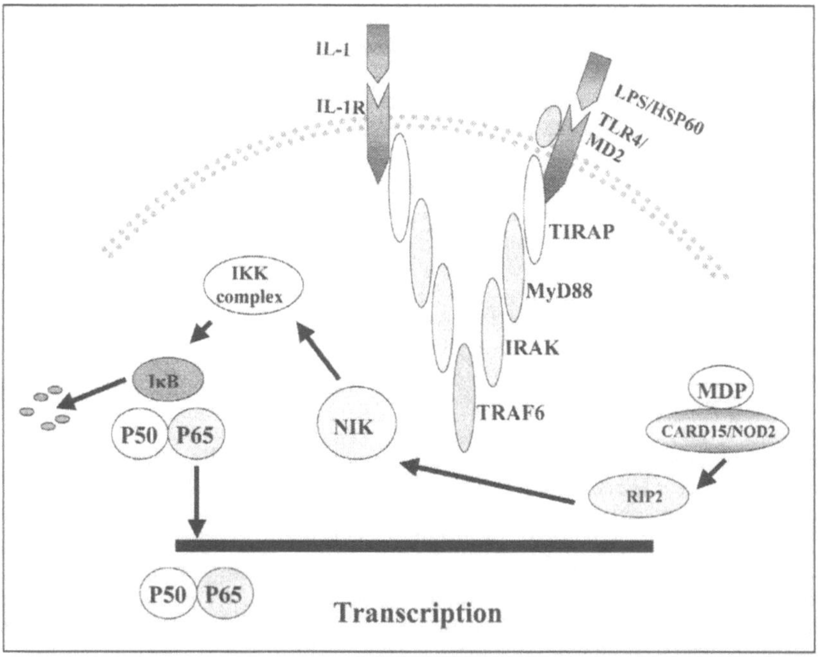

Figure 2. TLR and CARD/NOD signaling pathways. Toll-like receptors (TLR) use intracellular adapter proteins and kinases that are identical to those mediating the IL-1/IL-1 receptor signaling cascade. Ligation of membrane-bound TLR or intracellular CARD/NOD activate the IκB kinase (IKK) complex that phosphorylates IκBα to activate NFκB-mediated transcription of multiple proinflammatory and protective molecules.

production of inhibitory molecules, including COX-2, IκBα and PPARγ, which restore mucosal homeostasis and downregulate inflammatory responses. In addition to transcriptional regulation by products of the NFκB and MAPK signaling pathways, production and secretion of proinflammatory and inhibitory molecules is modulated by mRNA stabilization, protein translation and post-translational modification.

Commensal bacteria and purified bacterial adjuvants can stimulate innate immune responses in experimental intestinal inflammation. Intramural injection of purified peptidoglycan-polysaccharide complexes induces biphasic enterocolitis in susceptible rat strains; acute inflammation is mediated by macrophages and proinflammatory cytokines, while chronic granulomatous intestinal and systemic inflammation is T lymphocyte dependent.[84] Similar pathologic responses are induced by intramural injection of mycobacterial cord factor.[85] Acute enterocolitis induced by indomethacin is dramatically attenuated in germ-free rats and decreased by metronidazole.[28,86] Similarly, broad spectrum antibiotics prevent and treat acute colitis induced by DSS[20] and attenuate acute TNBS-induced colitis in rats.[87]

It is important to note that protective immune responses can also be stimulated by bacterial adjuvants. Rachmilewitz, Raz and colleagues have demonstrated that parenterally injected or oral CpG, either of synthetic, probiotic or native *E. coli* origin, can prevent and treat experimental colitis in several murine models.[88,89] This protection is mediated through TLR-9,[89] consistent with the observation that DNA from probiotic bacteria inhibits proinflammatory responses in colonic epithelial cells through TLR-9.[90] Probiotic bacterial species stimulate IL-10 secretion in rat splenocytes and downregulate TNF secretion induced by cecal bacterial lysates[91] and stimulate IL-10 secretion by ileal pouch mucosa.[92] Of relevance to mucosal homeostasis, *E. coli*

LPS or cecal bacterial lysates stimulate IL-10 secretion by murine mesenteric lymph node (MLN) cells.[93,94] Fractionation shows that the vast majority of IL-10 is produced by CD11c+ dendritic cells.[94] The importance of endogenous production of IL-10 by innate immune cells in maintaining mucosal homeostasis is illustrated by selective production of IL-12 p40 by MLN IL-10 deficient dendritic cells stimulated with cecal bacterial lysate,[94] and the selective ability of adoptive transfer of in vitro bacterial antigen-pulsed IL-10-/- dendritic cells (but not wild type dendritic cells) to accelerate the onset of colitis in gnotobiotic IL-10-/- recipient mice.[95] Additional mechanisms of protective homeostatic adaptive immune responses include PPARγ, HSP 25 and HSP 72.[96,97] In addition to its proinflammatory properties, NFκB appears to mediate protective effects, since NFκB-deficient Rag 2-/- mice have more aggressive colitis after *H. hepaticus* infection[98] and DSS induces more vigorous colitis in mice with an intestinal epithelial cell specific blockade of NFκB activation.[99]

Acquired Immune Responses

Experimental Rodent Models

Cellular and humoral immune responses to commensal bacteria have been documented in a number of mouse and rat models of experimental colitis (Table 6). The TH1-dominated immune responses in most models of chronic colitis are directed to enteric bacterial antigens[16,24,39,100-106] and precede the onset of clinical and histologic disease, suggesting that they are involved in the pathogenesis of disease rather than being secondarily activated.[39] TH1 cell lines or clones recognizing enteric bacterial antigens can induce colitis. Cecal bacterial antigen-responsive CD4+ T cells induced colitis in SCID recipients via CD40-CD40 ligand interactions.[102,107] *H. hepaticus*-responsive CD4+ T cells induced colitis in *H. hepaticus* infected but not in noninfected Rag 2-/- mice showing the importance of the continued presence of enteric antigen.[24] Of considerable interest, cotransfer of *H. hepaticus*-specific CD45RB^{low} CD4+ T cells (both CD25+ and CD25- populations) blocked onset of disease in a bacterial antigen-specific manner.[24] IL-10 but not TGFβ mediated these bacterial antigen-specific regulatory T cells. Moreover, CD4+ T cell clones recognizing a 15-mer peptide epitope of the flagellar hook protein (Flg E) of *H. hepaticus* presented by I-A^b could transfer colitis to T cell deficient mice.[105] Responsive T cells selectively recognized some, but not all *Helicobacter species* and a *Schistosona mansoni*-reactive clone did not induce disease. These results complement previous observations that the constant presence of enteric bacterial antigen (or antigens produced by bacteria) is required for induction of colitis after T cell transfer[16,25,108] as well as bacterial antigen specificity in experimental colitis. Monoassociation or dual association with *E. coli* and/or *E. faecalis* induces T cell-mediated colitis in gnotobiotic IL-10-/- mice.[39;40] In this model, MLN CD4+ T cells selectively respond to the bacterial species inducing disease, and *E. faecalis*-monoassociated mice develop an oligoclonal expansion of T cells in the colonic lamina propria and draining MLN.[109] Takahashi et al demonstrated restricted TCR Vβ 13 and 14 clonotypes in colonic lamina propria CD4+ T cells from IL-10-/- mice with colitis but not in noninflamed colons or in peripheral lymphoid tissues.[110] Veltkamp et al demonstrated in the CD3ε transgenic model that MLN CD4+ T cells selectively responded to cecal bacterial antigens but not to nonbacterial luminal or colonic epithelial antigens.[16] However, bacterial-specific TH1 responses are not necessarily pathogenic, since monoassociated IL-10-/- mice can develop detectible T cell responses without clinical or histologic evidence of colitis.[111] Of key importance, wild type mice exhibit protective, regulatory T cell responses and tolerance to commensal bacteria.[101] Bacterial antigen-specific regulatory T cells can mediate tolerance. Cotransfer of cecal bacterial antigen-responsive CD4+ T cell lines that secrete IL-10.[112] or *H. hepaticus*-specific CD4+ CD45RB^{low} T cells[105] prevent colitis in T cell-deficient mice. Lamina propria CD4+ T cells from normal mice suppress enteric bacterial antigen-induced proliferation of TH1 CD4+ T cells isolated from SCID mice with colitis induced by transfer of CD45RB^{high} cells.[113] This protection does not depend on in vitro stimulation by bacterial

Table 6. Immune responses to commensal bacteria in experimental colitis

Model	Antigen	Response
A. Cellular immune responses		
IL-10 (129 SvEv)	Cecal lysate	↑ IFNγ by MLN CD4+ T cells (94)
	E. coli, E. faecalis *Klebsiella pneumoniae*	↑ IFNγ by MLN CD4+ T cells (precedes histologic colitis) (39)
	Clostridium sordelii	↑ proliferation, but no disease (111)
IL-10-/- (C57Bl/6)	*H. hepaticus*	↑ IFNγ by MLN CD4+ T cell clones (24)
	H. hepaticus flagellar hook protein	↑ IFNγ by MLN CD4+ T cell clones (105)
C3H/HeJ Bir	Cecal lysate	↑ MLN CD4+ T cell proliferation (102)
	Cecal lysate	↑ IFNγ or IL-10 by T cell lines (112)
	Flagellin	↑ proliferation (139)
BM→$CD_3\varepsilon$ Tg	Cecal lysate	↑ IFNγ by MLN CD4+ T cells (16)
Mdr-1-/-	Cecal lysate	↑ proliferation LP and IE 6 cells (100)
TNBS	Fecal lysate	↑ proliferation spleen and LP CD4+ T cells (101)
HLA B27 Tg rat	Cecal lysate	↑ IFNγ by MLN CD4+ T cells (104)
Citrobacter freundii	*C. freundii* intimin	↑ IFNγ, IL-2 splenocytes (with ConA or αCD3 Ab stimulation) (103)
B. Humoral immune responses		
C3H/HeJ Bir	Cecal lysate	↑ IgG by Western blot (118)
	E. coli	↑ IgG by Western blot (118)
	Flagellin	↑ IgG by Western blot, ELISA (139)
HLA B27 Tg rats	B. vulgatus HSP 60	↑ IgG by Western blot (119)

MLN, mesenteric lymph node; LP, intestinal lamina propria; IEL, intraepithelial lymphocyte.

antigens. Bacterial antigen-responsive IL-10 secreting regulatory cells are found in the colons of IL-2-/- mice with colitis, but are functionally defective, since they do not suppress IFNγ production.[114] Although $CD45RB^{low}$ T cells contain CD25+ and CD25- regulatory T cells that prevent disease and downregulate bacterial-responsive TH1 cells, Asseman et al[115] recently demonstrated a colitogenic population of CD25- $CD45RB^{low}$ CD4+ T cells that could transfer colitis to T cell deficient mice. Their absence in germ-free mice suggests that these cells are stimulated by commensal bacteria. Although mice with experimental colitis display loss of tolerance, feeding *E. coli* lysates can attenuate colitis in IL-10-/- mice and wild type mice fed DSS by downregulating mucosal proinflammatory cytokine production and decreasing lymphocyte proliferation to cecal bacterial lysate.[116]

Cecal lymphoid aggregates appear to be the preferential site in which pathogenic T cells are activated. Removal of the cecal tip ("appendectomy"), which contains a large lymphoid aggregate, attenuates experimental colitis in TCRα-/- mice,[117] while cecal bypass diminishes distal colitis and even gastritis in HLA B27 transgenic rats despite not changing luminal bacterial concentrations in remote organs.[21] Similar to the epidemiologic reports of protection against ulcerative colitis by early appendectomy, delayed removal of the cecal tip after onset of colitis does not decrease disease.[117] Although not formally documented in these studies, naïve T cells are presumably stimulated by luminal bacteria in cecal lymphoid tissues, undergo clonal expansion and home back to the colonic lamina propria.

Serologic responses to luminal bacterial antigens, including unfractionated cecal bacterial lysates, multiple commensal species,[118] several *Bacteroides species*, the core glycolipid of LPS and heat shock protein 60 (HSP 60) of *B. vulgatus*[119] have been reported in multiple colitis models (Table 6). In several studies, these serologic responses crossreact with host antigens, as in the case of pANCA and HSP 60.[120,121] Antibodies to dietary and self antigens have been described in Gαi2-/- mice before and after onset of colitis.[122] However, B lymphocytes are not required for induction of experimental colitis, and actually mediate protection in several models.[123-125] Protection does not depend on antibody production and in some settings is IL-10 dependent.[125,126] Secretion of IL-10 and TGFβ in rat MLN B cells is stimulated by cecal bacterial lysate.[19] Furthermore, B lymphocytes are important in clearing experimental *Citrobacter rodentium* infection.[127]

Human IBD Patients

IBD patients exhibit loss of tolerance by mounting both serologic and cellular immune responses to commensal enteric bacteria. Duchmann et al demonstrated that T cells from patients with Crohn's disease or ulcerative colitis proliferated in response to both autologous and heterologous fecal bacteria, while T cells from normal patients only responded to heterologous bacteria.[128] Subsequent studies by the same group showed that T cell clones from both IBD patients and normals reacted to *Salmonella typhorium, Yersinia enterocolitis, H. pylori* and *E. coli*, but the lamina propria T cells from IBD patients responded more actively.[129] CD4+ T cell clones from IBD patients also react to *Bacteroides* and *Bifidobacteria species*.[130] Van den Bogaerde et al reported that Crohn's disease patients had peripheral blood T cell proliferative responses to a variety of bacteria (*Bacteroides, E. coli* and *Klebsiella*), yeast (*Saccharomyces*) and food antigens.[131] Rectal administration of yeast and citrus fruit antigens enhanced rectal blood flow in Crohn's disease patients to a greater degree than controls. These antigens also induced greater skin test and peripheral blood lymphocyte proliferative responses in Crohn's disease patients.[132] These results suggest that loss of tolerance in IBD patients is a generalized phenomenon to both bacterial and food antigens, although there is very little clinical supportive evidence that dietary antigens cause pathogenic T cell responses. Presumably anti-*Saccharomyces cerevisiae* antibodies (ASCA) in Crohn's patients are due to dietary baker's yeast.

Crohn's disease patients, and to a lesser degree those with ulcerative colitis, have increased systemic and mucosal antibody responses to a number of commensal bacterial species.[133-135] The most consistent antibody responses are against *E. coli, Peptostreptococci, Eubacterium* and *Bacteroides species*. Perinuclear antineutrophilic cytoplasmic antibodies (pANCA) crossreact with cecal bacteria and several enteric bacterial species, including *E. coli, Bacteroides caccae* and *Mycobacteria*.[103,136] More recent studies have demonstrated selective serologic responses to a variety of defined bacterial antigens in IBD patients (Table 7) that have been suggested to be useful to distinguish Crohn's disease from ulcerative colitis, IBD from irritable bowel syndrome, and identifying clinically important subsets of Crohn's disease patients.[137,138] The Cedars-Sinai group have demonstrated that approximately 80% of Crohn's disease patients react to either I2 (*Pseudomonas fluorescence*), OmpC (*E. coli*) or oligomannan (*Saccharomyces cerevisiae*) and that 25% of patients respond to all 3 microbial antigens.[138] Moreover, antibody responses to those antigens were associated with a small intestinal fibrostenosing phenotype and high antibody titers were associated with surgical resection. Although these provocative results need to be confirmed, serologic responses to bacteria may be helpful in predicting the natural history of disease and identifying clinical subgroups for antibiotic or probiotic intervention.

Conclusion

Convincing data support the concept that commensal enteric bacteria induce pathogenic T cell responses that cause chronic experimental intestinal inflammation in genetically susceptible mice and rats. Genetic susceptibility is a key determinant of immune responses to luminal bacteria; susceptible hosts mount pathogenic TH1, or less commonly, TH2 responses to their

Table 7. Serologic responses to specific microbial antigens by IBD patients and controls

	Frequency of Positive Serologic Responses			
Antigen	**Ulcerative Colitis, %**	**Crohn's Disease, %**	**Inflammatory Controls, %**	**Normal Controls, %**
pANCA	57-66	13-15	8	0-5
ASCA	6-12	55-61	11	1-5
Bacteroides vulgatus (26 kDa)	54	NR[1]	NR	9
Pseudomonas fluorescens (I_2)	10	54	19	4
Mycobacterium (HupB, 32 kDa)	10*	90	NR	0
Fusobacterium varium	61	13	13	3

*IgA antibody, N=10/group. ASCA, anti-saccharomyces cerevisiae antibody; NR, not reported; pANCA, perinuclear antineutrophil cytoplasmic antibody. Adapted from Sartor RB (Current Opinion 2001).

luminal bacteria, while normal hosts exhibit tolerance to their own bacteria mediated by regulatory T cells. Developing evidence suggests that commensal bacterial antigens stimulate these regulatory T cells. Gnotobiotic rodent studies indicate that different enteric commensal bacteria have selective abilities to induce intestinal inflammation and that specificity of bacterial induction of disease may be genetically determined. Some bacterial species, especially *Lactobacilli* and *Bifidobacterial species*, prevent colitis. Food antigens and antibody responses to either enteric bacterial or food antigens appear to have no pathogenic role in experimental enterocolitis. T cell responses to bacteria in human IBD patients are only beginning to be studied, but patients with Crohn's disease and ulcerative colitis appear to have defective mucosal tolerance. Crohn's disease patients exhibit serologic responses to a variety of commensal bacterial species and defined bacterial antigens; these antibody responses may have diagnostic and predictive value. IBD patients have immune responses to dietary antigens, but the pathogenic role of those responses is unknown. These combined results strongly suggest that selective manipulation of the complex microbial milieu by antibiotics, probiotics and prebiotics alone or in combination may have an important place in treating IBD patients.

References

1. Sartor RB. Animal models of intestinal inflammation. In: Sartor RB, Sandborn WJ, eds. Kirsner's Inflammatory Bowel Diseases. 6th ed. Elsevier Publishers, 2004:120-137.
2. Wilson KH. Natural biota of the human gastrointestinal tract. In: Blaser MJ, Smith PD, Ravdin JI et al, eds. Infections of the Gastrointestinal Tract. 2nd ed. Philadelphia: Lippincott Williams & Wilkins, 2002:45-56.
3. Sartor RB. Microbial influences in inflammatory bowel disease: Role in pathogenesis and clinical implications. In: Sartor RB, Sandborn WJ, eds. Kirsner's Inflammatory Bowel Diseases. 6th ed. Elsevier Publishers, 2004:138-162.
4. Tannock GW. Molecular assessment of intestinal microflora. Am J Clin Nutr 2001; 73:410S-414S.
5. Sartor RB. Microbial agents in the pathogenesis, differential diagnosis, and complications of inflammatory bowel diseases. In: Blaser MJ, Smith PD, Ravdin JI et al, eds. Infections of the GI Tract. 2nd ed. New York: Raven Press, 2002:383-413.
6. Sartor RB. Targeting enteric bacteria in treatment of inflammatory bowel diseases: Why, how and when. Current Opinion in Gastroenterology 2003; 19:358-365.
7. Schultz M, Rath HC. The possible role of probiotic therapy in inflammatory bowel disease. In: Tannock GW, ed. Probiotics and Prebiotics: Where are We Going? Wymondham: Caister Academic Press, 2002:175-237.
8. Elson CO, Cong Y. Understanding immune-microbial homeostasis in intestine. Immunol Res 2002; 26:87-94.

9. Elson CO. Genes, microbes, and T cells - New therapeutic targets in Crohn's disease. N Engl J Med 2002; 346:614-616.
10. Heller F, Duchmann R. Intestinal flora and mucosal immune responses. Int J Med Microbiol 2003; 293:77-86.
11. Sellon RK, Tonkonogy S, Schultz M et al. Resident enteric bacteria are necessary for development of spontaneous colitis and immune system activation in interleukin-10-deficient mice. Infect Immun 1998; 66:5224-5231.
12. Rath HC, Herfarth HH, Ikeda JS et al. Normal luminal bacteria, especially Bacteroides species, mediate chronic colitis, gastritis, and arthritis in HLA-B27/human beta2 microglobulin transgenic rats. J Clin Invest 1996; 98:945-953.
13. Dianda L, Hanby AM, Wright NA et al. T cell receptor-alpha beta-deficient mice fail to develop colitis in the absence of a microbial environment. Am J Pathol 1997; 150:91-97.
14. Schultz M, Tonkonogy SL, Sellon RK et al. IL-2-deficient mice raised under germfree conditions develop delayed mild focal intestinal inflammation. Am J Physiol 1999; 276:G1461-G1472
15. Contractor NV, Bassiri H, Reya T et al. Lymphoid hyperplasia, autoimmunity, and compromised intestinal intraepithelial lymphocyte development in colitis-free gnotobiotic IL-2-deficient mice. J Immunol 1998; 160:385-394.
16. Veltkamp C, Tonkonogy SL, de Jong YP et al. Continuous stimulation by normal luminal bacteria is essential for the development and perpetuation of colitis in Tg(epsilon26) mice. Gastroenterology 2001; 120:900-913.
17. Taurog JD, Richardson JA, Croft JT et al. The germfree state prevents development of gut and joint inflammatory disease in HLA-B27 transgenic rats. J Exp Med 1994; 180:2359-2364.
18. Kuhn R, Lohler J, Rennick D et al. Interleukin-10-deficient mice develop chronic enterocolitis. Cell 1993; 75:263-274.
19. Hoentjen F, Harmsen HJ, Braat H et al. Antibiotics with a selective aerobic or anaerobic spectrum have different therapeutic activities in various regions of the colon in interleukin-10 gene deficient mice. Gut 2003; 52:1721-1727.
20. Rath HC, Schultz M, Freitag R et al. Different subsets of enteric bacteria induce and perpetuate experimental colitis in rats and mice. Infect Immun 2001; 69:2277-2285.
21. Rath HC, Ikeda JS, Linde HJ et al. Varying cecal bacterial loads influences colitis and gastritis in HLA-B27 transgenic rats. Gastroenterology 1999; 116:310-319.
22. Kullberg MC, Ward JM, Gorelick P et al. Helicobacter hepaticus triggers colitis in specific-pathogen-free interleukin-10 (IL-10)-deficient mice through an IL-12 and gamma interferon-dependent mechanism. Infect Immun 1998; 66:5157-5166.
23. Dieleman LA, Arends A, Tonkonogy SL et al. Helicobacter hepaticus does not induce or potentiate colitis in interleukin-10-deficient mice. Infect Immun 2000; 68:5107-5113.
24. Kullberg MC, Jankovic D, Gorelick PL et al. Bacteria-triggered CD4(+) T regulatory cells suppress Helicobacter hepaticus-induced colitis. J Exp Med 2003; 196:505-515.
25. Iqbal N, Oliver JR, Wagner FH et al. T helper 1 and T helper 2 cells are pathogenic in an antigen-specific model of colitis. J Exp Med 2002; 195:71-84.
26. Aranda R, Sydora BC, McAllister PL et al. Analysis of intestinal lymphocytes in mouse colitis mediated by transfer of CD4+, CD45RBhigh T cells to SCID recipients. J Immunol 1997; 158:3464-3473.
27. Onderdonk AB, Franklin ML, Cisneros RL. Production of experimental ulcerative colitis in gnotobiotic guinea pigs with simplified microflora. Infect Immun 1981; 32:225-231.
28. Yamada T, Deitch E, Specian RD et al. Mechanisms of acute and chronic intestinal inflammation induced by indomethacin. Inflammation 1993; 17:641-662.
29. Wood JD, Peck OC, Tefend KS et al. Evidence that colitis is initiated by environmental stress and sustained by fecal factors in the cotton-top tamarin (Saguinus oedipus). Dig Dis Sci 2000; 45:385-393.
30. Axelsson LG, Landstrom E, Bylund-Fellenius AC. Experimental colitis induced by dextran sulphate sodium in mice: Beneficial effects of sulphasalazine and olsalazine. Aliment Pharmacol Ther 1998; 12:925-934.
31. Hudcovic T, Stepankova R, Cebra J et al. The role of microflora in the development of intestinal inflammation: Acute and chronic colitis induced by dextran sulfate in germ-free and conventionally reared immunocompetent and immunodeficient mice. Folia Microbiol 2001; 46:565-572.
32. Rath HC, Wilson KH, Sartor RB. Differential induction of colitis and gastritis in HLA-B27 transgenic rats selectively colonized with Bacteroides vulgatus and Escherichia coli. Infect Immun 1999; 67:2969-2974.
33. Waidmann M, Bechtold O, Frick JS et al. Bacteroides vulgatus protects against Escherichia coli-induced colitis in gnotobiotic interleukin-2-deficient mice. Gastroenterology 2003; 125:162-177.

34. Kishi D, Takahashi I, Kai Y et al. Alteration of V beta usage and cytokine production of CD4+ TCR beta beta homodimer T cells by elimination of Bacteroides vulgatus prevents colitis in TCR alpha-chain-deficient mice. J Immunol 2000; 165:5891-5899.
35. Higgins LM, Frankel G, Douce G et al. Citrobacter rodentium infection in mice elicits a mucosal Th1 cytokine response and lesions similar to those in murine inflammatory bowel disease. Infect Immun 1999; 67:3031-3039.
36. Higgins LM, Frankel G, Connerton I et al. Role of bacterial intimin in colonic hyperplasia and inflammation. Science 1999; 285:588-591.
37. Mann BA, Kim SC, Sartor RB. Selective induction of experimental colitis by monoassociation of HLA-B27 transgenic rats with various enteric Bacteroides species. Gastroenterology 2003; 124:A322, (abstract).
38. Maggio-Price L, Shows D, Waggie K et al. Helicobacter bilis infection accelerates and H. hepaticus infection delays the development of colitis in multiple drug resistance-deficient (mdr1a-/-) mice. Am J Pathol 2002; 160:739-751.
39. Kim SC, Tonkonogy SL, Albright CA et al. Variable phenotypes of enterocolitis in interleukin 10-deficient mice monoassociated with two different commensal bacteria. Gastroenterology 2005; 128(4):891-906.
40. Kim SC, Tonkongy SL, Bower M et al. Dual-association of gnotobiotic IL-10-/- mice with two nonpathogenic commensal bacterial species accelerates colitis. Gastroenterology 2004; 126:A291.
41. Tsang J, Brown RS, Andersen GL et al. Selective expansion of colitogenic commensal bacterial species in SPF IL-10-/- mice. Gastroenterology 2004; 126:A291.
42. Isaacs KL, Sartor RB. Antibiotics in IBD. Gastroenterol Clin North Am 2004; 33:335-345.
43. Gionchetti P, Rizzello F, Venturi A et al. Oral bacteriotherapy as maintenance treatment in patients with chronic pouchitis: A double-blind, placebo-controlled trial. Gastroenterology 2000; 119:305-309.
44. Kruis W, Schutz E, Fric P et al. Double-blind comparison of an oral Escherichia coli preparation and mesalazine in maintaining remission of ulcerative colitis. Aliment Pharmacol Ther 1997; 11:853-858.
45. Harper PH, Lee EC, Kettlewell MG et al. Role of the faecal stream in the maintenance of Crohn's colitis. Gut 1985; 26:279-284.
46. D'Haens GR, Geboes K, Peeters M et al. Early lesions of recurrent Crohn's disease caused by infusion of intestinal contents in excluded ileum. Gastroenterology 1998; 114:262-267.
47. Sandborn WJ. Pouchitis following ileal pouch-anal anastomosis: Definition, pathogenesis, and treatment. Gastroenterology 1994; 107:1856-1860.
48. Dieleman LA, Heizer WD. Nutritional issues in inflammatory bowel disease. Gastroenterol Clin North Am 1998; 27:435-451.
49. Ambrose NS, Johnson M, Burdon DW et al. Incidence of pathogenic bacteria from mesenteric lymph nodes and ileal serosa during Crohn's disease surgery. Br J Surg 1984; 71:623-625.
50. Liu Y, Van Kruiningen HJ, West AB et al. Immunocytochemical evidence of Listeria, Escherichia coli, and Streptococcus antigens in Crohn's disease. Gastroenterology 1995; 108:1396-1404.
51. Sutton CL, Kim J, Yamane A et al. Identification of a novel bacterial sequence associated with Crohn's disease. Gastroenterology 2000; 119:23-31.
52. Wei B, Huang T, Dalwadi HN et al. Pseudomonas fluorescens encodes the Crohn's disease-associated I2 sequence and T-cell superantigen. Infect Immun 2002; 70:6567-6575.
53. Meddings JB. Review article: Intestinal permeability in Crohn's disease. Aliment Pharmacol Ther 1997; 11(Suppl):47-53.
54. Thjodleifsson B, Sigthorsson G, Cariglia N et al. Subclinical intestinal inflammation: An inherited abnormality in Crohn's disease relatives? Gastroenterology 2003; 124:1728-1737.
55. Swidsinski A, Ladhoff A, Pernthaler A et al. Mucosal flora in inflammatory bowel disease. Gastroenterology 2002; 122:44-54.
56. Kleessen B, Kroesen AJ, Buhr HJ et al. Mucosal and invading bacteria in patients with inflammatory bowel disease compared with controls. Scand J Gastroenterol 2002; 37:1034-1041.
57. Wehkamp J, Harder J, Weichenthal M et al. Inducible and constitutive beta-defensins are differentially expressed in Crohn's disease and ulcerative colitis. Inflamm Bowel Dis 2003; 9:215-223.
58. Lala S, Ogura Y, Osborne C et al. Crohn's disease and the NOD2 gene: A role for paneth cells. Gastroenterology 2003; 125:47-57.
59. Darfeuille-Michaud A, Neut C, Barnich N et al. Presence of adherent Escherichia coli strains in ileal mucosa of patients with Crohn's disease. Gastroenterology 1998; 115:1405-1413.
60. Glasser AL, Boudeau J, Barnich N et al. Adherent invasive Escherichia coli strains from patients with Crohn's disease survive and replicate within macrophages without inducing host cell death. Infect Immun 2001; 69:5529-5537.
61. Rayment N, Mylonaki M, Rampton D et al. Colocalisation of E. coli with macrophages in lamina propria in patients with active inflammatory bowel disease (IBD). Gut 2003; in press,(abst).

62. Hisamatsu T, Suzuki M, Reinecker HC et al. CARD15/NOD2 functions as an anti-bacterial factor in human intestinal epithelial cells. Gastroenterology 2003; 124:993-1000.
63. Korzenik JR, Dieckgraefe BK. Is Crohn's disease an immunodeficiency? A hypothesis suggesting possible early events in the pathogenesis of Crohn's disease. Dig Dis Sci 2000; 45:1121-1129.
64. Dieckgraefe BK, Korzenik JR. Treatment of active Crohn's disease with recombinant human granulocyte-macrophage colony-stimulating factor. Lancet 2002; 360:1478-1480.
65. Takeda K, Takeuchi O, Akira S. Recognition of lipopeptides by Toll-like receptors. J Endotoxin Res 2002; 8:459-463.
66. Tapping RI, Tobias PS. Mycobacterial lipoarabinomannan mediates physical interactions between TLR1 and TLR2 to induce signaling. J Endotoxin Res 2003; 9:264-268.
67. Haller D, Jobin C. Interaction between resident luminal bacteria and the host: Can a healthy relationship turn sour? J Pediatr Gastroenterol Nutr 2004; 38:123-136.
68. Sartor RB, Hoentjen F. Proinflammatory cytokines and signaling pathways in intestinal innate immune cells. In: Mestecky MS, ed. Mucosal Immunol. 3rd ed. 2004: in press,
69. Uhlig HH, Powrie F. Dendritic cells and the intestinal bacterial flora: A role for localized mucosal immune responses. J Clin Invest 2003; 112:648-651.
70. Stagg AJ, Hart AL, Knight SC et al. The dendritic cell: Its role in intestinal inflammation and relationship with gut bacteria. Gut 2003; 52:1522-1529.
71. Werling D, Jungi TW. TOLL-like receptors linking innate and adaptive immune response. Vet Immunol Immunopathol 2003; 91:1-12.
72. Akira S, Hemmi H. Recognition of pathogen-associated molecular patterns by TLR family. Immunol Lett 2003; 85:85-95.
73. Girardin SE, Boneca IG, Viala J et al. Nod2 is a general sensor of peptidoglycan through muramyl dipeptide (MDP) detection. J Biol Chem 2003; 278:8869-8872.
74. Bonen DK, Ogura Y, Nicolae DL et al. Crohn's disease-associated NOD2 variants share a signaling defect in response to lipopolysaccharide and peptidoglycan. Gastroenterology 2003; 124:140-146.
75. Girardin SE, Boneca IG, Carneiro LA et al. Nod1 detects a unique muropeptide from gram-negative bacterial peptidoglycan. Science 2003; 300:1584-1587.
76. Harton JA, Linhoff MW, Zhang J et al. Cutting edge: CATERPILLER: A large family of mammalian genes containing CARD, pyrin, nucleotide-binding, and leucine-rich repeat domains. J Immunol 2002; 169:4088-4093.
77. Mahida YR. The key role of macrophages in the immunopathogenesis of inflammatory bowel disease. Inflamm Bowel Dis 2000; 6:21-33.
78. Haller D, Russo MP, Sartor RB, Jobin C. IKKbeta and phosphatidylinositol 3-kinase/Akt participate in nonpathogenic Gram-negative enteric bacteria-induced RelA phosphorylation and NF-kappa B activation in both primary and intestinal epithelial cell lines. J Biol Chem 2002; 277:38168-38178.
79. Haller D, Holt L, Schwabe RF et al. Transforming growth factor-{beta}1 inhibits nonpathogenic Gramnegative bacteria-induced NF-{kappa}B recruitment to the interleukin-6 gene promoter in intestinal epithelial cells through modulation of histone acetylation. J Biol Chem 2003; 278:23851-23860.
80. Becker C, Wirtz S, Blessing M et al. Constitutive p40 promoter activation and IL-23 production in the terminal ileum mediated by dendritic cells. J Clin Invest 2003; 112:693-706.
81. Rescigno M, Urbano M, Valzasina B et al. Dendritic cells express tight junction proteins and penetrate gut epithelial monolayers to sample bacteria. Nature Immunol 2001; 2:361-367.
82. Oshitani N, Hato F, Kitagawa S et al. Analysis of intestinal HLA-DR bound peptides and dysregulated immune responses to enteric flora in the pathogenesis of inflammatory bowel disease. Int J Mol Med 2003; 11:99-104.
83. Jobin C, Sartor RB. The I kappa B/NF-kappa B system: A key determinant of mucosal inflammation and protection. Am J Physiol Cell Physiol 2000; 278:C451-C462.
84. Sartor RB, De La Cadena RA, Green KD et al. Selective kallikrein-kinin system activation in inbred rats differentially susceptible to granulomatous enterocolitis. Gastroenterology 1996; 110:1467-1481.
85. Sogawa M, Matsumoto T, Yamagami H et al. A murine model of granulomatous colitis with mesenteric lymphadenitis induced by mycobacterial cord factor. Virchows Arch 2003; 442:151-158.
86. Robert A, Asano T. Resistance of germfree rats to indomethacin-induced intestinal lesions. Prostaglandins 1977; 14:333-341.
87. Videla S, Vilaseca J, Guarner F et al. Role of intestinal microflora in chronic inflammation and ulceration of the rat colon. Gut 1994; 35:1090-1097.
88. Rachmilewitz D, Karmeli F, Takabayashi K et al. Immunostimulatory DNA ameliorates experimental and spontaneous murine colitis. Gastroenterology 2002; 122:1428-1441.

89. Rachmilewitz D, Katakara K, Karmeli F et al. Toll-like receptor 9 signaling mediates the anti-inflammatory effects of probiotics in murine experimental colitis. Gastroenterology 2004; 126:520-528
90. Jijon H, Backer J, Diaz H et al. DNA from probiotic bacteria modulates murine and human epithelial and immune function. Gastroenterology 2004; 126:1358-1373.
91. Cender CJ, Haller D, Walters C et al. VSL #3 alters cytokine production of unfractionated splenocytes upon stimulation with cecal bacterial lysate: Immunomodulation by this probiotic combination. Gastroenterology 2002; 122:A145, (abst).
92. Ulisse S, Gionchetti P, D'Alo S et al. Expression of cytokines, inducible nitric oxide synthase, and matrix metalloproteinases in pouchitis: Effects of probiotic treatment. Am J Gastroenterol 2001; 96:2691-2699.
93. Kim SC, Tonkonogy SL, Sartor RB. Role of endogenous IL-10 in downregulating proinflammatory cytokine expression. Gastroenterology 2001; 120:A183, (abst).
94. Albright C, Tonkonogy SL, Sartor RB. Endogenous IL-10 inhibits APC stimulation of T lymphocyte responses to luminal bacteria. Gastroenterology 2002; 122:A270, (abst).
95. Albright C, Tonkonogy SL, Frelinger JA et al. Adoptive transfer of E. faecalis-pulsed dendritic cells accelerates colitis in IL-10 deficient mice. Gastroenterology 2003; 124:A73, (abst).
96. Dubuquoy L, Jansson EA, Deeb S et al. Impaired expression of peroxisome proliferator-activated receptor gamma in ulcerative colitis. Gastroenterology 2003; 124:1265-1276.
97. Kojima K, Musch MW, Ren H et al. Enteric flora and lymphocyte-derived cytokines determine expression of heat shock proteins in mouse colonic epithelial cells. Gastroenterology 2003; 124:1395-1407.
98. Tomczak MF, Erdman SE, Poutahidis T et al. NF-kappa B is required within the innate immune system to inhibit microflora-induced colitis and expression of IL-12 p40. J Immunol 2003; 171:1484-1492.
99. Russo MP, Boudreau F, Li F et al. NFkappaB blockade exacerbates experimental colitis in transgenic mice expressing an intestinal epithelial cells (IEC) specific IkappaB super-repressor. Gastroenterology 2001; 120:A70, (abst).
100. Panwala CM, Jones JC, Viney JL. A novel model of inflammatory bowel disease: Mice deficient for the multiple drug resistance gene, mdr1a, spontaneously develop colitis. J Immunol 1998; 161:5733-5744.
101. Duchmann R, Schmitt E, Knolle P et al. Tolerance towards resident intestinal flora in mice is abrogated in experimental colitis and restored by treatment with interleukin-10 or antibodies to interleukin-12. Eur J Immunol 1996; 26:934-938.
102. Cong Y, Brandwein SL, McCabe RP et al. CD4+ T cells reactive to enteric bacterial antigens in spontaneously colitic C3H/HeJBir mice: Increased T helper cell type 1 response and ability to transfer disease. J Exp Med 1998; 187:855-864.
103. Cohavy O, Harth G, Horwitz M et al. Identification of a novel mycobacterial histone H1 homologue (HupB) as an antigenic target of pANCA monoclonal antibody and serum immunoglobulin A from patients with Crohn's disease. Infect Immun 1999; 67:6510-6517.
104. Qian B-F, Hoentjen F, Dieleman LA et al. Dysregulated luminal bacterial antigen-specific T cell responses and antigen presenting cell function in HLA-B27 transgenic rats with chronic colitis. Gastroenterology 2003; 124:A487, (abst).
105. Kullberg MC, Andersen JF, Gorelick PL et al. Induction of colitis by a CD4+ T cell clone specific for a bacterial epitope. Proc Natl Acad Sci USA 2003; 100:15830-15835.
106. Brimnes J, Reimann J, Nissen M et al. Enteric bacterial antigens activate CD4(+) T cells from scid mice with inflammatory bowel disease. Eur J Immunol 2001; 31:23-31.
107. Cong Y, Weaver CT, Lazenby A et al. Colitis induced by enteric bacterial antigen-specific CD4+ T cells requires CD40-CD40 ligand interactions for a sustained increase in mucosal IL-12. J Immunol 2000; 165:2173-2182.
108. Yoshida M, Shirai Y, Watanabe T et al. Differential localization of colitogenic TH1 and TH2 cells monospecific to a micro flora-associated antigen in mice. Gastroenterology 2002; 123:1949-1961.
109. Brown J, Haddad W, Balish E et al. Defining the TCR repertoire in interleukin-10 deficient mouse colitis induced by monoassociation with E. faecalis. Gastroenterology 2002; 122:A262, (abst).
110. Takahashi I, Matsuda J, Gapin L et al. Colitis-related public T cells are selected in the colonic lamina propria of IL-10-deficient mice. Clin Immunol 2002; 102:237-248.
111. Sydora BC, Tavernini MM, Jewell LD et al. Effect of bacterial monoassociation on tolerance and intestinal inflammation in IL-10 gene-deficient mice. Gastroenterology 2001; 120:A517, (abst).
112. Cong Y, Weaver CT, Lazenby A et al. Bacterial-reactive T regulatory cells inhibit pathogenic immune responses to the enteric flora. J Immunol 2002; 169:6112-6119.

113. Gad M, Brimnes J, Claesson MH. CD4+ T regulatory cells from the colonic lamina propria of normal mice inhibit proliferation of enterobacteria-reactive, disease-inducing Th1-cells from scid mice with colitis. Clin Exp Immunol 2003; 131:34-40.
114. Waidmann M, Allemand Y, Lehmann J et al. Microflora reactive IL-10 producing regulatory T cells are present in the colon of IL-2 deficient mice but lack efficacious inhibition of IFN-gamma and TNF-alpha production. Gut 2002; 50:170-179.
115. Asseman C, Read S, Powrie F. Colitogenic Th1 cells are present in the antigen-experienced T cell pool in normal mice: Control by CD4+ regulatory T cells and IL-10. J Immunol 2003; 171:971-978.
116. Konrad A, Mahler M, Flogerzi B et al. Amelioration of murine colitis by feeding a solution of lysed Escherichia coli. Scand J Gastroenterol 2003; 38:172-179.
117. Mizoguchi A, Mizoguchi E, Chiba C et al. Role of appendix in the development of inflammatory bowel disease in TCR-alpha mutant mice. J Exp Med 1996; 184:707-715.
118. Brandwein SL, McCabe RP, Cong Y et al. Spontaneously colitic C3H/HeJBir mice demonstrate selective antibody reactivity to antigens of the enteric bacterial flora. J Immunol 1997; 159:44-52.
119. Mann BA, Rocha ER, Smith CJ et al. Determining the role of heat shock protein 60 of Bacteroides vulgatus in the induction of colitis in the HLA B27 transgenic rat model. Abst Am Soc Microbiol 2003; B-285, (abst).
120. Seibold F, Brandwein S, Simpson S et al. pANCA represents a cross-reactivity to enteric bacterial antigens. J Clin Immunol 1998; 18:153-160.
121. Leung FW, Heng MC, Allen S et al. Involvement of luminal bacteria, heat shock protein 60, macrophages and gammadelta T cells in dextran sulfate sodium-induced colitis in rats. Dig Dis Sci 2000; 45:1472-1479.
122. Uhlig HH, Hultgren HE, Ohman BL et al. Antibody response to dietary and autoantigens in G alpha i2-deficient mice. Eur J Gastroenterol Hepatol 2001; 13:1421-1429.
123. Ma A, Datta M, Margosian E et al. T cells, but not B cells, are required for bowel inflammation in interleukin 2-deficient mice. J Exp Med 1995; 182:1567-1572.
124. Davidson NJ, Leach MW, Fort MM et al. T helper cell 1-type CD4+ T cells, but not B cells, mediate colitis in interleukin 10-deficient mice. J Exp Med 1996; 184:241-251.
125. Mizoguchi A, Mizoguchi E, Smith RN et al. Suppressive role of B cells in chronic colitis of T cell receptor alpha mutant mice. J Exp Med 1997; 186:1749-1756.
126. Dalwadi H, Wei B, Schrage M et al. B cell developmental requirement for the G alpha i2 gene. J Immunol 2003; 170:1707-1715.
127. Simmons CP, Clare S, Ghaem-Maghami M et al. Central role for B lymphocytes and CD4+ T cells in immunity to infection by the attaching and effacing pathogen Citrobacter rodentium. Infect Immun 2003; 71:5077-5086.
128. Duchmann R, Kaiser I, Hermann E et al. Tolerance exists towards resident intestinal flora but is broken in active inflammatory bowel disease (IBD). Clin Exp Immunol 1995; 102:448-455.
129. Duchmann R, Marker-Hermann E, Meyer zum Buschenfelde KH. Bacteria-specific T-cell clones are selective in their reactivity towards different enterobacteria or H. pylori and increased in inflammatory bowel disease. Scand J Immunol. 1996; 44:71-79.
130. Duchmann R, May E, Heike M et al. T cell specificity and cross reactivity towards Enterobacteria, Bacteroides, Bifidobacterium, and antigens from resident intestinal flora in humans. Gut 1999; 44:812-818.
131. van den Bogaerde J, Kamm MA, Knight SC. Immune sensitization to food, yeast and bacteria in Crohn's disease. Aliment Pharmacol Ther 2001; 15:1647-1653.
132. Van Den Bogaerde J, Cahill J, Emmanuel AV et al. Gut mucosal response to food antigens in Crohn's disease. Aliment Pharmacol Ther 2002; 16:1903-1915.
133. Macpherson A, Khoo UY, Forgacs I et al. Mucosal antibodies in inflammatory bowel disease are directed against intestinal bacteria. Gut 1996; 38:365-375.
134. Auer IO, Roder A, Wensinck F et al. Selected bacterial antibodies in Crohn's disease and ulcerative colitis. Scand J Gastroenterol 1983; 18:217-223.
135. Tabaqchali S, O'Donoghue DP, Bettelheim KA. Escherichia coli antibodies in patients with inflammatory bowel disease. Gut 1978; 19:108-113.
136. Cohavy O, Bruckner D, Gordon LK et al. Colonic bacteria express an ulcerative colitis pANCA-related protein epitope. Infect Immun 2000; 68:1542-1548.
137. Landers CJ, Cohavy O, Misra R et al. Selected loss of tolerance evidenced by Crohn's disease-associated immune responses to auto- and microbial antigens. Gastroenterology 2002; 123:689-699.
138. Mow WS, Vasiliauskas EA, Lin YC et al. Association of antibody responses to microbial antigens and complications of small bowel Crohn's disease. Gastroenterology 2004; 126;414-424.
139. Lodes MJ, Cong Y, Elson CO et al. Bacterial flagellin is a dominant antigen in Crohn's disease. J Clin Invest 2004; 113:1296-1306.

CHAPTER 5

Experimental Models of Mucosal Inflammation

Warren Strober and Ivan J. Fuss

Introduction

While studies of models of mucosal inflammation has been a mainstay of IBD research for the past half century, it is only in the last 10-15 years that this kind of study has taken its place as *primus inter pares* among the many approaches to studying these diseases. The reason this has come about is both simple and complex. It is simple because this period has seen the advent of a myriad of new models that individually and collectively allow the exhaustive exploration of many aspects of IBD that were formerly impossible in any other way. This includes the study of the possible effects of various types of genetic and immunologic abnormalities on disease pathogenesis and disease therapy. It is complex because the full and adequate exploration of these models inevitably requires an ability to apply a sophisticated knowledge of current molecular and cellular methodology to the study of living organisms. As attested to by the material reviewed here, the explosion in our knowledge of IBD derived from models is spectacular. With their use it has been possible to construct the main features of IBD on a detailed immunologic level, to explore the role of the bacterial microflora on disease pathogenesis, and to begin to understand the genetic underpinnings of the human disease. In addition, it has been possible to predict the efficacy of new treatments of the disease.

In this chapter we will review the main features of models of mucosal inflammation in three ways. First, we discuss the features of models that characterize the models as a whole and that therefore give rise to over-arching principles of mucosal inflammation that provide a sound basis of the human counterpart diseases. Second, we will consider two types of models (the cell-transfer model and the hapten-induced models) in considerable detail. These models are among those subjected to persistent in-depth study and have thus yielded a substantial fraction of the sum total of knowledge that has been gleaned from the study of all of the models. In addition, these models are among those most extensively studied and together they yield the majority of the insights that can be gleaned from the study of all models. Third and last, we have compiled a comprehensive table of all models of mucosal inflammation so that the models can be appropriately grouped and special insights that can be derived from each model can be briefly stated. While this tripartite approach should be very adequate in giving the reader a strong inkling of the models field, the reader is also encouraged to consult other recent reviews of this area including: *Annual Review of Immunology*[191] and *Nature Immunology Review*.[16]

Immune Mechanisms in Inflammatory Bowel Disease, edited by Richard S. Blumberg and Markus F. Neurath. ©2006 Eurekah.com and Springer Science+Business Media.

General Features of Models of Mucosal Inflammation

Final Common Pathways of Experimental Mucosal Inflammation

Although experimental models of mucosal inflammation arise from a myriad of causes, they are ruled by the Th1/Th2 paradigm in that they usually resolve into a Th1 or a Th2 final common pathway of responsiveness.[191] Furthermore, for reasons that are only partly understood, the Th1 pathway is by far the predominant one, since most models are characterized by excessive IFN-γ responses and little or no increase in IL-4 responses. One possible reason for this "bias" is that, as discussed below, the inflammation is driven by the antigens and other substances in the mucosal microflora, which means that it may be influenced in its earliest stages by innate immune responses mediated by toll-like receptors (TLR's). This creates a Th1 bias because TLR's generally initiate a signaling pathway that leads to activation of NF-κB and IL-12 production.[200] Perhaps the best example of this is the Th1 inflammation occurring spontaneously in Stat3-deficient mice in whom inflammation is extinguished by inter-breeding with TLR4-deficient and thus LPS-nonresponsive mice.[95] Another example is TNBS-colitis which, as discussed in greater detail below, exhibit a genetically-determined increase in IL-12(p70) production following LPS administration;[15] in this case, however, such overproduction could conceivably result from stimulation of multiple TLR ligands not just LPS.

The strong tendency of mucosal inflammations to exhibit a Th1 response raises the question of what conditions are in fact necessary for the occurrence of a Th2-mediated inflammation. One factor is simply the genetic background of the mouse in whom the inflammation is established. This is seen in the fact that TNBS-colitis, a colitis induced by per-rectal administration of a haptenating agent that is discussed at length below, is a pure Th1 disease in SJL/J mice, but a mixed Th1/Th2 disease in BALB/c mice[48,145] shows that the same inflammation-inducing agent can induce either Th1 or Th1/Th2 disease presumable because in some mouse strains there is sufficient Th2 directing cytokines (?IL-2 and IL-4) produced early on to overcome innate or regulatory responses that would ordinarily move the system in the direction of a Th1 response. Another factor may be the nature of the antigen or antigens that are driving the inflammation. Thus, it has been shown that low affinity interactions between antigen and TCR favor Th2 T cell differentiation;[157] this, in turn, may relate to the fact that such interactions allow the cell to produce the small amounts of IL-4 necessary to initiate Th2 responses.[74] In any case, this mechanism could explain the fact that mice with TCR α-chain deficiency develop a Th2 colitis because these mice recognize antigens using a TCR composed of dimeric receptors composed of two β-chains that necessarily recognize antigen with low affinity.[86,135] Yet another reason a mucosal inflammation may be related to a Th2 response relates to a unique model of mucosal inflammation, the oxazolone-colitis model, in which it has been shown that the Th2 response is due to the induction of NKT cells that produce IL-13.[79] This model is discussed in detail below. In the present context, however, it should be noted that the development of these cells may require particular cytokines, including cytokines produced by epithelial cells, that direct Th2 NKT cell differentiation.

Once a pathway of Th differentiation develops in a given model that pathway is usually, but not always, fixed. In the latter regard, there are at least two models in which an initial Th1 inflammation later becomes a mixed Th1/Th2 inflammation, the IL-10 knock-out model and the Samp1/Yit model.[5,190] The reason for this partial "conversion" is not clear in either of these models, but in relation to IL-10 deficiency one might speculate that lack of IL-10 negative regulation leads to excessive Th2 responses as well as excessive Th1 responses, although the latter is noted first because it more easily developes in the mucosal microenvironment.

The possible significance of dividing models of mucosal inflammation into Th1 and Th2 models lies in the fact that these models may relate to specific forms of human IBD. Th1 models, for instance, invariably give rise to transmural inflammations that generally mimic Crohn's disease that is now recognized as a Th1 inflammation as well.[63,70,138] The latter has been more or less definitively established by the fact that patients with Crohn's disease can be effectively treated with anti-IL-12 and such therapy results in normalization of Th1 cytokines without effects on Th2 cytokines.[129] This equivalence of murine Th1 models to human Crohn's disease is not mirrored by a parallel equivalence of murine Th2 models to human ulcerative colitis. In two separate studies relating to this issue, it was shown that repletion of SCID or RAG-2-deficient mice with Th1 or Th2 T cells bearing T cell receptors (TCRs) specific for albumin peptides and then infected with recombinant *E. coli* expressing ovalbumin (ECOVA) resulted in a colitis developing in both cases that was somewhat different, although in neither case did the Th2 cells give rise to a histologic picture resembling ulcerative colitis.[87,220] Thus, in one study repletion of cells from IFN-γ-deficient mice (Th2 cells) gave rise to an inflammation that was more superficial and associated with a greater neutrophil infiltration as compared to the inflammation seen with repletion of cells from IL-4-deficient mice (Th1 cells); in addition, the Th2 cell-driven inflammation was associated with a follicular pattern of cell infiltration whereas the Th1 cell-driven inflammation was associated with a diffuse pattern of cell infiltration.[219] However, the degree of crypt loss was similar in both inflammations. Similarly, another study, polarized Th1 and Th2 cell lines were used for repletion and in this case the only difference in the pattern of inflammation noted was that the Th2 cell-driven colitis was characterized by a lesser T cell infiltration and a greater neutrophil and eosinophil infiltration than the Th1 cell-driven model.[87] From these studies, one might conclude that all other factors being equal, a Th2 polarity of the T cells driving an inflammation does not per se lead to a distinctive inflammation similar to ulcerative colitis. However, a different picture emerges with respect to oxazolone colitis. As will be discussed in considerable detail below, in this case a Th2-inflammation driven by IL-13-producing NKT cells does in fact lead to an lesion similar to ulcerative colitis and this similarity is mirrored by the fact that ulcerative colitis lesions also contain IL-13-producing NKT cells.[68] Thus, the distinction between models mimicking Crohn's disease and ulcerative colitis lies not only in the Th1/Th2 polarity of the T cells, but also in the type of T cell involved in these diseases.

Effector Cell vs. Regulatory Cell Defects as a Cause of Experimental Mucosal Inflammation

One of the major characteristics of the mucosal immune system is its capacity to respond vigorously to mucosal pathogens and yet to respond poorly, if at all, to commensal bacteria comprising the mucosal microflora or to food antigens.

This Janus-like quality allows the mucosal immune system to reconcile its host defense function with its need to establish a blissful tolerance to noninjurious components of the mucosal environment that therefore preempts the development of mucosal autoimmunity. Responsiveness vs. nonresponsiveness to components of the mucosal microflora is also reflected in the models of mucosal inflammation—as one might expect from the fact discussed in detail below that the mucosal microflora drive the inflammatory response. The fact is that such models of inflammation can either be the result of an excessively robust response to components of the mucosal microflora or of a poorly developed tolerogenic response to these same components.

The mechanisms governing mucosal unresponsiveness (also known as oral tolerance) are only slowly revealing themselves and indeed the study of models of mucosal inflammation has contributed to this understanding. It is now clear that tolerance is a two component process involving both the establishment of anergy/deletion of reactive cells and the induction of regulatory cells that suppress responses via their production of suppressive cytokines or negative cell-cell interactions.[107,192] Anergy/deletion is the major mechanism of oral tolerance when the antigen load is high and occurs in the mucosal lymphoid tissue itself or in systemic lymphoid tissue as a result of the dissemination of mucosal antigens (review reference 192). With respect to antigens in the mucosal microflora, it seems likely that these are functionally equivalent to "true" self-antigens since they are present in the body for the life of the organism and there is now firm evidence that these antigens have ready access to the internal milieu.[8,119] Thus, these antigens could conceivably induce tolerance in the thymus (as do self-antigens) as well as in the periphery. Regulatory cell induction is the second mechanism of tolerance and is most obviously occurring at low antigen loads.[107,192] However, this may be more apparent than real, since it is likely that regulatory cells are produced at all concentration of antigens but are more difficult to discern in the face of a deletion process. In this context, it is important to mention that induction of regulatory cells may function cooperatively with induction of anergy/deletion in that the latter process is unlikely to eliminate all cells responsive to mucosal antigens and the cells that escape this process require control by regulatory cells.

Initial studies of regulatory cells induced in the mucosa indicated that they were TGF-β-producing cells, termed Th3 cells in recognition of the fact that they were distinct in their cytokine profile from either Th1 or Th2 cells.[26,28,213] More recent studies, however, have provided evidence that they may also be CD25+ regulatory T cells that develop in the thymus in response to positive selection by self-antigens.[62,196] Since the latter could in part consist of mucosal antigens as noted above, it is reasonable to postulate that CD25+ regulatory cells induced in the thymus ultimately return to the mucosa and are expanded at the latter site by exposure to the same mucosal antigens. Additional support for the idea that Th3 cells and CD25+ cells are over-lapping cell populations is that CD25+ cells have been shown to mediate suppression via the expression of cell-surface or secreted TGF-β.[1,141,142,170] Finally, it is important to note that Th3 cells may also contain a population of IL-10-producing cells, so that this population of regulatory cells may be somewhat heterogeneous.[212,213]

The ready appearance of regulatory T cells following oral antigen exposure has led investigators to question what conditions in the mucosa are present that favors the induction of such cells. One factor could be that, as mentioned, these cells develop in the thymus in relation to selection by mucosal (self) antigens so that the mere presence of such antigens in the mucosal milieu is sufficient reason for their predominance in this lymphoid site. Another possibility is that mucosal T cells are induced by antigen presenting dendritic cells that have a unique capacity to induce (or expand preexisting populations of) regulatory cells.[12] Evidence for this comes from the fact that dendritic cells in the mucosal tissues have a substantially different cytokine profile than dendritic cells in the spleen.[89] In particular, Peyer's patches contain CD11c+ subpopulations that produce less IL-12 and more IL-10 than their splenic counterparts. Whether these differences originate from unique interactions with adjacent epithelial cells or from other factors in the mucosal milieu remains to be seen. Recently it has been shown that dendritic cells in the terminal ileum are cells that harbor antigens derived from bacteria in the mucosal microflora and constitutively produce IL-23.[8] This appears to be a homeostatic mechanism of antigen uptake that may paradoxically lead to the induction or expansion of regulatory T cells.

So far in our discussion we have been concerned with the induction of mucosal unresponsiveness or oral tolerance. The obverse of this, mucosal responsiveness, depends on the presence of mucosal adjuvants that are either intrinsic to the mucosal antigen or accompany the mucosal antigen and that are defined by their ability to induce positive Th1 or Th2 responses one way or another. For the most part this again involves antigen-presenting dendritic cells and/or other cells in that adjuvants induce the latter to produce the requisite amount of IL-12 or IL-4 to bring about the differentiation of Th1 or Th2 effector cells.[18] For Th1 responses, this would most likely involve (but is not limited to) substances that interact with TLRs and thereby induce IL-12 or TNF-α production. On the other hand, for Th2 responses this would involve substance that, like cholera toxin, interacts with G-protein-linked receptors that ultimately lead to inhibition of IL-12 production and the enhancement of IL-10 production.[18,19]

These mechanisms of mucosal unresponseness and responsiveness are evident in the various murine models of inflammation and indeed allow us to broadly classify the models on the basis of defects leading to deficient regulatory response or excessive effector cell response. The latter, or "Type I models" are well represented by the hapten-induced colitides, either TNBS-colitis or oxazolone colitis which is discussed at length below. Suffice to say here that TNBS-colitis induction (at least in certain mouse strains) appears to depend on a genetically-determined IL-12 hyper-responsiveness which, in turn, creates conditions leading to a Th1 response induced by TNBS.[15] A second example of this type of model is mice bearing a Stat4 transgene. These mice mount increased Th1 responses when exposed to DNP-KLH, presumably because even normal IL-12 responses evoke increased Stat4 signaling; in addition, when exposed to antigens in autologous mucosal microflora in vitro they develop cells that induce colitis in a SCID recipient, indicating that they are effector cells that overcome any protective effects of the regulatory cells normally in the cell population.[214] The latter observation illustrates the relationship of the flora to colitis induction in a mouse with an aberrant effector cell response. Yet another example of a Type I model are mice with NF-κB defects due to homozygous deletion of p50 and heterozygous deletion of p65 (p50-/-p65+/- mice).[58] Here, one again observes a heightened IL-12 response, in this case probably due to a qualitative change in NF-κB signaling that results in activated NF-κB components that specifically induce the IL-12 promotor.

The Type II models on the other hand, are well represented by colitis occurring in mice with IL-10 deficiency or foxp3 deficiency, the former directly affecting the secretion of an important regulatory cytokine, the latter a key intra-cellular component necessary for the development of CD25+ regulatory cells.[41,60,83] In addition, the cell transfer model discussed at length below is an excellent example of a Type II model, in this case revealed by the transfer of naïve T cells to immunodeficient mice alone or accompanied by mature T cells.[160,161] As we shall see, in the former case the cell population do not contain and cannot develop into regulatory cells and the mice develop colitis, whereas in the latter case, a population of cells is provided that does contain potentially expandable regulatory T cells. Finally, the existence of a number of models of mucosal inflammation that are clearly Type II models and thus originate from a regulatory T cell defect naturally raises the question of whether human inflammatory bowel disease can be due to a Type II defect. In reality, this has been difficult to prove in humans, absent secure ways of quantitating regulatory T cells (i.e., distinguishing them from activated cells that bear similar markers) and measuring their primary functional effects. One group of investigators has championed the existence of such a defect in patients (in this case, the loss of CD8+ regulatory T cells,[133,204] but this finding needs to be verified by other investigators and further substantiated as a primary disease defect.

The Role of the Bacterial Mucosal Microflora in Models of Mucosal Inflammation

Regardless of whether a mucosal inflammation occurs spontaneously or is induced and regardless of whether it is a Th1 or Th2 effector cell or a form of regulatory cell disorder, the inflammation is initiated and sustained by the resident mucosal microflora. This key fact is shown quite simply and definitively by the data in Table 1 that shows that the vast majority of experimental mucosal inflammations do not occur if the susceptible mouse is maintained in a gnotobiotic (germ-free) environment. Two seeming exceptions to this rule provide further substantiation of this point after careful analysis. The first is the IL-2 KO mouse that develops severe and aggressive gastritis, duodenitis and colitis under conventional conditions but still can manifest mild, focal gastrointestinal inflammation under germ-free conditions (at least in one study).[35,176,183] However, the latter inflammation is usually associated with peri-portal hepatic inflammation, hemolytic anemia and generalized lymphoid hyperplasia and is thus part of a generalized inflammatory state that is best characterized as an autoimmune diathesis that is independent of an inflammation in mucosal tissues (and thus the mucosal microflora). This view is strengthened by the fact that IL-2 deficient mice have poor regulatory T cell development.[115] The second exception is dextran sulfate sodium-induced colitis (DSS-colitis), a colitis that can also be observed under germ-free conditions, but only in an attenuated form and after repeated doses of dextran sulfate to achieve a "chronic" form of inflammation.[4,167,202] However, since DSS-colitis can be induced in SCID mice and thus is caused by the direct activation of macrophages by dextran sulfate occurring independently of T cell activation, it seems likely that the inflammation is not, strictly speaking, an immunologic event that is

Table 1. Comparison between the degree of microbial environment and onset of intestinal inflammation

	SPF	Germ-Free
Regulatory cell defect models		
SCID-transfer colitis	+	0
IL-10 deficiency colitis	++	0
IL-2 deficiency colitis	++	Mild, focal
Tgε26	+	0
Increased effector cell response models		
SAMP-1/Yit mice	+	0
TCR-α chain colitis	+	0
DSS colitis	+	0/+
Indomethacin colitis	+	0
HLA B_{27} TG rat	+	0

dependent on stimulation by antigens in the microflora.[46] This explains its' relative independence of the mucosal microflora.

Additional evidence that the mucosal microflora drives mucosal inflammation in experimental colidites comes from studies that demonstrate that whereas lymphocytes of mice do not proliferate when stimulated by antigens derived from bacteria in their own flora, they do proliferate when stimulated by antigens derived from bacteria in the flora of other mice, even when the latter are cage littermates and are of the same strain.[50-52] It thus appears that mice develop oral tolerance to "autologous flora" that is every bit as specific as tolerance to "self antigens".

A related observation of interest is that oral tolerance for bacterial antigens in the flora is lost in the inflamed intestine of mice with TNBS-colitis and regained once the inflammation has subsided.[52] Furthermore, tolerance to autologous flora could be restored and the colitis abrogated in the colitic mice treated with IL-10 or anti-IL-12. As discussed more completely below, these findings suggest that TNBS-colitis is, at least in part, initially driven by an innate immune response to antigens in the mucosal microflora and only later by an adaptive immune response to antigens in the mucosal microflora that cross-react with TNBS-modified proteins. A parallel situation obtains in IL-2 KO mice or Stat4 transgenic mice that are subjected to systemic immunization with TNP-KLH or other TNP-substituted proteins. Such mice exhibit rapid onset of a colitis that is (in the case of the IL-2 KO mouse) identical to the spontaneously occurring colitis.[56,214] Such colitis appears to be driven by mucosal responses to antigens (TNP-substituted proteins) that cross-react with antigens in the mucosal microflora that are not adequately restrained by tolerogenic mechanisms because of the lack of regulatory cells (IL-2 KO mice) or excessive effector cell drive (Stat4 transgenic mice).[214] Finally, it is important to note that loss of tolerance to antigens in the mucosal microflora is also a feature of human IBD, in that lymphocytes from Crohn's disease patients but not control patients can be shown to respond to antigens in their mucosal microflora with the production of cytokines.[51,52]

Yet another support for the idea that the mucosal microflora is a central factor in experimental mucosal inflammation derives from the fact that in most cases of spontaneously occurring intestinal inflammation the inflammation is most severe in the colon, the intestinal site harboring the highest concentration of intestinal bacteria. Thus, even in cases where the inflammation can also affect the small bowel under some conditions, e.g., IL-10 KO mice, the major and sometimes only intestinal area involved is the colon. This relationship is so strong that it raises the question of why experimental colitis mainly limited to the small intestine, such as in the $TNF^{\Delta ARE}$ mouse that over-produce TNF-α or the SAMP1/Yit mouse, ever occurs.[96,97] One, at the moment, purely speculative possibility is that the underlying lesions in these mice lead to changes in small intestinal epithelial barrier function so that, in effect, the intestinal site most in contact with the mucosal microflora (and thus most subject to stimulation by the mucosal microflora) is the small intestine despite the relative paucity of organisms at this site. A second possibility, also one that is poorly supported by data, is that certain underlying immune abnormalities are better compensated in the large intestine than the small intestine so that, for instance, TNF-α overproduction may be better tolerated in the large intestine than in the small intestine. This question is not trivial because it relates to the differential localization of IBD in humans that is still not explained.

Having established that antigens in the mucosal microflora drive experimental mucosal inflammations, it becomes of interest to define whether these antigens are few or many in number. If few, i.e., related to a single organism, then strategies of IBD treatment can involve antigen-specific tolerization schemes or defined antibiotic regimens. If many, and related to many classes of organisms, treatment of an underlying disorder of immunologic reactivity is more appropriate.

The first question that can be asked in this context is whether experimental inflammation can be caused by an immunologic response to single antigen. Three very carefully defined models indicate that this is in fact the case. The first is the so-called ECOVA model developed independently by two groups.[87,219] As alluded to previously, in this model SCID mice are reconstituted with TCR transgenic mice with a TCR specific for an OVA peptide and then infected with a recombinant *E. coli* (ECOVA) that expresses this peptide. It is found that infection with ECOVA but not with an *E. coli* expressing an irrelevant antigen evokes colitis in the reconstituted mice, thereby showing that a response to an antigenic peptide is sufficient to induce inflammation. The second model is the IL-10 KO mouse infected with *H. hepaticus* that develops inflammation because of a lack of a regulatory cell response.[103,104] Here, it can be shown that clonal CD4+ T cells specific for a peptide antigen derived from *H. hepaticus* can induce disease in recipient RAG2 KO mice infected with *H. hepaticus*, but not an irrelevant organism.[102] Yet a third model indicating that a single antigen can induce colitis is the CD8+ T cell-driven model of colitis in which mice expressing an OVA peptide under a epithelial cell specific promotor (and thus expressed only in epithelial cells) is injected with CD8+ T cells bearing a TCR transgene specific for this antigen and the animal is then infected with a vesicular stomatitis virus encoding OVA or simply the latter virus not encoding OVA plus cholera toxin or OVA plus anti-CD40, two dendritic cell activators.[207] Taken together, the above data definitively establish that T cell activation by antigen is a critical step in the evolution of various kinds of colitis. In addition, they indicate without question that a single antigen (or, more precisely, a single antigen peptide) can induce full-blown intestinal inflammation under various circumstances. Nevertheless, they fall short of proving that a single antigen is usually operative in experimental colitis, since the single antigen causing disease in these models may in less contrived settings be accompanied by numerous other antigens with similar disease causing capabilities.

Evidence that this latter assertion is in fact the case is inherent in the extensive analysis of the spontaneous colitis occurring in a substrain of the LPS-nonresponsive C3H/HeJ mouse, known as the C3H/HeJBir mouse. With this model it has been shown that CD4+ T cells from C3H/HeJBir mice stimulated in vitro by lysates derived from the mucosal microflora react more vigorously than similar cells from ordinary C3H/HeJ mice yet were nonreactive to either food antigens or epithelial cell-derived antigens.[33] Moreover, both whole CD4+ T and clonal CD4+ T cells from C3H/HeJBir mice transfer disease to naïve disease-free SCID recipients. Thus the disease is likely due to CD4+ T cells reactive with antigens in the microbial microflora and that individual antigens can induce disease (as in the studies discussed above).

In further studies the nature of the bacteria to which the C3H/HeJBir mouse is responding was investigated. Here it was found that while antisera from the colitic mice reacted with a multitude of bacterial antigens the great majority of these antigens were derived from *Enterobacteriacae* and *Enterococcus spp.*, which together comprise less than 1% of the total flora.[17]

Thus, it could be inferred that only a small number of organisms relative to the total number of organisms in the microbial flora were responsible for production of antigens causing disease. Finally, analysis of T cell receptors in T cell lines derived from colitic mice by determination of CDR1 length distribution revealed a pauci-clonality in the individual lines which, however, was qualitatively different in the various lines.[17,34] Taken together, this body of work indicates that while the antigens that induce effector T cells that lead to colitis constitute a small part of the universe of antigens in the microbial microflora and are derived from relatively few types of organisms, the antigens manifest very considerable multiplicity.

In a recent update of the above work, bacterial lysate DNA from C3H/HeJBir mice was inserted into a phage library which was then expressed in an *E. coli* culture and screened by antisera from the colitic mice to identify bacterial antigens recognized by the antisera.[114] About one quarter of the phage clones expressed a flagellin associated with various organisms, the others expressed a variety of other intracellular and surface proteins. Further studies revealed that other types of colitic mice such as IL-10 KO mice and MD1a KO mice also had antibodies to flagellin and that flagellin-specific T cells derived from C3H/HeJBir mice could cause disease in SCID mice. In addition, a subgroup of patients with Crohn's disease, but not with ulcerative colitis had high titres of anti-flagellin antibodies. These data strongly suggest that flagellin antigens are eliciting responses in colitic animals and humans. However, as in the case of other studies of single antigens causing disease this in itself does not prove that this is the only antigen or even the major antigen causing disease. Further studies in which the antigens previously identified as those capable of eliciting immune responses in colitic C3H/HeJBir mice need to be done to determine how truly dominant flagellin antigens are in the pathogenesis of colitis.

Another approach to the study of the range of antigens in the mucosal microflora inducing disease and one already mentioned in relation to the C3H/HeJBir model, is to determine the clonality of the T cells in the inflamed tissue. Obviously, a high degree of clonality would suggest that the mouse is reacting to a particular antigen (or set of related antigens). One example of this approach is the extensive study of T cell clonality in the TCR α-chain KO mouse wherein it was shown that the T cells in the inflamed tissue exhibit considerable T cell clonality characterized by the presence of a common (public) motif in the CDR3 region of the TCR.[135] However, as already mentioned, in the absence of the α-chain the cells express an aberrant TCR composed of two β-chains and this in itself could limit clonality. A more realistic test of this approach was conducted in relation to the cell transfer model.[2] Here again, one sees evidence of TCR oligoclonality; however, the oligoclonality is "private", i.e., it refers to oligoclonality within a given mouse and not common to all the mice. This implies that each mouse is responding to a restricted set of antigens, but that this set differs from mouse to mouse. Thus, the TCR repertoire studies do not support the idea that a very narrow group of antigens cause colitis. Of interest, a similar conclusion can be drawn from several studies of oligoclonality in T cells from human IBD lesions where again oligoclonality is present, but is private, and thus consistent with the concept that antigens eliciting responses vary from patient to patient.[164]

Studies of whether bacterial monoassociation can cause disease in certain mouse models of mucosal inflammation have also been used to assess whether such inflammation can be caused by a single antigen or at least a very limited number of antigens. Using this approach, HLA-B27 transgenic rats maintained in a germ-free environment and thus free of disease were colonized with a cocktail of five different bacterial strains.[168] The rats developed moderate colitis and gastritis if *Bacteroides vulgatus* was included in the cocktail but not if this organism was excluded. In addition, exposure to *B. vulgatus* alone led to the same degree of colonic inflammation as exposure to the whole cocktail including *B. vulgatus*, but not to gastric inflammation, suggesting that some bacteria, although not involved in the induction of disease at a primary intestinal site (in this case, the colon) may account for inflammation at secondary intestinal sites. Similar findings were found in studies of monoassociation of the TCR-α-chain deficient mouse and in mice with carageenan-induced colitis.[20,93,151] Finally, while monoassociation of germ-free mice with *Enterococcus fecalis* exhibited mild colitis, association of these mice with *E. fecalis* and *Bacterioides* species led to a severe colitis, suggesting a synergistic effect between the two strains.[172,181] Taken together, these results suggest that anaerobic bacteria, such as *B. vulgatus*, can themselves initiate colitis in HLA-B27 transgenic rats and in certain mouse models of colitis whereas other bacterial species, although unable to initiate colitis independently, have an important role in mediating inflammation in other parts of the GI tract and/or can modulate the intensity of the colitis if given along with the anaerobic bacteria. However, as in the case of studies discussed above in which single antigens cause colitis, these findings should not be taken as proof that only a single organism causes colitis in mice exposed to a conventional mucosal microflora.

A related point of interest is that specific bacterial flora present in a particular niche of the intestine may have increased importance in eliciting inflammation in that specific area. This idea is supported by recent studies in which the effect of the administration of narrow or broad spectrum antibiotics was assessed on the onset and prognosis of colitis in various regions of the intestine of IL-10 deficient mice colonized with specific-pathogen free flora was evaluated.[81] Thus, in a prevention study relating to this phenomenon, all antibiotics tested, including metronidazole, ciprofloxacin, or a vancomycin-imipenem combination, decreased inflammation in the caecum and colon. However, in a treatment study, ciprofloxacin and vancomycin-imipenem were found to be more effective in decreasing caecal inflammation. This correlated with a reduced *E. coli* and *Enterococcus faecalis* concentration in the cecum after antibiotic administration. In contrast, metronidazole and the vancomycin-imipenem reduced colonic injury and correlated with an elimination of anaerobic bacteria, including *Bacteroides* species.

Yet another observation that relates to the fact that antibiotics manifest regional differences in their ability to treat colitis arises from the treatment of patients with Crohn's disease. In particular, it has been shown that treatment of these patients with metronidazole is more effective in those with colonic versus ileal involvement whereas the opposite is true with respect to treatment with ciprofloxacin. This may reflect the fact that the ileal mucosa from Crohn's disease patients (with post-operative reoccurrence of disease) is associated with increased numbers of adherent/invasive *E. coli* strains that are more effectively treated by ciprofloxacin than metronidazole.[38]

The ability of organisms to associate with a particular intestinal niche may also potentiate its ability to cause inflammation. This is shown by the fact that in genetically susceptible Lewis or Wistar rats the creation of a jejunal self-filling blind loop can lead to hepatobiliary inflammation resembling sclerosing cholangitis and can cause reactivation of quiescent arthritis.[110] In addition, in some patients undergoing surgical treatment for morbid obesity creation of bypassed jejunoileal segments leads to the occurrence of arthritis as well as hepatic and skin inflammation.[49] In both cases, metronidazole or broad-spectrum antibiotics with activity against anaerobic organisms can reverse these "remote" inflammatory manifestations. Other examples of bacteria in a particular intestinal niche causing inflammation is the HLA-B27 transgenic rat with a cecal self-filling blind loop.[168] This surgical manipulation of the gut leads to significantly increased cecal concentrations of *Bacterioides spp.* and an increased aerobic bacteria ratio and causes markedly enhanced cecal inflammation; furthermore, exclusion of the cecum from the fecal stream decreases the total bacterial load in the cecum and results in almost complete healing of the cecal inflammation. A very similar situation obtains in TCR-α chain-deficient mice, where early removal of the tip of the cecum containing a large lympoid aggregate leads to attenuation of subsequent colitis.[134] Overall, these data paint a convincing picture that bacteria occupying a particular niches in the intestine are of increased importance in generating effector cells that ultimately cause disease in that niche and suggest that subsets of aerobic or anaerobic bacteria show regional differences in their capacity to mediate mucosal inflammation.

While certain organisms appear to be involved in the induction of mucosal inflammation, other organisms (called probiotic organisms (or, simply, probiotics)) appear to have a special capacity to quell inflammation.[180] A number of mechanisms have been put forward to explain the anti-inflammatory effect of probiotics, none of them entirely satisfactory. One mechanism is based on the idea that probiotics act by replacing bacteria in the mucosal microflora that are capable of causing mucosal inflammation. This mechanism, however, appears to be unlikely since it is clear from the preceding discussion that if this were so, probiotics would have to replace bacterial species occupying many different niches. In addition, probiotics would have to cause a massive and easily measurable alteration in the gut flora and this is manifestly not the case.

Another possible mechanism of probiotic activity is that while probiotics do not displace other bacteria, they do limit the latter's growth by producing anti-microbial peptides or by inhibiting the first step of bacterial colonization, bacterial adhesion to epithelial cells. This is possibly exemplified by the fact that one probiotic organism, *Lactobacillus salivarius* to produce a broad spectrum *bacteriocin* that exhibits anti-microbial activity against a range of microorganisms such as *Bacillus*, *Staphylococcus*, *Enterococcus*, and *Listeria* species without at the same time having activity against itself.[65] In addition, *Lactobacillus plantarum* 299v has been shown to enhance epithelial cell mucin gene expression (MUC2, MUC3), thereby inhibiting the adherence of enteropathogen *E. coli* to HT-29 epithelial cells.[117] As for the ability of probiotics to interfere with bacterial cell adherence it has been shown that certain Lactobacillus species, such as *lactobacillus GG*, exhibit excellent adhesion to epithelial cell in vitro and such adhesion inhibits cell attachment and invasion by both pathogenic bacteria such as *S. typhimurium* and *Yersinia* as well now by nonpathogenic *E. coli*.[106] This theory of probiotic activity, however, has the same problem as the displacement theory: probiotic anti-microbials do not appear to be potent enough to suppress the vast majority of the organisms in the mucosal microflora.

Another theory of probiotic activity is that probiotics enhance epithelial cell barrier function. Evidence for this concept is that the intestinal permeability of 14-day-old suckling rats is greatly enhanced by daily gavage with cow's milk, but such enhancement was counteracted by the addition of *Lactobacillus GG* to the milk.[88] In addition, while the oral administration of *Lactobacillus* plantarum 299v to mice that develop spontaneous bacterial peritonitis due to portal hypertension did not inhibit the bacterial translocation of intestinal *E. coli* and therefore did not change the course of the disease, such administration did reduce bacterial translocation in rats with methotrexate-induced colitis and in humans after abdominal surgery.[128,130,169,] Studies of IL-10 KO mice corroborate these findings, in that it was initially shown that IL-10 KO mice exhibit decreased IgA secretion and significantly lower levels of resident luminal *Lactobacillus sp*; it was then shown that the administration of oral lactulose to nursing mothers, a maneuver that increases *Lactobacillus* levels in fecal samples led to normalized levels of IgA.[120] In a related study, it was shown that spontaneously developing colitis in IL-10 KO mice could be inhibited by administration of a cocktail of probiotic organisms (called VSL #3) and that this was associated with enhanced epithelial barrier function documented by increased resistance to *Salmonella* invasion of epithelial monolayers.[121,122] Finally, in studies germane to the onset of inflammation in IL-10 KO mice, it was shown that colitis occurring after transfer of the mice from a germfree to a SPF environment was ameliorated if germ-free animals were first colonized with *Lactobacillus plantarum 299v* before transfer.[184]

A final theory of the mechanism for probiotics is that probiotics affect mucosal inflammation by down-regulating chemokine and/or proinflammatory cytokine responses through the induction of regulatory T cells. This theory is, in principle, quite attractive because it does not require the probiotics to have a "mass effect", i.e., they do not have to affect every cell, but rather a small subpopulation of cells that intrinsically operate in low number. Some evidence in favor of this theory, comes from studies showing that probiotic administration is associated with decreased inflammatory cytokine responses and increased regulatory cytokine responses. Thus, stimulation of epithelial cells in vitro by an inflammatory stimulus (*Salmonella*) leading to IL-8 production and NF-κB activation is inhibited in the presence of the aforementioned probiotic cocktail, VSL #3.[121] In addition, in vivo administration of VSL#3 leads to decreased TNF-α/IFN-γ secretion in colitic IL-10 KO mice.[122] Finally, the subcutaneous administration of *Lactobacillus salivarius* to

IL-10 KO mice or to mice with collagen-induced arthritis leads to attenuation of disease and is associated with decrease production of inflammatory mediators such as IL-12 and TNF-α associated with the appearance of increased numbers of TGF-β secreting cells.[186]

In a related series of studies indirectly bearing on the role of suppressor cells in probiotic function, it was shown initially that immunostimulatory DNA sequences (DNA containing unmethylated CpG sequences) can prevent the development and/or treat established experimental colitis of various forms.[149,166] Later, it was shown that both *E. coli* DNA and probiotic DNA were able to inhibit DSS-colitis in wild-type but not in TLR9 KO mice, but was able to do so in both TLR2 and TLR4 KO mice.[165] These data led to the suggestion that immunostimulatory sequences act by signaling TLR-9 on the surface of immunoregulatory dendritic cells, but leave unexplained the question of why sequences in all bacteria in the mucosal microflora don't ordinarily act as a probiotic. One possibility comes from the observation that *Lactobacillus reuteri*, a poor inducer of IL-12, is capable of inhibiting DC activation by other *Lactobacillus* species.[31] This suggests that different species of *lactobacillus* (and, by extension, other members of the mucosal microflora) exert different dendritic cell activation patterns.

From the above discussion of regulatory cells in relation to probiotic activity, it is clear that much additional work will be necessary to substantiate that such cells play a major role and to show that that role is mediated by immunoregulatory dendritic cells.

The Genetic Basis of Models of Mucosal Inflammation

Whether a model of mucosal inflammation is due to an obvious primary genetic defect such as IL-10 deficiency or is induced by a primary environmental factor such as administration of TNBS, the severity of the inflammation or even whether the inflammation occurs at all depends on the mouse strain in which it is expressed. This inevitably points to the fact that genetic factors present in the various mouse strains influence the inflammation. In recent years it has become possible to pinpoint the chromosomal locus or loci that encode these factors since strains of low and high susceptibility can be interbred and the precise location of loci can be identified by correlating susceptibility with inheritance of chromosomal (micro-satellite) markers associated with one or the other strain. Once a chromosomal locus has been identified by a such a "genome-wide search" it is then at least theoretically possible to zero in on the actual gene involved by creating congenic strains and correlating inheritance of disease susceptibility with informative cross-over events within the isolated chromosomal area. Assuming that these genetic factors also operate in human IBD, these techniques also offer a unique approach to identifying human disease genes.

Initial studies of genetic factors in murine models involved analysis of dextran-sulfate (DSS)-colitis, perhaps because of the ease of inducing this form of experimental inflammation in various mouse strains. It should be noted, however, that this model of mucosal inflammation is not strictly dependent on the function of activated T cells and thus the results of genetic studies using this model may not be applicable to other mucosal inflammations.[46] In any case, in the relevant genetic studies of DSS-colitis, it was found that different mouse strains manifest different levels of susceptibility to induction of inflammation and, indeed, a strain known to be susceptible to spontaneously developing colitis, the C_3H/HeJBir strain, or to be susceptible to autoimmune disease generally, the NOD strain, displays increased susceptibility to DSS-colitis.[123] In addition, it was shown that the NON/Lt mouse is resistant to DSS-colitis even though it bears the same MHC genes as the NOD strain; this and subsequent studies indicate that MHC genes are not involved as susceptibility genes in the various forms of mucosal inflammation even though there is some evidence that MHC genes do function as risk factors in human IBD.[14,154]

The above genetic studies of DSS-colitis were prelude to a genome-wide search for chromosomal susceptibility areas as described above. In this case F2 progeny of crosses between susceptible C3H/HeJ mice and resistant C57BL/6 mice were studied and definite susceptibility loci on chromosomes 2 and 5 were identified (designated *Dssc2* and *Dssc1*, respectively) along with suggestive loci on chromosomes 18, 1 and 11.[124] In addition, analysis of congenic substrains derived from the highly susceptible NOD/Lt strain carrying chromosomal intervals on chromosomes 2 and 9 from the resistant NON/Lt strain corresponding to *Dssc2* and a locus on chromosome 9 only weakly associated with disease in the genome wide search, showed that these regions led to reduced susceptibility to colitis. Taken as a whole these data indicate that multiple genetic factors influence DSS-colitis and thereby establish the precedent that even in relation to a relatively straight-forward form of experimental colitis, the underlying genetic factors are complex.

Additional studies of various murine models of inflammation have involved genome-wide screens of TNBS-colitis, IL-10-deficient mice and SAMP1/Yit mice. The screen of TNBS-colitis susceptible SJL/J mice and resistant C57BL/6 mice intercrosses led to the discovery of two susceptibility loci, one on chromosome 9 (*Tnbs1*) and one on chromosome 11(*Tnbs2*).[15] Of interest, the locus on chromosome 9 was highly significant in male mice, not in female mice whereas the locus on chromosome 11 was, conversely, highly significant in female mice and not in male mice. This apparent association with gender may be more apparent than real, however, since an analysis of male mice discordant for the association with the chromosome 9 locus did show an association with the chromosome 11 locus. The association of susceptibility with the chromosome 11 locus was corroborated by the finding that C57BL/10 mice, whose genome is 99% homologous with resistant CD57BL/6 mice, have a nonhomologous area on chromosome 11 that is coextensive with the susceptibility region in SJL/J mice and, presumably for this reason, are susceptible to TNBS-colitis. Another significant finding from this study was that susceptible SJL/J mice exhibit a greatly increased serum IL-12p40 response following IP injection of LPS as compared with C57BL6/mice and this response is a genetically determined phenomenon also linked to the chromosome 11 susceptibility locus. Thus, it is likely that the IL-12 response and susceptibility are related and that, as discussed further below, an "aberrant" IL-12 response is the immunologic basis of susceptibility to TNBS-colitis in SJL/J mice. The chromosome 11 locus is also of interest because it is a susceptibility locus for experimental asthma, and harbors a number of genes of immunologic interest, including the IL-4 and IL-13 genes as well as a newly discovered gene family known as the "TIM" genes that may influence T cell differentiation into Th1 and Th2 cells.[101] This may explain the fact that SJL/J mice are also susceptible to oxazolone-colitis—a Th2 colitis.

Additional genetic studies in this case focusing on the SAMP1/YitFc subline of the SAMP/Yit mice have yielded data that is in some ways parallel to the data derived from the analysis of TNBS-colitis. The SAMP1/Yit mouse strain was derived originally from AKR mice that were bred to study premature aging.[131] However, the breeding was not precise and some genetic contamination occurred that ultimately (and fortuitously) led to mice that spontaneously develop mucosal inflammation. Interest in this model is high because the inflammation closely resembles human Crohn's disease both in its macroscopic appearance as skip lesions in the small intestine and in its microscopic appearance as a granulomatous infiltrate affecting the full thickness of the bowel wall. However, it is not yet clear that SAMP1/Yit disease is immunologically equivalent to Crohn's disease.

In the genetic studies of SAMP1/Yit mice, genetic analysis of intercrosses between SAMP1/YitFc mice (by definition susceptible mice) and resistant C57BL/6 mice led to identification of a susceptibility locus (*Ibdq1*) on the same chromosome 9 region identified in the TNBS-colitis model noted above as well as less well defined loci on chromosome 6 and the X chromosome.[98] It was noted that *Ibdq1* contains the IL-10R (α chain) and IL-18 genes, but no mutations of these genes were found in sequencing studies. In addition, no functional abnormalities involving IL-10 or its receptor was found, and with the SAMP1/yit mice displaying increased expression of IL-18 prior to disease onset, it remains to be seen if this abnormality is genetically controlled or is merely a part of the SAMP1/Yit inflammation. The analysis of coassociation of particular loci with disease was conducted using various features of the inflammation and the chromosome 9 locus most closely correlated with the appearance of epithelial cell injury. This suggests that this region contains a disease gene affecting epithelial integrity which may in fact be the same as the disease gene affecting DSS-colitis and TNBS-colitis. If so, this could be a more general disease gene affecting several forms of experimental mucosal inflammation that could conceivably be operative in human disease as well.

Finally, a study of the genetic factors affecting the occurrence of mucosal inflammation associated with IL-10 deficiency was conducted by analysis of intercrosses of susceptible C3H/HeJBir and resistant C57BL46 mice.[61] This analysis also led to a complex genetic picture in that loci on chromosomes 3 (Cdcs1), 1 (Cdcs2), 2 (Cdcs3), 8 (Cdcs4), 17 (Cdcs5) and 18 Cdcs6) were identified. These susceptibility locus were contributed by the C3H (Cdcs1-3) or the B6 (Cdcs4-6) parent and correlated with somewhat different pathologic markers. The fact that even the resistant parent contributed susceptibility genes (albeit with relatively weak associations) underscores the difficulty of isolating susceptibility genes in human outbred populations.

By far the most significant association was with the locus on chromosome 3. This locus is quite broad (i.e., spans a large genetic region) and thus contains a large number of genes of potential interest.[61] Perhaps the most relevant is *NFκb1*, a gene encoding the NF-κB p105 transcription factor. Interestingly, in an as yet unpublished study this locus was also found to influence susceptibility to colitis occurring in mice with *Gi2α* deficiency (G. Bouma, Vrije University, the Netherlands, personal communication), suggesting that, as in the case of the susceptibility loci on chromosome 9 and 11, it may influence susceptibility to inflammation in many models of inflammation and have broad significance.

From the studies summarized above it is clear that the data so far obtained from the study of genetic factors in murine models of mucosal inflammation only scratches the surface of the possible knowledge to be gained in this area. In particular, the field must move from discovery of susceptibility loci to the characterization of specific genes within loci that are in fact responsible for disease susceptibility. Given the complexity of genetic factors in human disease, this may indeed be the only way to discover candidate genes affecting human disease.

Finally, regarding the relation of the data described above to human disease it should be pointed out that none of the disease susceptibility loci so far discovered are syntenic with the locus of the homozygous NOD2 gene mutation that has been shown to be the cause of disease in perhaps 15% of patients with Crohn's disease. Recently, however, mice with targeted deletions in this gene have been developed and it has been shown that in the absence of normal NOD2 function bacterial peptidoglycan activation of NF-κB via TLR2 is dysregulated and one sees increased activation and subsequent IL-12 and Th1 cytokine production.[211] Evidently, NOD2, via its ligand muramyl dipeptide (a substance derived from the intracellular digestion of peptidoglycan) provides a break on peptidoglycan activation of NF-κB and thus prevents excessive Th1 responses to a ubiquitous bacterial component. Thus, it now appears that the basis of disease in the presence of a functional NOD2 mutation is the facilitation of an excessive Th1 response rather than a reduced Th1 response. Interestingly, this validates and preserves the notion that Crohn's disease arises from an excessive cytokine response to components in the mucosal microflora. Further studies to corroborate this conclusion in patients are now awaited.

The Role of the Epithelial Cell Barrier in Models of Mucosal Inflammation

In recent years it has become increasingly apparent that epithelial cells play a crucial role in the maintenance of mucosal homeostasis. This role takes two broad forms: first, epithelial cells recognize and respond to a variety of microbial constituents derived from both members of the mucosal microflora or from invading pathogens and thus constitute a forward host defense line that prevents microbial colonization; second, the epithelial cells by virtue of their ability to form tight junctions constitute physical impediment to the trans-epithelial penetration of mucosal organisms. It is fair to say that the properties and functions grouped under these epithelial cell characteristics comprise the active and passive components of an "epithelial cell barrier" that regulates the ability of intact luminal organisms to gain entry into the lamina propria proper and thus limits the extent and modifies the nature of the exposure of mucosal lymphoid cells to the many potential antigens (and other stimulants) in the mucosal milieu.[158]

Perhaps the most important way by which epithelial cells recognize mucosal microbial constituents is through the expression of a range of Toll-like receptors (TLRs) that specifically bind a wide variety of microbial components. These include TLRs that bind peptidoglycans (TLR2), viral double stranded RNA(TLR3), lipopolysacharides(TLR4), flagellins(TLR5) and bacteria CpG's(TLR9). TLR signaling of epithelial cells via these components results in the production of both pro- and anti-inflammatory cytokines and cyclo-oxygenase/lipoxygenase components;[200] in addition, it results in the elaboration of chemokines that lead to the influx of both acute and chronic inflammatory cells into the lamina propria that deal with potentially invasive organisms at the mucosal surface.[200] It follows that a defect in this inflammatory response of epithelial cells could conceivably comprise the initial phase of the chronic inflammation characteristic of IBD. However, whether or not this is true is inevitably clouded by the fact that professional APCs such as dendritic cells (epithelial cells comprise a class of nonprofessional APCs) also express TLRs so that a defect in TLR expression or function could be acting through a professional APC abnormality, not an epithelial cell abnormality. With this caveat in mind, it is possible that the spontaneous colitis seen in C3H/HeJBir mice (that lack TLR4 function and that thus display LPS unresponsiveness) is a direct effect of the inability of epithelial cells to respond normally to LPS; this could lead to colitis if one assumes that TLR4 dysfunction is accompanied by compensatory overactivity of other TLRs that then cause excessive Th1 responses. In addition, as already mentioned, as discussed above, mice lacking NOD2, an intra-cellular protein recognizing a component of peptidogyclan exhibit greatly increased Th1 responses when stimulated by peptidoglycan via TLR2 because NOD2 normally down-regulates TLR2 induction of IL-12.[211] Thus, in this case, a defect in an intra-cellular protein that regulates a TLR function appears to be a factor in disease. Correlating with these data in mice, there is recent evidence that TLR expression on epithelial cells is altered in Crohn's disease.[25] If this can be proven to be a primary effect than the role of TLRs or epithelial cells in mucosal inflammation will assume considerable importance.

While defects in epithelial cell responsiveness as a manifestation of epithelial barrier dysfunction is only a potential cause of mucosal inflammation, physical disruption of the barrier is a very real, tangible cause of such inflammation. This is beautifully exemplified by mice that express a dominant-negative form of N-cadherin in epithelial cells in a random pattern leading to areas of normal epithelial cell-cell adhesion and areas of abnormal or severly disrupted cell-cell adhesion.[80] Such mice are remarkable in that they exhibit severe mucosal inflammation, but only in areas subjacent to epithelial cells having abnormal cell-cell adhesion, not in areas subjacent to epithelial cells with normal adhesion. This model shows that even in the presence of a completely normal mucosal immune system, the mere over-exposure of mucosal immune elements to the mucosal microflora leads to disease. Thus, the capacity of the epithelial cells to form a physical separation between the mucosal milieu and the mucosal immune elements is a critical component in the prevention of mucosal inflammation. It should be added that there is now excellent evidence that bacteria in the mucosal microflora do gain access to the mucosal immune system in some form and thereby stimulate tonic (innate) mucosal immune responses.[8,119] However, in ways that are still, poorly understood, such entry occurs under conditions in which frank inflammation does not occur, possibly because the organisms are no longer in an immunogenic form that can stimulate inflammation.

Representative Models of Mucosal Inflammation

The Cell-Transfer Model

One of the most important of the murine models to be intensively studied over the past decade has been the cell transfer model (also called the SCID-transfer model), wherein immunodeficient RAG2 KO or SCID mice are reconstituted with naïve CD4+ T cells expressing the CD45RB^{hi} (naïve) T cell marker or both these naïve CD4+T cells and memory CD4+T cells expressing the CD45RB^{lo} (memory) T cell marker.[161] In the former case, the mice invariably

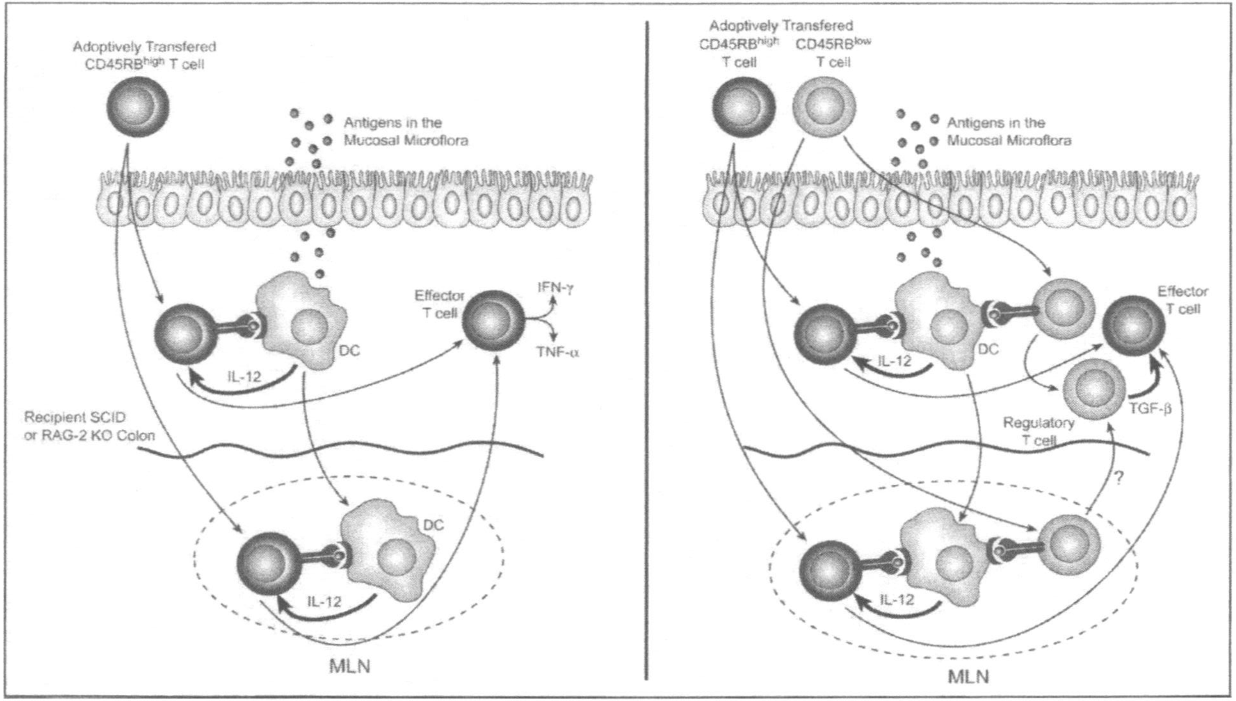

Figure 1. Cellular mechanisms of cell-transfer colitis. Left panel) Introduction of (adoptive transfer) of CD45RB^{high} (naïve) CD4+ T cells to a SCID or RAG-2 KO mouse results in activation of CD4+ effector cells in either the lamina propria or the mesenteric lymph node by APCs at these locations. Activated cells under the influence of IL-12 become Th1 polarized T cells that then traffic to the lamina propria where they produce Th1 cytokines and mediate inflammation of the colon. Right panel) Introduction (adoptive transfer) of both CD45RB^{high} (naïve) CD4+ T cells and CD45RB^{low} (mature) CD4 cells leads to activation of both effector T cells in the naïve T cell population and regulatory T cells in the mature T cell population, again in either the lamina propria or mesenteric lymph nodes. The regulatory T cells dominate the response and the mouse remains free of inflammation. Regulatory cells in the mature population are CD25+ cells whose function requires TGF-β, IL-10, and activation via CTLA-4. The antigens stimulating the cell population are either self-antigens or antigens in the mucosal microflora.

develop a transmural Th1 colitis whereas in the latter case they remain free of colitis (See Fig. 1). The particular power of this model is that it allows one to separately characterize the effector T cells that develop in the naïve T cell population in response to exposure to the intestinal microflora antigens and the regulatory cells in the memory T cell population.

Cell transfer colitis is characterized by a greatly thickened mucosa due to a dense infiltration with mononuclear cell in a pattern resembling that seen in human Crohn's disease. The basic immunologic effector mechanism underlying this inflammation is the development among the transferred naïve (CD45RB^{hi} cells) of a large population of Th1 cells producing IFN-γ and TNF-α.[163] That this is an IL-12-driven process is shown by the fact that the inflammation can be prevented by administration of anti-IL-12 and (less effectively) by anti-TNF-α.[113,163] However, the transfer of naïve cells from mice unable to produce the STAT4 transcription factor still results in disease even though STAT4 is necessary for good IFN-γ responses, probably because the cells can still elaborate TNF-α.[32] Recently, some evidence has come to light indicating that whereas IL-12 may be necessary to initiate the Th1 inflammation in transfer colitis, IL-23 may be necessary to maintain the inflammation; however, this thesis still requires addition study before it can be fully accepted.[99,153]

As in the case of other mucosal models of colitis, the colitis in the cell-transfer model requires the presence of a normal intestinal microflora. This is shown by the fact that transfer of cells to mice being reared in a germ-free facility do not develop disease and also by the fact that CD4+ T cells from colitic mice produce Th1 cytokines when stimulated by APC's pulsed with antigens in fecal extracts of normal but not germ-free mice.[22,162] As mentioned previously, while the T cells in mice with colitis manifest a restricted T cell receptor repertoire, the restriction differs from mouse to mouse (despite the fact that the mice bear the same MHC antigens and are genetically identical). This is compatible with the view that although the antigens causing the disease in any given mouse is somewhat circumscribed, a larger universe of antigens may be causing disease in the mouse population as a whole.

Turning now to the characterization of the regulatory cells in the CD45RB^{low} cell population, it has been established that the sub-population of cells involved are the naturally occurring mature (CD45RB^{low}) CD25+ regulatory cells that have been shown to be involved in the induction of peripheral tolerance and the prevention of autoimmunity in many other systems.[127,170,177] It is now known that such cells develop in the thymus as a result of positive selection by self antigens but it is still unclear why they are not deleted as are other CD4+ T cells with self-antigen specificity. One possibility is that these cells have special properties by virtue of their expression of unique genes that encode for survival and/or suppressor capability. Evidence that this is so comes from the recent finding that CD25+ cells express a molecule known as foxp3 whose absence is marked by the development of autoimmunity and colitis in both mice and humans.[83] Of particular significance to the development of colitis is that CD25+ T cells may recognize antigens in the bacterial microflora as well as self-antigens. This possibility arises from the fact that antigens in the bacterial flora have access to the internal milieu and can therefore be expressed by APC's in the thymus to nascent regulatory cells. Thus, one scenario for the development of mucosal CD25+ T cells is that the latter initially develop in the thymus in relation to antigens in the mucosal microflora and then migrate to the mucosa after leaving the thymus where they reencounter stimulatory antigens and are expanded. They are then in a position to regulate mucosal responses to antigens in the microflora that would otherwise induce inflammation. This explanation of the origin of CD25+ regulatory cells that prevent colitis explains the fact that the naïve cell population do not contain these cells and cannot serve as a source for these cells since CD45RB^{high} cells do not contain a subpopulation of CD25+ cells (which are all mature cells); in addition, they have a set of specificities for exogenous antigens that do not include either self-antigens or antigens in the mucosal microflora.

Having established that CD25+ cells are responsible for the regulatory effects of the CD45RB^{low} cells the question immediately presents itself as to how these cells mediate protection from inflammation. Since administration of antibodies to the inhibitory cytokine TGF-β

abolishes CD45RB^{low} cell protection,[160] one factor that is clearly involved is TGF-β, either that produced by CD25+ cells themselves or that produced by cells induced to produce TGF-β by CD25+ cells. Information derived from several independent studies of CD25+ T cell function strongly favors the former possibility. These studies have established that: 1) CD25+ cells bear cell surface TGF-β in a latent form (TGF-β linked with latency-associated protein (LAP) that is activated upon cell-cell contact; 2) CD25+ cell suppression in vitro is inhibited by soluble recombinant LAP, a molecule that blocks recognition of cell surface latent TGF-β; and, most importantly, 3) cells bearing TGF-β mediate suppression of inflammation in the cell-transfer model and cells from TGF-β KO mice fail to protect from colitis in this model.[142] These powerful data supporting the concept that suppression by CD25+ cells is mediated by TGF-β produced by these cells, including TGF-β bound to the cell surface, has recently been corroborated in other models of inflammation involving the development of autoimmune diabetes.[10,156] In one such study, for instance, it was shown that cells bearing surface TGF-β are present at lesional sites but cannot affect amelioration of inflammation in mice whose T cells do not respond to TGF-β because they bear a dominant-negative TGF-β receptor transgene that prevents expression of the TGF-β receptor.[71]

Another molecule implicated in the regulatory effect of CD25+ cells in cell-transfer colitis is the T cell costimulatory molecule CTLA-4, a cell-surface molecule that is preferentially expressed on CD25+ cells. Thus, in studies where purified CD25+ regulatory T cells were coadministered with naïve T cells, protection from the development of colitis was inhibited not only by TGF-β (as indicated above) but also by anti-CTLA-4.[170] If indeed two molecules are implicated in CD25+ suppression, how do the mechanisms of suppression underlying their suppressor effects relate to one another. One possibility is that these mechanisms are inter-dependent; this is favored by studies showing that stimulation of T cells via CTLA-4 enhances TGF-β production,[27] so that during regulatory events T regulatory cells may require CTLA-4 signaling to produce sufficient TGF-β to achieve suppression. Another possibility is that TGF-β and CTLA-4 subtend independent and additive mechanisms of suppression; this finds support in the well-known fact that CTLA-4 (as well as TGF-β) can be shown to mediate down-regulation of immune responses independently of other known factors.[29]

IL-10 is yet another factor implicated in the regulation of mucosal inflammation. In initial studies it was shown that neither SCID mice reconstituted with normal CD45RB^{high} and IL-10-deficient CD45RB^{low} cells nor SCID mice reconstituted with normal CD45RB^{high} and CD45RB^{low} cells and administered anti-IL10R antibody are protected from colitis.[3b] This finding strongly suggests that IL-10 produced by CD25+ cells can function as a second inhibitory cytokine that prevents the development of colitis independently of TGF-β (and CTLA-4). Since IL-10 has clear-cut suppressive effects, it is possible that it functions independently of other regulatory factors, particularly when it is produced in large amounts. This is the situation that probably obtains in two cases: 1) when RAG-2 KO mice that produce excess IL-10 due to a IL-10 transgene expressed in epithelial cells;[75] or 2) when SCID mice that are administered so-called Tr1 regulatory cells that produce high levels of IL-10 (and very little TGF-β) are reconstituted with CD45RB^{high} cells (and that are thus expected to develop colitis).[73] However, when IL-10 is not produced in high amounts, as in the case of ordinarily cell transfer colitis, its' regulatory function may relate to its role in supporting the expansion of TGF-β-producing CD25+ cells or in the function of such cells. This view is favored by findings in the TNBS-colitis model to be discussed below that show that TGF-β-producing cells do not expand in the absence of IL-10.[67]

In more recent studies, the role of IL-10 in the regulation of inflammation has been refined by studies of its relation to inflammation caused by colitogenic T cells in CD45RB^{high} and CD45RB^{low} cells separately. Here it has been found that CD45RB^{low} cells obtained from normal mice, as expected, do not cause colitis whereas those obtained from IL-10 deficient mice cause colitis; in addition, transfer of CD45RB^{low} cells from normal mice to RAG-2 recipients cause colitis if the mice are simultaneously treated with anti-IL-10R.[3,127] Together, these studies suggest that colitogenic T cells exist within the CD45RB^{low} cell population that are ordinarily inhibited by IL-10 from functioning as such by IL10 in the RAG-2 KO host. Indeed, even treatment of normal mice with anti-IL-10R results in the development of colitis suggesting that such regulation also occurs in normal mice. Finally, it has been shown that while anti-IL-10R treatment prevents the protective effect of CD25+ T cells on inflammation caused by CD45RB^{low} CD25+ Tcells from IL-10 deficient mice, i.e., colitogenic T cells in the mature T cell population, it does not prevent the protective effect of these cells on inflammation caused by (splenic) CD45RB^{high} T cells, a naïve cell population that does contain a subpopulation of preformed colitogenic T cells.[3] The reason for this discrepancy is uncertain, but it has been suggested that it relates to the affinities of naïve and mature T cells for mucosal antigens, and thus their susceptibility to be regulated. Alternatively, it relates to the ability of the two populations to rapidly produce Th1 cytokines which act as countervailing factors in the development of regulatory responses.

Another point that should be noted in relation to IL-10 is that in contrast to TGF-β-producing CD25+ regulatory cells, IL-10-producing CD25$^-$ Tr1 regulatory cells develop in post-thymic T cell populations in relation to exposure to exogenous antigens.[173] Nevertheless, it is evident that the latter cells do not appear to play a major role in cell-transfer colitis since inflammation is only partially affected by transfer of CD25$^-$ T cells in the CD45RB^{low} population (and then only at the highest cell dose) despite the fact that a sub-population of naïve CD45RB^{high} T cells transferred to SCID or RAG-2 KO mice might be expected to develop into these cells.[3a] Moreover, this failure of Tr1 cells to serve as regulatory cells in cell-transfer colitis cannot be attributed to the fact that the stimulating antigens are arising from normal flora rather than from a mucosal pathogen. This is evident from a recent study that shows that RAG-2 KO mice reconstituted with CD45RB^{high} cells and then infected with *H.hepaticus* developed inflammation when cotransferred CD25+ cells but not when cotransferred CD25- cells. In addition, the protection afforded by the CD25+ cells was typical of CD25+ cells in that it was blocked by administration of anti-TGF-β or anti-IL-10R.[127] In contrast, infection of normal mice with *H. hepaticus* led to development of CD25+ regulatory cells that when transferred to RAG-2 recipient repleted with cells from IL-10 deficient mice prevented inflammation following infection with *H. hepaticus*.[103] In this case the regulatory cells were IL-10 producing Tr1-like cells whose activity was blocked by anti-IL-10R but not by anti-TGF-β. These contrasting results suggest that development of cells in immunodeficient/lymphopenic RAG-2 KO mice favors the emergence of CD25+ regulatory cells whereas the same infection in normal mice favors the development of Tr1 regulatory cells. Thus, it may be that the immunodeficient/lymphopenic host offers a permissive milieu for the development of CD25+ cells which may not obtain in animals with intact immune systems. This introduces the possibility the studies of cell-transfer colitis may lead to an over-emphasis of the role of CD25+ T cells mucosal homeostasis.

A very important recent development relating to the role of regulatory cells in cell transfer colitis is that the administration of CD25+ T cells to mice with previously established colitis can cure the on-going inflammation.[140] Examination of tissues in mice with colitis administered the regulatory cells revealed that the latter were present both in the lamina propria and in the mesenteric lymph nodes and that at both sites the CD25+ cells were found in close association with clusters of CD11c+ dendritic cells and T cells. These findings suggest that CD25+ regulatory cells can operate at the level of APC's or T cells to affect inhibition of an inflammatory response. A similar result was obtained in a model of colitis induced in SCID mice repleted with CD45RB^{high} cells and then infected with *Leishmania major* organisms.[111] In this case it was shown that the therapeutic effect was reversed by anti-TGF-β, anti-IL-10 and/or CTLA-4, the factors mentioned above as being involved in CD25+ T cell regulation. However, while each of these antibodies reversed inflammation, anti-IL-10R and anti-TGF-β did not reestablish and anti-CTLA-4 did reestablish IL-4 and IFN-γ secretion. This interesting result suggests that the mechanism of suppression for the cytokines involves a complex interaction of these factors.

As mentioned at the outset of this discussion of the cell-transfer model, antigens in the bacterial flora are the driving force of the inflammation, just as it is in other models of mucosal inflammation. This, in turn, implies that the inflammatory process involves, at its inception, the uptake of such antigens by antigen-presenting cells (APC's) in the Peyer's patches or the lamina propria followed by the presentation of processed antigen to T cells either in the lamina propria or in the mesenteric lymph nodes (following migration of APC's to the latter sites). That such processing and migration is indeed occurring during the inflammation seen in the cell-transfer model is shown in recent work documenting that in this model increased numbers of CD134+ cells (i.e., cells bearing an surface molecule associated with APC's) appear in the mesenteric lymph nodes and depletion of these cells with systemically administered anti-CD134 leads to reversal of colitis.[126] In this context, it is also important to note that cell transfer colitis does not occur in mice deficient in migration-inhibitor factor, a cytokine important for the function and activation of APCs in tissues.[42]

In summary of this discussion of cell transfer colitis, the major "take-home" lesson is that antigens in the mucosal microflora can induce both effector cell and regulatory cell responses in the mucosal immune system and it is the balance achieved by these responses that determines whether or not inflammation will occur.

Hapten-Induced Experimental Colitides

A second and equally revealing class of models studied in recent years are the hapten-induced models of inflammation, i.e., the colitides induced by the intra-rectal application of the haptenating agents trinitrobenzene sulphonic acid (TNBS) or oxazolone in the presence of ethanol.[13,145] These models have proven to be useful in the elucidation of the mechanisms of mucosal inflammation because they allow easy study of the events governing the onset of the inflammation, and, in common with the cell-transfer model, the events governing the regulation of the inflammation. In addition, while TNBS-colitis is a Th1-mediated inflammation, oxazalone-colitis is a Th2-mediated inflammation; this provides an opportunity to compare and contrast the inflammations induced in similar ways but governed by different types of T cells. Finally, the hapten-induced models are to some extent strain specific and therefore allow the study of genetic factors in mucosal inflammation.

TNBS-Colitis

We shall begin our discussion of hapten-induced models of inflammation with a consideration of TNBS-colitis, a colitis characterized by a transmural inflammation that on histologic grounds is considered a model of Crohn's disease. While this form of experimental inflammation has been known and studied for at least two decades,[139,217] it was not until a mouse model of the colitis in SJL/J mice was studied in the mid-1990's and shown to be a Th1-mediated disease replete with over-production of IL-12, IFN-γ and TNF-α that the model gained wide currency as a vehicle for the study of experimental mucosal inflammation.[145] Later it was appreciated that to some extent TNBS-colitis was strain specific in that certain strains (e.g., C57BL/6 mice) were highly resistant to the development of colitis whereas other strains (e.g., SJL/J mice) were highly susceptible to the development of colitis.[57] In addition, while a "pure" Th1 inflammation obtained in the SJL/J strain, in other strains (e.g., Balb/c mice) a mixed Th1/Th2 inflammation was induced by hapten administration.[48] These strain differences allow study of haptene-induced colitis mediated by both Th1 or mixed Th1/Th2 responses and, in addition, are the basis of genetic studies of TNBS-colitis. Finally, since mice of many strains don't develop colitis following administration of TNBS, the strain differences unequivocably establish that TNBS-colitis is not simply due to a chemical injury to the colonic mucosa.

An initial dividend accruing from early studies of this model was that the inflammation required antigen presentation of inducing antigens in that it was blocked by administration of anti-CD40L, an antibody that interferes with APC-T cell signaling.[193] This observation was later reiterated in studies of other models and established that antigen processing by APC's is a critical early step in the establishment of the inflammation.[112,113] A second dividend was that administration of monoclonal anti-IL12 antibody (anti-p40 chain antibody) not only prevented TNBS-colitis when given concomitantly with the TNBS but also was a highly effective treatment for TNBS-colitis when given after colitis was established.[145] These observations established that TNBS-colitis was indeed an IL-12-driven inflammation or else one driven by IL-23 (a related cytokine that also contains a p40 chain). The efficacy of anti-IL-12 p40 in the treatment of TNBS-colitis was later replicated in other Th1-mediated experimental colitides and ultimately led to a trial of anti-IL-12 as a therapeutic agent in humans with Crohn's disease.[129] This trial has recently been completed and has provided initial evidence that the human inflammation is also an IL-12-driven inflammation and that anti-IL-12 administration is an effective treatment for this disease. It should be noted that studies of the possible mechanisms by which anti-IL-12 led to amelioration of TNBS-colitis showed that the most likely mechanism was not abatement of IFN-γ secretion, but rather the induction of apoptosis of effector T cells mediated, at least in part, by Fas/Fas ligand interactions.[69] This bodes well for the durability of the anti-IL-12 treatment of patients.

Given the prominent role of TNF-α in Crohn's disease it was of interest, early on, to determine the role of this cytokine in TNBS-colitis. Indeed, it was found that this colitis could not be induced in TNF-α KO mice and was particularly severe in mice over-expressing TNF-α due to the presence of a TNF-α transgene.[146] Since TNBS-colitis can be induced in an IFN-γ-R -deficient mouse this means that TNF-α is even more important than IFN-γ to the immunopathologic process of this inflammation.[24] This may be because TNF-α has a greater role in the down-stream tissue damage occurring in TNBS-colitis than does IFN-γ. Many other cytokines (and chemokines) are produced in the course of TNBS-colitis, including IL-1, IL-6, IL-17 and IL-18.[66,201] These cytokines add to the pathogenic milieu of TNBS-colitis but are probably not essential for the development of the inflammation.

Surprisingly, TNBS-induced colitis is similar to the spontaneous colitides occurring in mice with immunologic abnormalities, in that it is driven by antigens in the mucosal microflora. Thus, while TNP-haptenated proteins may be the initial antigenic stimulus leading to inflammation in TNBS-colitis, antigens in the mucosal microflora become secondary antigenic stimuli. This is shown by the fact that TNBS-colitis is characterized by cells that react with antigens in the microflora which then disappear with anti-IL-12 treatment.[52] In addition, intra-rectal administration of ethanol to mice with the intention of breaking the mucosal barrier and inducing heightened tolerance (unresponsiveness) to mucosal microflora antigens, is followed by increased resistance to the development of TNBS-colitis. A further relationship between the mucosal microflora and TNBS-colitis arises from the observation that mouse strains that are susceptible to the induction of TNBS-colitis are also strains that exhibit high in vivo IL-12 p70 responses following LPS administration; in addition, as described above, the inheritance of both of these characteristics can be mapped to the same chromosomal region on chromosome 11.[15] Thus it appears that the tendency to develop TNBS-colitis is a genetic trait that depends on an ability to react to a bacterial product such as LPS, or perhaps other bacterial products that stimulate the innate immune system via TLRs and/or other sensing molecules. (See further examination of this observation in the discussion above of the genetic aspects of mouse models of inflammation).

As in the case of cell-transfer colitis, TNBS-colitis can be influenced by the provision of regulatory T cells. However, in this case the latter are not provided by direct administration of cell populations containing regulatory cells (as when $CD45RB^{low}$ cells are coadministered this $CD45RB^{high}$ cells in the cell transfer model) but in a more physiologic way involving the feeding to TNP-haptenated colonic proteins (a complex mixture of proteins obtained by homogenization of colonic specimens and then subjecting the mixture to TNP-haptenization by TNBS) at the same time as TNBS is given by the intra-rectal route to induce colitis.[144] Such feeding of TNP-haptenated protein has the effect of inducing regulatory T cells in the same manner as does the feeding of soluble protein in the induction of oral tolerance. Thus, as a result of such feeding, T cells producing TGF-β appear in the lamina propria that suppress TNBS-colitis in the same mouse or in a second mouse to which the cells are transferred. Since this protective effect of feeding is reversed by systemic administration of anti-TGF-β, it can be safely inferred that, as in the cell-transfer model, it is in fact mediated by TGF-β. Interestingly, the TGF-β-producing T cells necessary for the protective effect did not have to be induced by the same antigen as that responsible for the colitis: feeding oxazolone, another haptenating agent (see discussion below) in a soluble form also prevented TNBS-colitis (I. Fuss, personal observation). This cross-protection undoubtably arises from the fact that the regulatory cells are antigen nonspecific in their suppressor function. These feeding studies show very vividly that TNBS-colitis results from a Th1 response that is unopposed by regulatory T cell mechanisms normally limiting responses to mucosal antigens capable of causing inflammation. The same conclusion can be derived from studies of the cell-transfer model, but in this case it is clearer that the regulatory response is identical to that normally occurring in the feeding of antigens and oral tolerance induction. As such, they highlight the fact that IBD may sometimes result from a dysregulated response to mucosal antigens.

As noted in the discussion of cell-transfer colitis, IL-10 as well as TGF-β plays an important role in the regulation of mucosal inflammation. This is also the case in TNBS-colitis as indicated in extensive studies in which the role of these regulatory cytokines were explored using cell transfer techniques in combination with the administration of antibodies to TGF-β and IL-10. In initial studies of this type it was shown that CD4+ T cells extracted from the lamina propria of mice fed TNP-haptenated colonic protein (TNP-HCP) and then administered TNBS per rectum prevented TNBS colitis in a recipient mouse whereas similar cells extracted from the lamina propria of mice only administered TNBS per rectum did not prevent colitis .[67] In addition, the cells from fed mice produced much higher amounts of TGF-β in vitro than cells from unfed mice but the amounts of IL-10 produced by the two cell populations were about the same. This was the first indication that TGF-β rather than IL-10 was the primary regulatory cytokine in this system.

In further studies, mice fed TNP-HCP were subsequently given TNBS per rectum along with anti-IL-10 or anti-TGF-β.[67] While in both cases the antibodies prevented the protective effect of the feeding, the antibodies had very different and revealing effects on cytokine secretion. Whereas anti-TGF-β administration led to greatly reduced TGF-β secretion, it had no effect on IL-10 secretion; in contrast, anti-IL-10 administration led to both reduced IL-10 secretion and reduced TGF-β secretion. Additional studies involving transfer of cells and administration of anti-IL10 to donor mice or to recipient mice undergoing TNBS-colitis established that the effect of anti-IL-10 was not on the induction of TGF-β-producing cells in the donor mice, but rather on their ability to either proliferate or to exert regulatory function, in the recipient mice. Taken together, these studies strongly suggest that the protective effect of IL-10 secretion in the TNBS model is an indirect consequence of its effect on the ability to generate cells producing TGF-β. In addition, they show that this supportive effect of IL10 is manifest, not during the induction of TGF-β-producing cells, but later, during the period these cells would be exerting regulatory effects. At present, it is unclear how IL-10 is providing this supportive effect. One possibility is that it modulates the Th1 response, so that the latter does not inhibit the expansion of TGF-β producing cells; another is that it facilitates TGF-β suppressor function by preventing potential target cells from becoming resistant to TGF-β.

In the cell transfer studies described above the regulatory cells contained within the lamina propria cell populations following TNP-HCP feeding were always administered at the time of initiation of TNBS-colitis, not after the latter had been well established. In fact, in unpublished studies it has not been possible to treat (reverse) established TNBS-colitis by administration of this source of regulatory cells (I. Fuss, personal observation). Whether this is due to the presence of relatively low number of regulatory cells in the lamina propria of fed mice or because regulatory cells in this system do not expand or cannot function in the face of a full-blown Th1 response is not known. If the latter is the case, than regulatory cell function in TNBS-colitis will differ from that in cell-transfer colitis since, as we have seen, administration of CD25+ cells to mice with established cell-transfer colitis is an effective means of reversing inflammation in the latter model.

While it is not currently possible to treat TNBS-colitis using cells derived from TNP-HCP-fed mice, it is possible to treat this colitis with TGF-β-producing cells induced by a form of gene therapy. In particular, it has been shown that intra-nasal administration of a plasmid encoding active TGF-β (TGF-β not bound to LAP) under a CMV promotor leads to the appearance of cells in the spleen and lamina propria (both T cells and macrophages) that are producing TGF-β.[92] This observation was followed by the demonstration that administration of the plasmid not only prevents TNBS-colitis, it very effectively treats this colitis. A second form of the TGF-β plasmid consists of a TGF-β gene under the control of a promotor whose activity is regulated by the presence of the antibiotic, doxycycline.[91] This plasmid allows one to control TGF-β production by doxycycline coadministration and, with its use, it has been possible to show that established TNBS-colitis is treatable by plasmid administration, but only when doxycycline is administered. This relates the therapeutic effect of the plasmid to TGF-β production in a precise and definitive way.

An observation of great interest to arise from the TGF-β plasmid administration studies just described is that the TGF-β production was always accompanied by IL-10 production.[91] Investigation of this fact disclosed that the IL-10 promotor has a Smad binding site and that a luciferase reporter gene linked to a TGF-β promotor produces a luciferase signal in cells exposed to TGF-β.[91] In addition, it was found that TGF-β-producing cells both synthesize IL-10 themselves and induce other cells to synthesize IL-10. These observations underscore the close relation of TGF-β and IL-10 and suggest that TGF-β and IL-10 production are coordinated events. In addition, they suggest that TGF-β requires IL-10 for its suppressor function as predicted by the in vivo studies described above.

As mentioned at the outset of this discussion of hapten-induced colitis, TNBS-colitis is strain-specific and therefore allows genetic studies of the factors that lead to susceptibility to disease induction. As described above, these studies led to the identification of susceptibility regions on chromosomes 9 and 11, the latter subsequently verified by the fact that C57BL/10 mice, a strain that is identical with the C57BL/6 strain save for the susceptibility region on chromosome 11 that it shares with the SJL/J strain, is also susceptible to TNBS-colitis.[15] This genetic area is of great potential interest because it harbors the IL-12 p40 gene as well as other genes that relate to cytokine production by T cells. In addition, as also discussed above, independent studies of systemic IL-12 p70 responses of SJL/J mice and C57BL/6 mice following LPS administration revealed that SJL/J mice exhibit far higher responses than C57BL/6 mice and this capacity is a genetic trait that maps to the susceptibility region on chromosome 11. Thus, the picture that emerges is that one cause of the susceptibility to TNBS-colitis in SJL/J mice is that such mice have a tendency to mount high LPS-induced IL-12 responses, or, alternatively, high IL-12 responses to other TLR ligands.

These genetic findings allow us to create a rational scenario that describes the immunological processes that underlie the development of TNBS-colitis in SJL/J mice. In this scenario, the introduction of TNBS/ethanol into the rectum of mice results in a compromise in the epithelial cell barrier and leads to the entry of bacteria in the mucosal microflora into the gut wall. This, in turn, leads to the stimulation of macrophages by bacterial lipopolysaccharide via TLR4 or by other bacteria-derived TLR ligands. While in most mouse strains, such stimulation leads to only modest IL-12 production, in the colitis-susceptible SJL/J strain it evokes a robust IL-12 response that prepares the mucosal immune system for the induction of a vigorous TNP-specific Th1 response to TNP-haptenated proteins formed by the exposure to TNBS. A second phase of the inflammation occurs when the mouse with nascent colitis begins to respond to antigens in the microflora as well as to the TNP-haptenated protein. This response to antigens to which the mouse was formerly tolerant occurs because regulatory cells that normally maintain such tolerance are compromised by the infection/inflammation. Only by induction of a more vigorous regulatory response via antigen feeding does one get a regulatory cell response sufficient to prevent inflammation. Thus, in this scenario, TNBS-colitis in SJL/J mice is seen as an amalgam of an innate and adaptive immune response involving both effector and regulatory elements.

Finally, as mentioned above, TNBS-colitis can also be induced in BALB/c mice, although in this case the characteristics of the colitis and its underlying cytokine basis is somewhat different from that in SJL/J mice.[48] In BALB/c mice, more prolonged administration of TNBS is necessary to induce colitis so that here it is reasonable to assume that colitis induction does not require the jump-start provided by the genetically determined elevated innate IL-12 response. Of considerable interest is the fact that in recent studies it has been shown that TNBS-colitis can be induced in both SCID and RAG2 KO BALB/c mice that is comparable to that seen in wild-type BALB/c mice.[64] This finding suggests that even in the absence of a gene affecting responses to LPS or other TLR ligands, the initiating event of the colitis is an innate immune response. One caveat concerning these data is that since SCID and RAG2 KO mice lack regulatory T cells it seems likely that while TNBS-olitis can develop in the absence of T cells, it can only do so because of the concomitant absence of regulatory T cells.

Oxazolone Colitis

A second and very different form of hapten-induced colitis is that induced by another haptenating agent, oxazolone, and is thus called oxazolone-colitis. This colitis is also induced in SJL/J mice or C57BL/10 mice, i.e., mice with a region on chromosome 11 that predisposes to TNBS-colitis, but the susceptibility of various mouse strains to this form of colitis is less well studied than for TNBS-colitis and it is still unclear what genetic factors predispose to this form of hapten-induced colitis. In initial studies, it was found that administration of oxazolone led to a short-lived colitis (lasting about 5 days in surviving mice) that was characterized by superficial inflammation of the gut wall, associated with edema, increased goblet cell activity and mucous formation, ulceration of the epithelial cell layer and neutrophil accummulation; in addition, the inflammation was limited to the distal half of the colon.[13] This pathological picture differed greatly from the densely packed, transmural lesion of TNBS-colitis and while the latter colitis resembles Crohn's disease, oxazolone colitis was more reminiscent of ulcerative colitis.

A second major difference between TNBS-colitis and oxazolone-colitis emerged from studies of the cytokine secretion patterns in the latter inflammation. These studies revealed that CD4+ T cells present in oxazolone-colitis lesions produce large amounts of IL-4 and IL-5, and normal or decreased amounts of IFN-γ.[13] Furthermore, it became evident that this Th2 cytokine response was in fact the immediate cause of oxazolone-colitis since administration of anti-IL-4 to mice at the time of disease induction prevented the development of colitis, whereas anti-IL-12 administration led to a severe pancolitis affecting both distal and proximal segments of colon. Finally, it is important to mention that TGF-β secretion was greatly increased in oxazolone colitis and seemed to account for the distal distribution of the disease since secretion of this cytokine was higher in the proximal colon than in the distal colon and anti-TGF-β led to severe colitis affecting the entire length of the colon.[13]

In further studies a more prolonged form of oxazolone-colitis, lasting 11-14 days was examined.[79] This was achieved using a presensitization regimen in which the skin was painted with oxazolone about five days before intra-rectal challenge and the latter was performed with a low concentration of oxazolone. With this "chronic" model it was observed that the initial IL-4 response gave way after about 4-5 days to an increasing IL-13 response and the latter was also intrinsic to disease pathogenesis since the colitis was prevented by administration of IL-13R2α-Fc, a soluble receptor for IL-13 that blocks IL-13 interaction with its signaling receptor.

These findings led to studies seeking the cellular origin of the IL-13. An important clue came from the observation that cell populations passed through columns with bound cell-specific antibodies were depleted of IL-13-producing cells. This suggested that the latter bore a high affinity Fc receptor and drew attention to the possibility that the IL-13 was being produced by an NKT cell. This, in fact, proved to be the case, since subsequent study showed that oxazolone colitis was prevented by treatment of mice with anti-NK1.1 antibodies i.e., antibodies that deplete NKT cells. In addition,it was shown that oxazolone colitis was prevented by treatment of mice with an anti-CD1 antibody that blocks NKT cell activation and that CD1 KO mice or Jα281 KO mice (the latter mice that cannot express a component of the invariant TCR that is expressed by most NKT cells) are highly resistant to the development of oxazolone-colitis. Finally, stimulation of cells from the lesional tissue of oxazolone-colitis with the glycolipid antigen, α-galactosyl-ceramide, an antigen that is a specific stimulant of NKT cells bearing an invariant TCR and that requires CD1-restricted presentation was shown to induce IL-13.[79] Taken together, these studies left little doubt that the IL-13-producing cell in oxazolone-colitis was in fact an NKT cell.

In order to understand the significance of the fact that NKT cells are intrinsic to the inflammation of oxazolone-colitis, we need to digress for a moment to discuss the nature and function of these cells. An NKT cells is a kind of hybrid cell that has characteristics of both an NK cell and a T cell in that it is a cell with T cell receptors (TCRs) that nevertheless bears NK markers (such as NK1.1). In addition to its distinctive markers, the NKT cell can be defined by the fact that it recognizes glycolipid antigens presented to it by a highly conserved MHC class I-like molecule, (CD1 in mice and CD1d in humans) and in most cases it expresses a so-called invariant TCR that is made up of a particular TCRα chain (Vα14-Jα281 in mice and Vα24-Jα18 in humans) linked to a restricted repertoire of Vβ chains, hence the fact that it can be stimulated by a particular prototype glycolipid, the aforementioned α-galactosyl-ceramide and can be recognized by α-galactosyl-ceramide-loaded tetramers that bind to the invariant TCR.[100] However, it is important to recognize that while the great majority of NKT cells have these attributes, a minority of NKT cells ("nonclassical" NKT cells), while still recognizing antigen restricted by CD1, bear noninvariant TCRs.[21] Other relevant characteristics of NKT cells are that in mice they are either CD4+ cells or CD4-/CD8- cells and in humans they can be CD8α-bearing cells as well.[100] There is some evidence that these phenotypic differences define cells with different effector functions and different capacities to produce Th1 or Th2 cytokines; however, the factors that control their differentiation into cells with these different properties are poorly understood. Yet another characteristic of NK cells is their ability to act as cytotoxic T cells and it is this property that accounts for the fact that NKT cells have anti-tumor cell activity and act as effector cells in certain inflammatory states such as experimental hepatitis.[6,205] In the latter regard, it has been shown that NKT cells in hepatic lesions secrete perforin and granzymes that mediate cellular injury.[37] Finally, it is important to emphasize that NKT can be activated by CD1-bearing antigen presenting cells in the absence of added antigen.[23] This indicates that NKT cells probably recognize self-antigens present in the groove of CD1 derived from the antigen-presenting cell; in other words NKT cells are self-antigen specific. In recent studies it has been shown that the ability of NKT cells to respond to self-antigens may be the key to understanding their role in host defense; in particular, it has been shown that NKT cells stimulated by self-antigen (in the context of CD1) produce large amounts of IFN-γ if also exposed to IL-12 derived from macrophages/dendritic cells that have taken up an invading organism.[39] In this view, they are an important component of the innate immune system since they do not require exogenous antigens to begin secreting cytokines.

With this information on NKT cells in hand, we can begin to construct a possible mechanism of how these cells mediate oxazolone-colitis (see Fig. 2). The starting point is that oxazolone acts as a unique kind of haptenating agent that has the particular ability to induce NKT cell development, presumably by giving rise to a lipid moiety that is presented to a small but expandable population of NKT cells in the context of CD1 on antigen presenting cells. In the

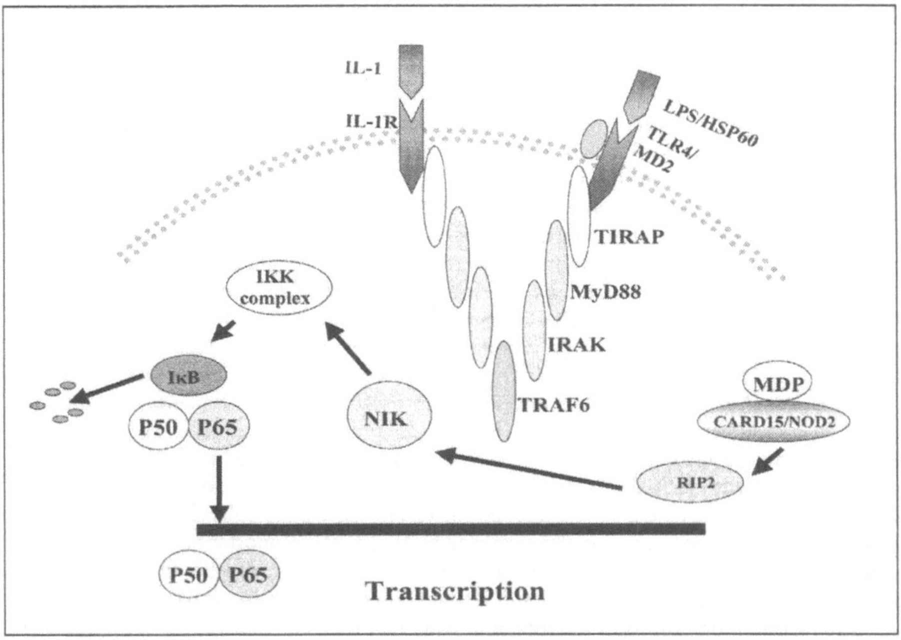

Figure 2. Cellular mechanisms of oxazolone colitis. The introduction of oxazalone per rectum to SJL/J or C57BL/10 mice results in the generation of a glycolipid antigen that is taken up and presented by lamina propria dendritic cells bearing CD1 to NKT cells. NKT cells differentiate into cells producing IL-4 and then IL-13 under the influence of various cytokines including cytokines possibly produced by epithelial cells. Inflammation is caused by NKT cells that can either attack epithelial cells directly as cytotoxic cells or indirectly via the production of IL-13. IL-13 may also potentiate NKT cell cytotoxic function. Oxazolone colitis is also negatively regulated by induction of TGF-β secretion by an as yet poorly defined cell population.

intestinal tract, the latter would include not only "professional" antigen presenting cells, but also gut epithelial cells which have been shown to express CD1.[206] These NKT cells then proliferate and differentiate into Th2 type cells possibly because they are under the influence of inducing cytokines uniquely present in the mucosal immune system. In any case, they begin to produce first IL-4 and then IL-13, the latter a cytokine which may act as an autocrine factor that causes the NKT cells to become cytotoxic effector cells with the capacity to cause tissue injury. The immediate target of possible NKT cell activity in oxazolone colitis is in fact the epithelial cell since, as mentioned, this cell bears CD1 and can present antigens to NKT cells. Such cytotoxicity would be responsible for ulceration of the mucosa and alteration of the epithelial barrier. Once this occurs there is the possibility of penetration of the mucosa by bacterial in the mucosal microflora and thus the possibility of further NKT cell stimulation by antigens in the flora that are cross-reactive with self antigens. Finally, it should be noted that changes in epithelial barrier function could be brought about by the IL-13 itself (F. Heller, Charite Hospital, Berlin, Germany, manuscript in preparation), indicating that this cytokine may have a dual pathogenetic role: one directly affecting epithelial cells and a second as a stimulant of NKT cell cytotoxicity.

As mentioned above, the factors regulating the differentiation or indeed the proliferation of NKT cells into various kinds of effector cells is still not well understood. One possibility is that NKT cell are regulated by cytokines (including novel cytokines) whose secretion determines NKT cell development. Evidence that this is indeed the case comes from numerous studies showing that NKT cell proliferation is amplified by IL-12, IL-18 and other cytokines.[54] In addition, it has recently been shown that mice that are deficient in EBI3, a component of a newly described cytokine known as IL-27 (which is made up of EBI3 linked to p28) has greatly decreased numbers of invariant NKT cells defined by α-galactosyl-ceramide tetramers and produced decreased amount of cytokines upon both in vitro and in vivo stimulation by α-galactosyl-ceramide, particularly IL-4 production.[148] Finally, and most importantly in relation to the present discussion, EBI3-deficient mice were resistant to the development of oxazolone-colitis. These studies suggest that EBI3, perhaps in the form of IL-27 plays an important role in the pathogenesis of oxazolone-colitis. Of interest, in EBI3 production; a link between oxazolone-colitis and ulcerative colitis is also increased in ulcerative colitis, suggesting that this molecule may also play an important role in human inflammation as well.[30]

The above analysis of the oxazolone-colitis model has led to an extensive study of human ulcerative colitis, the form of IBD that is histologically mimicked by the model. In extensive studies it was shown that cells from the lamina propria of patients do produce greatly increased amounts of IL-13 whereas similar cells from patients with Crohn's disease produce only a minor increase in this cytokine.[68] In addition, as in oxazolone colitis, evidence was obtained that the IL-13 was being produced by an NKT cell: the producing cell was CD161+ (NK1.1+) prior to stimulation and could be stimulated to produce IL-13 by antigen-presenting cells that were transfected with CD1d. Finally, and most importantly, it was shown that NKT cells extracted from patient tissues were cytotoxic for CD1d-expressing epithelial cell targets and that such cytotoxicity was substantially enhanced by IL-13. One important difference between the humans with ulcerative colitis and mice with oxazolone colitis was that the human NKT cells did not bear invariant chain TCRs and thus could not be stimulated to produce IL-13 by α-galactocyl-ceramide. We can thus conclude that the NKT cell that may be causing tissue injury in ulcerative colitis is a nonclassical, noninvariant NKT cell.

Overall, these studies of ulcerative colitis do two things concurrently. First, they validate the concept that oxazolone colitis is indeed a model of this form of human IBD and, secondly, they allow one to postulate a mechanism of disease pathogenesis for ulcerative colitis that is similar to that described above for oxazolone colitis. One additional fact concerning this mechanism to be gleaned from the human disease is that NKT cells do in fact have the capacity to be cytotoxic for epithelial cells and such cytotoxicity is enhanced by IL-13. Thus, the findings in the human disease go beyond what has yet been found in the model. The final proof that ulcerative colitis is due to IL-13-producing NKT cells awaits studies in which these components are blocked/depleted by administration of relevant inhibits such as antibodies to anti-IL-13. Such studies would provide additional insights into disease pathogenesis as well as provide a new and possibly potent treatment of ulcerative colitis.

Tabular Summary of Models of Mucosal Inflammation

In the above discussion of models of mucosal inflammation we have not conducted an individual analysis of the vast majority of models that have come "on-line" in the past several years. To remedy this situation, we now provide a summary of each of these models in tabular form, so as to provide a brief synopsis of the basic causes and chief characteristics of each model as well as any unique features that may be present (Tables 2-5). In addition, realizing that each model has its own story to tell and consequently its own set of insights to add, we have created a "take-home" lesson for each model that we think/hope captures the special contribution of that model. As will become apparent, the tables are arranged to reflect their most likely mechanism of pathogenesis and the reader is provided a reference or set of references that will allow him/her to learn more about each model.

Conclusion

As implied at the beginning of our discussion of models of mucosal inflammation, the discovery and subsequent immunologic/genetic analysis of these models was a watershed in the study of human inflammatory bowel disease as well as the study of autoimmunity in

Table 2. Inflammation due to increased effector T cell responses

Model	Underlying Defect or Precondition	Major Features	Unique Features	Take Home Message	References
TNBS-colitis; colitis induced by intra-rectal TNBS administration to SJL/J/or C57Bl/10 mice	Increased IL-12p70 response to LPS.	Th1 colitis transmural cell infiltration; Occasional granulomas	Genetic susceptibility; regions on Chrs 9 and 11	See text; colitis arises from innate and adaptive immune responses; it can be prevented by induction of oral tolerance to the hapten.	Neurath 1995 Neurath 1996 Fuss 1999 Bouma 2002
TNBS-colitis; colitis induced by intra-rectal TNBS administration to Balb/c mice		Mixed Th1/Th2 colitis. Follicular cell infiltrates.		Type of colitis induced by a common stimulus (TNBS) is dependent on the genetic background of the mouse.	Dohi 1999 Dohi 2000
Oxazolone colitis; colitis induced by intra-rectal oxazolone administration to SJL/J or C57Bl/10 mice		Th2 colitis; IL-4→IL-13 secretion. Secreted cytokines produced by NKT cells	Distal half of large bowel inflamed with single dose administration. Resolution of disease depends on TGF-β secretion.	See text. Th2, UC-like colitis can be caused by NKT cells producing IL-13. Epithelial cell damage may be due to NKT cell cytotoxicity.	Boirivant 1998 Heller 2002 Nieuwenhuis 2002
TCRα-chain deficiency	Inability to produce the α chain of the T cell receptor	Th2 colitis emerging as a response to cecal bacteria.	T cell display restricted Vβ repertoire. IL-10 producing CD1-bearing B cells ameliorate disease.	Th2 colitis can result from low affinity Ag-TCR interactions arising from the presence of an abnormal TCR. B cells can act as regulatory cells.	Mizoguchi 2000 Mizoguchi 1997 Mizoguchi 2002 Iijima 1999

continued on next page

Table 2. Continued

Model	Underlying Defect or Precondition	Major Features	Unique Features	Take Home Message	References
IL-7 Tg mice (SRα promotor)	Overexpression of IL-7 in epithelial cells, IELs and LPLs. ↓IL-7 in epithelial cells in areas of inflammation.	Th1 colitis associated with neutrophil and eosinophil infiltration. CD4+ T cells bearing αβ and γδ TCRs predominate.	Some autologous resemblance to UC: crypt abscesses and goblet cell loss.	IL-7 can act as a growth factor for T cells and lead to Th1 inflammation.	Yamazaki 2003 Watanabe 1998
Stat4 transgenic mice	Production of excessive amounts of Stat4	Th1 colitis develops when the mice are challenged with DNP-KLH or when their CD4+ cells are stimulated with autologous microflora and then transferred to SCID mice.		An increased capacity to activate the IL-12 signaling pathways can lead to colitis.	Wirtz 1999
SCID mice reconstituted with T-bet deficient or T-bet transgenic T cells.	T cells express abnormal amounts of T-bet.	T-bet deficient cells fail to induce Th1 colitis and T-bet overproducing cells are susceptible to Th1 colitis in the cell-transfer model.	T-bet deficient cells exhibit enhanced protective function and exhibit increased TGF-β signaling in the cell-transfer model.	Increased or decreased capacity to mediate Th1 responses have corresponding effects on the ability to induce colitis.	Neurath, J Exp Med, 2002
Conditional Stat-3 KO mice; deficiency due to TiE2-driven CRE expressed in b.m. cells	Inability to produce important intracellular signaling molecules.		Enhanced LPS (TLR 4)-induced NF-κB activation. Myeloid cell hyperplasia. Th1 inflammation in small and large intestine as well as in liver. Decreased NADPH oxidase induction.	Stat3 is an important miscellaneous regulator of innate and adaptive immune responses.	Kobayashi 2003 Takeda 1999

continued on next page

Table 2. Continued

Model	Underlying Defect or Precondition	Major Features	Unique Features	Take Home Message	References
G-protein subunit αi2-deficient mice.	Inability to produce Gαi2	Th1 colitis; altered G protein signaling leading to increased IL-12 production; abnormal B cell development and decreased LPS induction of IL-10; increased TCR-mediated T cell proliferation.	Background colitis susceptibility gene on Chr 3	G proteins regulate IL-12 synthesis.	Dalwadi 2003 Hornquist 1997 Rudolph 1995 He 2000
NF-κB p50 deficient mice	Inability to produce NF-κB p50 in NF-κB $p65^{+/-}$ mice.	Th1 colitis following *H. hepaticus* infection. Increased IL-12 p40 production but no increase in p35 production. ? IL-23 production.		Increased susceptibility to colitis may occur due to qualitative changes in NF-κB signaling.	Erdman 2001 Tomczak 2003
A20 deficient mice	Inability to produce A20, a zinc finger protein that acts as a de-ubiquitinating enzyme for NF-κB	Lack of de-ubiquitination leads to excessive NF-κB activation to various stimuli.	Inflammation of many organs. Transmural colitis. ↑TNF-α production.	Dysregulation of NF-κB activation leads to multi-focal inflammation.	Lee 2000 Evans 2004 Beg 1995
$TNF^{\Delta ARE}$ mice	↑TNF production due to deletion of AU-rich elements regulating TNF mRNA stability and translation.	Mice develop arthritis and a mainly small intestinal Th1 inflammation closely resembling Crohn's disease.	Gut inflammation dependent of TNF-RII. Arthritis dependent on TNF-RI.	TNF-α can induce a Crohn's-like inflammation.	Kontoyiannis 1999

continued on next page

Table 2. Continued

Model	Underlying Defect or Precondition	Major Features	Unique Features	Take Home Message	References
ECOVA colitis	GI infection with recombinant *E. coli* expressing OVA in SCID/RAG-2-/- mice bearing OVA-TCR Tg T cells.	Th1 or Th2 colitis depending on type of cells used to reconstitute mice. Th1 cell reconstitution: mononuclear infiltrates; granulomas; diffuse infiltrates. Th2 cell reconstitution: PMN + eosinophilic infiltrate; granulomas; follicles and more superficial infiltration.	Requires OVA associated with *E. coli* to induce disease.	Colitis can develop in response to a specific organism.	Yoshida 2002 Yoshida 2001 Iqbal 2002
OVA-Tg expressed in epithelial cells	Colitis induced in mice infected with VSV-OVA or in mice transferred OT-1 T cells and wild type VSV, cholera toxin, or CD40.	CD8+ T cell infiltration and increased epithelial cell cytotoxicity.		Epithelial cell presentation of antigen to CD8+ cells normally leads to tolerance but the latter can be broken with immunologic stimuli.	Vezys 2002 Vezys 2000
Rats bearing B7 transgene	Overexpression of a MHC class I antigen associated with arthritis in humans.	Th1 colitis as well as joint skin and heart inflammation.		Certain Class I molecules can present antigen to T cells that act as auto-immune effector cells.	Dieleman 2004 Rath 2001 Sartor 2000
"Light" transgenic mouse	Overexpression of "Light," a TNF superfamily member (TNFSFN) expressed on T cells. Light is a co-stimulatory model and Light transgenic mice display a hyperactive T cell population.	Th1 Colitis associated with lymphoproliferation and inflammation in many organs.	Glomerulitis; Autoantibodies	Increased T cell activation can lead to Th1 colitis.	Shaikh 2001

Table 3. Inflammation due to regulatory T cell defects

Model	Underlying Defect or Predisposing Condition	Major Consequences	Unique Features	Take Home Lesson	References
Cell transfer model	Transfer of cells (CD45RB^{hi} cells) into SCID or RAG2-/- mice that lack regulatory	Th1 transmural colitis.	Inflammation prevented by co-administration of regulatory cells in CD45RB^{low} cells.	See text; shows that lack of regulatory cells can lead to colitis.	Asseman 2003 Mottet 2003 Maloy 2003 Malmstrom
2001	T cells.				Read 2000 Asseman 1999 Powrie 1996 Powrie 1994 Kullberg 2003 Kullberg 2001 Kullberg 1998
IL-10 knock-out mice	Lack of ability to produce IL-10, a regulatory cytokine.	Th1 transmural colitis.	Genetic susceptibility region on Chr 3. An excessive Th2 response may occur in older mice.	IL-10 is a major regulatory cytokine that regulates mucosal responses.	Mahler 2002 Spencer 2002 Farmer 2001 Davidson 1998 Davidson 1996
CRFB2-4 KO mice	Lack of ability to produce a component of the IL-10 receptor (? IL-10 R2) and thus respond to IL-10.	Th2 colitis; same phenotype as IL-10 KO mice.		Same as above.	Spencer 1998
CIS 3/SOC 3 dominant negative transgenic mice	Decreased ability to downregulate STAT3.	Severe Th1 colitis after DSS administration due to STAT3-mediated overproduction of IL-6.		STAT3 dysregulation can lead to colitis if it is positive or negative.	Suzuki 2001 Hanada 2001

continued on next page

Table 3. Continued

Model	Underlying Defect or Predisposing Condition	Major Consequences	Unique Features	Take Home Lesson	References
IL-2 KO mice	Inability to produce IL-2.	Th2 colitis due to deficient regulatory cell development.	Apoptosis abnormalities; increase epithelial cell APC function; autoimmune hemolytic anemia. Lymphoproliferation and bone marrow infiltration.	IL-2 necessary for development of regulatory T cells.	Xiao 2003 Schultz 1999 Ludviksson 1999 Contractor 1998 Ludviksson 1997 Ehrhardt 1997 Sadlock 1993
IL-2R α or β chain KO mice	Inability to respond to IL-2 or IL-2/IL-15.	↓ Intestinal T cell development; both small and large intestinal inflammation.	β chain KO: granulocytopenia and bone marrow infiltration.		Poussier 2000
$Tg_{\epsilon}26$ mice	Mice bear CD3 transgene leading to disruption of intra-thymic T cell development.	Th1 colitis in mice reconstituted with normal bone marrow due to defect in thymic regulatory T cell development.	Colitis prevented by neonatal thymus but not adult thymus transplantation in mice reconstituted with normal bone marrow.	Decreased regulatory cell generation due to an intrathymic defect may lead to colitis.	De Jong 2000 MacKay 1998 Simpson 1998 Hollander 1995
Foxp3 KO mice	Mice lack scurfin, a product of the foxp3 gene.	Multi-organ autoimmune diseases including colitis; mice lack functional CD25+ regulatory cells.	Model of IPEX signaling in humans.	CD25+ regulatory cells expressing Foxp3 necessary to prevent colitis and autoimmunity.	Hori 2003 Bennett 2001
Mice bearing TGF-β RII dominant negative transgene (CD4 promotor)	Mice lack ability to respond to TGF-β	Th1 colitis; inflammation also observed in lungs and other organs; ↑ T cell activation.	Ig deposits glomeruli; susceptibility to experimental diabetes.	TGF-β signaling probably by CD25+ regulatory cells necessary to prevent colitis.	Gorelik 2000 Beck 2003 Hahm 2001

Table 4. Models due to defective barrier function

Model	Underlying Defect or Predisposing Condition	Major Consequences	Unique Features	Take Home Message	References
Mice with over-expression of dominant-negative (DN) N-cadherin in epithelial cells (fatty acid binding protein promotor).	DN N-cadherin over-expression leads to E-cadherin deficiency.	Epithelium with areas exhibiting poor cell-cell adhesion. Inflammation and ulcers occur in lamina propria subjacent to areas of such poor cell-cell adhesion.	Mice develop adenomas.	There is a direct correlation between barrier function and inflammation.	Hermiston 1995
DSS colitis	Colitis caused by injection of dextran sulfate.	Inflammation due to epithelial cell disruption and macrophage activation. Initial innate immune responses followed by adaptive Th1/Th2 responses to mucosal antigens.		Innate immune responses can lead to severe inflammation.	Dieleman 1994 Ohkosa 1985 Kitajima 1999 Onderdonk 1978 Rath 2001 Axelsson 1996 Egger 2000 Dieleman 1998 Mashimo 1996
Mdr1a deficient mice	Lack of production of epithelial cell transporter proteins leading to epithelial cell disruption.	Spontaneous development of Th1 colitis.		Model suggests that inability to export mucosal antigens may lead to abnormalities of epithelial cell barrier function.	Panwala 1998 Gottesman 1993 Leveille-Webster 1995 Satsangi 1996 Yacyshyn 1999

Table 5. Models due to an as yet undefined mechanism

Model	Underlying Defect or Predisposing Condition	Major Consequences	Unique Features	Take Home Message	References
C3H/HeJ Bir mouse	TLR4 deficit leading to LPS unresponsiveness	Th1 colitis usually localized in the cecum. Colitis spontaneously remits due to development of IL-10 producing regulatory T cells.	Extensive analysis of the immune response to antigens in the microflora indicate responses occur in only a relatively limited subpopulation of organisms.	Defective TLR4 responses may predispose to excessive innate or adaptive responses to gut organisms.	Lodes 2004 Cong 2002 Cong 2000 Cong 1998 Brandwein 1997 Sundberg 1994
SAMP1/Yit mouse	AKR substrain initially bred to achieve accelerated senescence and then bred to express colitis.	Th1 inflammation affects mainly the small intestine and causes skip lesions. Striking macroscopic and microscopic similarity to Crohn's disease.	Th2 responses may develop later in disease. Genetic susceptibility locus on Chr 11.		Bamias 2004 Matsumoto 1998 Kosiewicz 2001 Kozaiwa 2003 Rivera-Nieves 2003
Wiskott-Aldrich Syndrome Protein (WASP) deficiency	Inability to produce WASP, a cause of immunodeficiency in humans.	Superficial Th2 inflammation similar to UC. Mild immunodeficiency.	Mice develop osteoporotic bone disease.	?WASP defect leads to low affinity TCR interaction and Th2 responses.	Snapper 1998 Klein 2003

general. If we were to point to the major insights derived from these models we would mention first the elucidation of the essential role of the mucosal microflora as the inductive force of chronic mucosal inflammation and thus the conclusion that commensal bacteria rather than pathogens provide the antigens that induce pathogenic mucosal responses. As a related point, it now appears that while mucosal inflammation can be caused by a single organism, the data favor the view that many organisms (i.e., many antigens) are typically involved, with some more important than others in any given individual. These observations relative to the mucosal flora are best coupled with the second major fact to emerge from the models, namely that many different immune defects can lead to mucosal inflammation and thus the inductive antigens in the microflora play upon a multitude of abnormalities to cause chronic inflammation. This insight prepares us for the multigenic reality of human IBD and establishes that chronic mucosal inflammation results from one or another endogenous immunologic (or epithelial) defect that in some manner disturbs the normally tolerogenic response of the mucosal immune system to the endogenous microflora.

A third insight to be derived from the models and one that directly relates to this last point is the notion that regulatory T cell defects can cause mucosal inflammatory disease. This point emphasizes the exquisite balance of response and nonresponse in the mucosal immune system and that inflammation is an immediate consequence of imbalance. In the process of defining the mechanics of such regulatory cell abnormalities much has been learned concerning the origin and types of regulatory cytokines that might be involved in such defects and thus it is now possible to consider treatment of mucosal inflammatory disease by provision of these cytokines. A fourth key fact derived from the models and one that is perhaps the most useful from a therapeutic standpoint, is the establishment of the Th1/Th2 duality of mucosal inflammation and the discovery that Th1 inflammations resembling Crohn's disease are due to excess production of IL-12 (and/or IL-23). This insight has led directly to a trial of anti-IL-12 administration in the treatment of Crohn's disease that has already yielded highly promising results. In addition, the Th1/Th2 duality has led to the identification of an atypical Th2 (IL-13) response in ulcerative colitis and thus has pointed the way to a possible treatment of this form of IBD based on administration of IL-13 inhibitors.

These insights are, of course, not the only ones that will emerge from the study of models of mucosal inflammation. On the immediate horizon is the use of models to elucidate the nature of genetic abnormalities causing such inflammation and, by implication, causing inflammation in human disease. One possible example for such use is the study of the cause of inflammation due to the newly discovered NOD2 mutation in a subset of patients with Crohn's disease. Such studies could well lead to molecularly oriented treatments for IBD that would provide permanent prevention and relief from this very serious condition.

Acknowledgements

Thank you to the assistance of Joan Yang in the laboratory and Sara Kaul in typing and formatting this manuscript.

References

1. Annunziato F, Cosmi L, Liotta F et al. Phenotype, localization, and mechanism of suppression of CD4+CD25+ human thymocytes. J Exp Med 2002; 196(3):379-387.
2. Aranda R, Sydora BC, McAllister PL et al. Analysis of intestinal lymphocytes in mouse colitis mediated by transfer of CD4+, CD45RB^{high} T cells to SCID recipients. J Immunol 1997; 158:3464-3473.

3a. Asseman C, Read S, Powrie F. Colitogenic Th1 cells are present in the antigen-experienced T cell pool in normal mice: Control by CD4+ regulatory T cells and IL-10. J Immunol 2003; 171(2):971-978.

3b. Asseman C, Mauze S, Leach MW et al. An essential role for interleukin 10 in the function of regulatory T cells that inhibit intestinal inflammation. J Exp Med 1999; 190(7):995-1004.

4. Axelsson LG, Landstrom E, Goldschmitt TJ et al. Dextran sulfate sodium (DSS) induced experimental colitis in immunodeficient mice: Effects in CD4+ cell depleted, athymic and NK-cell depleted SCID mice. Inflamm Res 45:181-191.

5. Bamias G, Mishina M, Martin III CT et al. Differential role of Th1 and Th2 Cytokines in the pathogenesis of SAMP1/YitFc ileitis. Gastroenterology 2004; 126(4):A119.
6. Baron JL, Gardiner L, Nishimura S et al. Activation of a nonclassical NKT cell subset in a transgenic mouse model of hepatitis B virus infection. Immunity 2002; 16(4):583-594.
7. Beck PL, Rosenberg IM, Xavier RJ et al. Transforming growth factor-beta mediates intestinal healing and susceptibility to injury in vitro and in vivo through epithelial cells. Am J Pathol 2003; 162(2):597-608.
8. Becker C, Wirtz S, Blessing M et al. Constitutive p40 promotor activation and IL-23 production in the terminal ileum mediated by dendritic cells. J Clin Invest 2003; 112(5):693-706.
9. Beg AA, Sha WC Bronson RT et al. Embryonic lethality and liver degeneration in mice lacking the RelA component of NF-kappaB. Nature 1995; 376:167-170.
10. Belghith M, Bluestone JA, Barriot S et al. TGF-β-dependent mechanisms mediate restoration of self-tolerance induced by antibodies to CD3 in overt autoimmune diabetes. Nat Med 2003; 9(9):1202-1208.
11. Bennett CL, Christie J, Ramsdell F et al. The immune dysregulation, polyendocrinopathy, enteropathy, X-linked syndrome (IPEX) is caused by mutations of FOXP3. Nat Genet 2001; 27(1):20-21.
12. Bilsborough J, George TC, Norment A et al. Mucosal CD8α+DC, with a plasmacytoid phenotype, induce differentiation and support function of T cells with regulatory properties. Immunology 2003; 108(4):481-492.
13. Boirivant M, Fuss IJ, Chu A et al. Oxazalone colitis: A murine model of T helper cell type 2 colitis treatable with antibodies to interleukin 4. J Exp Med 1998; 188(1):1929-1939.
14. Bouma G, Crusius JB, Odkerk P et al. Secretion of tumour necrosis factor alpha and lymphotoxin alpha in relation to polymorphisms in the TNF genes and HLA-DR alleles. Relevance for inflammatory bowel disease. Scand J Immunol 1996; 43(4):456-463.
15. Bouma G, Kaushiva A, Strober W. Experimental murine colitis is regulated by two genetic loci, including one on chromosome 11 that regulates IL-12 responses. Gastroenterology 2002; 123(2):554-565.
16. Bouma G, Strober W. The immunological and genetic basis of inflammatory bowel disease. Nat Rev Immunol 2003; 3(7):521-533.
17. Brandwein SL, McCabe RP, Cong Y et al. Spontaneously colitic C3H/HeJBir mice demonstrate selective antibody reactivity to antigens of the enteric bacterial flora. J Immunol 1997; 159(1):44-52.
18. Braun MC, He J, Wu CY et al. Cholera toxin suppresses interleukin (IL)-12 production and IL-12 receptor β1 and β2 chain expression. J Exp Med 1999; 189(3):541-552.
19. Braun MC, Kelsall BL. Regulation of interleukin-12 production by G-protein-coupled receptors. Microbes Infect 2001; 3(2):99-107.
20. Breeling JL, Onderdonk AB, Cisneros RL et al. Bacterioides vulgatus outer membrane antigens associated with carageenan-induced colitis in guinea pigs. Infect Immunol 1998; 56(7):1754-1759.
21. Brigl M, Brenner MB. CD1: Antigen presentation and T cell function. Annu Rev Immunol 2004; 22:817-890.
22. Brimnes J, Reimann J, Nissen M et al. Enteric bacterial antigens activate CD4+ T cells from scid mice with inflammatory bowel disease. Eur J Immunol 2001; 31(2):23-31.
23. Brutkiewicz RR, Bennink JR, Yewdell JW et al. TAP-independent, beta 2-microglobulin-dependent surface expression of functional mouse CD1.1. J Exp Med 1995; 182(6):1913-1919.
24. Camoglio L, te Velde AA, de Boer A et al. Hapten-induced colitis associated with maintained Th1 and inflammatory responses in IFN-γ receptor-deficient mice. Eur J Immunol 2000; 30(5):1486-1495.
25. Cario E, Podolsky DK. Differential alteration in intestinal epithelial cell expression of toll-like receptor 3 (TLR3) and TLR4 in inflammatory bowel disease. Infect Immun 2000; 68(12):7010-7017.
26. Chen Y, Inobe J, Kuchroo VK et al. Oral tolerance in myelin basic protein T-cell receptor transgenic mice: Suppression of autoimmune encephalomyelitis and dose-dependent induction of regulatory cells. Proc Natl Acad Sci USA 1996; 93(1):388-391.
27. Chen W, Jin W, Wahl SM. Engagement of cytotoxic T lymphocyte-associated antigen 4 (CTLA-4) induces transforming growth factor beta (TGF-β) production by murine CD4+ T cells. J Exp Med 1998; 188(10):1849-1857.
28. Chen Y, Kuchroo VK, Inobe J et al. Regulatory T cell clones induced by oral tolerance: Suppression of autoimmune encephalomyelitis. Science 1994; 265(5176):1237-1240.
29. Chen W, Wahl SM. TGF-β: The missing link in CD4+CD25+ regulatory T cell-mediated immunosuppression. Cytokine Growth Factor Rev 2003; 14(2):85-89.
30. Christ AD, Stevens AC, Koeppen H et al. An interleukin 12-related cytokine is up-regulated in ulcerative colitis but not in Crohn's disease. Gastroenterology 1998; 115(2):307-313.
31. Christensen HR, Frokiaer H, Pestka JJ. Lactobacilli differentially modulate expression of cytokines and maturation surface markers in murine dendritic cells. J Immunol 2002; 168(1):171-178.

32. Claesson MH, Bregenholt S, Bonhagen K et al. Colitis-inducing potency of CD4+ T cells in immunodeficient, adoptive hosts depends on their state of activation, IL-12 responsiveness, and CD45RB surface phenotype. J Immunol 1999; 162(6):3702-3710.
33. Cong Y, Brandwein SL, McCabe RP et al. CD4+ T cells reactive to enteric bacterial antigens in spontaneously colitis C3H/HeJBir mice: Increased T helper cell type 1 response and ability to transfer disease. J Exp Med 1998; 187(6):855-864.
34. Cong Y, Weaver CT, Lazenby A et al. Colitis induced by enteric bacterial antigen-specific CD4+ T cells requires CD40-CD40 ligand interaction for a sustained increase in mucosal IL-12. J Immunol 2000; 165(4):2173-2182.
35. Contractor NV, Bassir H, Reya T et al. Lymphoid hyperplasia, autoimmunity, and compromised intestinal intraepithelial lymphocyte development in colitis-free gnotobiotic IL-2-deficient mice. J Immunol 1998; 160(1):385-394.
36. Contractor NV, Bassirir H, Reya T et al. Lymphoid hyperplasia, autoimmunity, and compromised intestinal intraepithelial lymphocyte development in colitis-free gnotobiotic IL-2-deficient mice. J Immunol 1998; 75:253-61.
37. Dao T, Heal WZ, Crispe IN. IL-18 augments perforin-dependent cytotoxicity of liver NK-T cells. J Immunol 1998; 161(5):2217-2222.
38. Darfeuille-Michaud A, Neut C, Barnich N et al. Presence of adherent Escherichia coli strains in ileal mucosa of patients with Crohn's disease. Gastroenterology 1998; 115(6):1405-1413.
39. Dascher CC, Brenner MB. CD1 antigen presentation and infectious disease. Contrib Microbiol 2003; 10:164-182.
40. Davidson NJ, Hudak SA, Lesley RE et al. IL-12, but not IFN-γ, plays a major role in sustaining the chronic phase of colitis in IL-10-deficient mice. J Immunol 1998; 161(6):3143-3149.
41. Davidson NJ, Leach MW, Fort MM et al. T helper cell 1-type CD4+ T cells, but not B cells, mediate colitis in interleukin 10-deficient mice. J Exp Med 1996; 184(1):241-251.
42. De Jong YP, Abadia-Molina AC, Satoskar AR et al. Development of chronic colitis is dependent on the cytokine MIF. Nat Immunol 2001; 2(11):1061-1066.
43. De Jong YP, Comiskey M, Kalled SL et al. Chronic murine colitis is dependent on the CD154/CD40 pathway and can be attenuated by anti-CD154 administration. Gastroenterology 2000; 119(3):715-723.
44. Dieleman LA, Hoentjen F, Quian BF et al. Reduced ratio of protective versus proinflammatory cytokine responses to commensal bacteria in HLA-B27 transgenic rats. Clin Exp Immunol 2004; 136(1):30-39.
45. Dieleman LA, Palmen MJ, Akoh H et al. Chronic experimental colitis induced by dextran sulphate sodium (DSS) is characterized by Th1 and Th2 cytokines. Clin Exp Immunol 114:385-391.
46. Dieleman LA, Ridwan BU, Tennyson GS et al. Dextran sulfate sodium-induced colitis occurs in severe combined immunodeficient mice. Gastroenterology 1994; 107(6):1643-1652.
47. Dohi T, Fujihashi K, Kiyono H et al. Mice deficient in Th1- and Th2-type cytokines develop distinct forms of hapten-induced colitis. Gastroenterology 2000; 119(3):724-733.
48. Dohi T, Fujihashi K, Rennert PD et al. Hapten-induced colitis is associated with colonic path hypertrophy and T helper cell 2-type responses. J Exp Med 1999; 189(8):1169-1180.
49. Drenick EJ, Roslyn JJ. Cure of arthritis-dermatitis syndrome due to intestinal bypass by resection of nonfunctional segment of blind loop. Dig Dis Sci 1990; 35(5):656-660.
50. Duchmann R, May E, Heike M et al. T celll specificity and cross reactivity towards enterobacteria, bacteriodes, bifidobacterium, and antigens from resident intestinal flora in humans. Gut 1999; 44(6):812-818.
51. Duchmann R, Neurath MF, Meyer zum Buschenfelde KH. Responses to self and nonself intestinal microflora in health and inflammatory bowel disease. Res Immunol 1999; 148(8-9):589-594.
52. Duchmann R, Schmitt E, Knolle P et al. Tolerance towards resident intestinal flora in mice is abrogated in experimental colitis and restored by treatment with interleukin-10 or antibodies to interleukin-12. Eur J Immunol 1996; 26(4):934-938.
53. Dalwadi H, Wei B, Schrage M et al. B cell developmental requirement for the G alpha i2 gene. J Immunol 2003; 170(4):1707-1715.
54. Eberl G, MacDonald HR. Selective induction of NK cell proliferation and cytotoxicity by activated NKT cells. Eur J Immunol 2000; 30(4):985-992.
55. Egger B, Bajaj-Elliott M, MacDonald MW et al. Characterization of acute murine dextran sodium sulphate colitis: Cytokine profile and dose dependency. Digestion 2000; 62:240-248.
56. Ehrhardt RO, Ludviksson BR, Gray B et al. Induction and prevention of colonic inflammation in IL-2-deficient mice. J Immunol 1997; 158(2):566-573.
57. Elson CO, Beagley KW, Sharmanov AT et al. Hapten-induced model of murine inflammatory bowel disease: Mucosa immune responses and protection by tolerance. J Immunol 1996; 157(5):2174-2185.

58. Erdman S, Fox JG, Dangler CA et al. Typhlocolitis in NF-kappa B-deficient mice. J Immunol 2001; 166(3):1443-1447.
59. Evans PC, Ovaa H, Hamon M et al. Zinc-finger protein A20, a regulator of inflammation and cell survival, has de-ubiquitinating activity. Biochem J 2004; 378(Pt 3):727-734.
60. Fantini MC, Becker C, Monteleone G et al. Cutting edge: TGF-β induces a regulatory phenotype in CD4+CD25- T cells through Foxp3 induction and down-regulation of Smad7. J Immunol 2004; 172(9):5149-5153.
61. Farmer MA, Sundberg JP, Bristol IJ et al. A major quantitative trait locus on chromosome 3 controls colitis severity in IL-10-deficient mice. Proc Natl Acad Sci USA 2001; 98(24):13820-13825.
62. Fehervari Z, Sakaguchi S. Development and function of CD25+CD4+ regulatory T cells. Curr Opin Immunol 2004; 16(2):203-208.
63. Fiocchi C. From immune activation to gut tissue injury: The pieces of the puzzle are coming together. Gastroenterology 1999; 117(5):1238-1241.
64. Fiorucci S, Mencarelli A, Palazzetti B et al. Importance of innate immunity and collagen binding integrin alpha1beta1 in TNBS-induced colitis. Immunity 2002; 17(6):769-780.
65. Flynn S, van Sinderen D, Thornton GM et al. Characterization of the genetic locus responsible for the production of ABP-118, a novel bacteriocin produced by the probiotic bacterium Lactobacillus salivarus subsp. Salivarus UCC118. Microbiology 2002; 148(Pt 4):973-984.
66. Franchimont D, Bouma G, Galon J et al. Adrenal cortical activation in murine colitis. Gastroenterology 2000; 119(6):1560-1568.
67. Fuss IJ, Boirivant M, Lacy B et al. The interrelated roles of TGF-β and IL-10 in the regulation of experimental colitis. J Immunol 2002; 168(2):900-908.
68. Fuss IJ, Heller F, Boirivant M et al. Nonclassical CD1d-restricted NK T cells that produce IL-13 characterize an atypical Th2 response in ulcerative colitis. J Clin Invest 2004; 113(10):1490-1497.
69. Fuss IJ, Marth T, Neurath MF et al. Anti-interleukin 12 treatment regulates apoptosis of Th1 T cells in experimental colitis in mice. Gastroenterology 1999; 117(5):1078-1088.
70. Fuss IJ, Neurath M, Boirivant M et al. Disparate CD4+ lamina propria (LP) lymphokine secretion profiles in inflammatory bowel disease. Crohn's disease LP cells manifest increased secretion of IFN-gamma, whereas ulcerative colitis LP cells manifest increased secretion of IL-5. J Immunol 1996; 157(3):1261-1270.
71. Gorelik L, Flavell RA. Abrogation of TGF-β signaling in T cells leads to spontaneous T cell differentiation and autoimmune disease. Immunity 2000; 12(2):171-81.
72. Gottesman MM, Pastan I. Biochemistry of multidrug resistance mediated by the multidrug transporter. Ann Rev Biochem 1993; 62:385-391.
73. Groux H, O'Garra A, Bigler M et al. A CD4+ T-cell subset inhibits antigen-specific T-cell responses and prevents colitis. Nature 1997; 389(6652):737-742.
74. Guo L, Hu-Li J, Paul WE. Probabilistic regulation of IL-4 production in Th2 cells: Accessibility at the IL4 locus. Immunity 2004; 20(2):193-203.
75. Hagenbaugh A, Sharma S, Dubinett SM et al. Altered immune responses in interleukin 10 transgenic mice. J Exp Med 1997; 185(12):2101-2110.
76. Hahm KB, Im YH, Parks TW et al. Loss of transforming growth factor beta signaling in the intestine contributes to tissue injury in inflammatory bowel disease. Gut 2001; 49:164-165.
77. Hanada T, Yoshida T, Kinjyo I et al. A mutant form of JAB/SOCS1 augments the cytokine-induced JAK/STAT pathway by accelerating degradation of wild-type JAB/CIS family proteins through the SOCS-box. J Biol Chem 2001; 276(44):40746-40754.
78. He J, Gurunathan S, Iwasaki A et al. Primary role for Gi protein signaling in the regulation of interleukin 12 production and the induction of T helper cell type 1 responses. J Exp Med 2000; 191(9):1605-1610.
79. Heller F, Fuss IJ, Nieuwenhuis EE et al. Oxazalone colitis, a Th2 colitis model resembling ulcerative colitis, is mediated by IL-13-producing NK-T cells. Immunity 2002; 17(5):629-638.
80. Hermiston ML, Gordon JL. Inflammatory bowel disease and adenomas in mice expressing a dominant negative N-cadherin. Science 1995; 270(2539):1203-1207.
81. Hoentjen F, Harmsen HJ, Braat H et al. Antibiotics with a selective aerobic or anaerobic spectrum have different therapeutic activities in various regions of the colon in interleukin 10 gene deficient mice. Gut 2003; 52(12):1721-1727.
82. Hollander GA, Simpson SJ, Mizoguchi E et al. Severe colitis in mie with aberrant thymic selection. Immunity 1995; 3(1):27-38.
83. Control of regulatory T cell development by the transcription factor Foxp3. Science 2003; 299(5609):1057-1061.
84. Hori S, Takahashi T, Sakaguchi S. Control of autoimmunity by naturally arising regulatory CD4+ T cells. Adv Immunol 2003; 81:331-371.

85. Hornquist CE, Lu X, Rogers-Fani PM et al. G(alpha)i2-deficient mice with colitis exhibit a local increase in memory CD4+ T cells and proinflammatory Th1-type cytokines. J Immunol 1999; 158(3):1068-1077.
86. Iijima H, Takahashi I, Kishi D et al. Alteration of interleukin 4 production results in the inhibition of T helper type 2 cell-dominated inflammatory bowel disease in T cell receptor alpha chain-deficient mice. J Exp Med 1999; 190(5):607-615.
87. Iqbal N, Oliver JR, Wagner FH et al. T helper 1 and T helper 2 cells are pathogenic in an antigen-specific model of colitis. J Exp Med 2002; 195(1):71-84.
88. Isolauri E, Majamaa H, Arvola T et al. Lactobacillus casei strain GG reverses increased intestinal permeability induced by cow milk in suckling rats. Gastroenterology 1993; 105(6):1643-1650.
89. Iwasaki A, Kelsall BL. Freshly isolated peyer's patch, but not spleen, dendritic cells produce interleukin 10 and induce the differentation of T helper type 2 cells. J Exp Med 1999; 190(2):229-239.
90. Kitajima S, Takoma S, Morimoto M. Changes in colonic mucosal permeability mouse colitis induced with dextran sulfate sodium. Exp Animal 1999; 48:137-143.
91. Kitani A, Fuss I, Nakamura K et al. Transforming growth factor (TGF)-β1-producing regulatory T cells induce Smad-mediated interleukin 10 secretion that facilitiates coordinated immunoregulatory activity and amelioration of TGF-β1-mediated fibrosis. J Exp Med 2003; 198(8):1179-1188.
92. Kitani A, Fuss IJ, Nakamura K et al. Treatment of experimental (Trinitrobenzene sulfonic acid) colitis by intranasal administration of transforming growth factor (TGF)-β1 plasmid: TGF-β1-mediated suppression of T helper cell type 1 response occurs by interleukin (IL)-10 induction and IL-12 receptor β2 chain downregulation. J Exp Med 2000; 192(1):41-52.
93. Kishi D, Takahashi I, Kai Y et al. Alteration of V beta usage and cytokine production of CD4+ TCR beta beta homodimers T cells by elimination of Bacterioides vulgatus prevents colitis in TCR alpha-chain-deficient mice. J Immunol 165:5891-5899.
94. Klein C, Nguyen D, Liu CH et al. Gene therapy for Wiskott-Aldrich syndrome: Rescue of T-cell signaling and amelioration of colitis upon transplantation of retrovirally transduced hematopoietic stem cells in mice. Blood 2003; 101(6):2159-2166.
95. Kobayashi M, Kweon MN, Kuwata H et al. Toll-like receptor-dependent production of IL-12p40 causes chronic enterocolitis in myeloid cell-specific Stat3-deficient mice. J Clin Invest 2003; 111(9):1297-308.
96. Kontoyiannis D, Pasparakis M, Pizarro TT et al. Impaired on/off regulation of TNF biosynthesis in mice lacking TNF AU-rich elements: Implications for joint and gut-associated immunopathologies. Immunity 1999; 10(3):387-398.
97. Kosiewicz MM, Nast CC, Krishnan A et al. Th1-type responses mediate spontaneous ileitis in a novel murine model of Crohn's disease. J Clin Invest 2001; 107(6):695-702.
98. Kozaiwa K, Sugawara K, Smith Jr MF et al. Identification of a quantitative trait locus for ileitis in a spontaneous mouse model of Crohn's disease: SAMP1/YitFc. Gastroenterology 2003; 125(2):477-490.
99. Krajina T, Leithauser F, Moller P et al. Colonic lamina propria dendritic cells in mice with CD4+ T cell-induced colitis. Eur J Immunol 2003; 33(4):1073-1083.
100. Kronenberg M, Gapin L. The unconventional lifestyle of NKT cells. Nat Rev Immunol 2002; 2(8):57-568.
101. Kuchroo VK, Umetsu DT, DeKruff RH et al. The TIM gene family: Emerging roles in immunity and disease. Nat Rev Immunol 2003; 3(6):454-62.
102. Kullberg MC, Anderson JF, Gorelick PL et al. Induction of colitis by a CD4+ T cell clone specific for a bacterial epitope. Proc Natl Acad Sci USA 2003; 100(26):15830-15835.
103 Kullberg MC, Jankovic D, Gorelick PL et al. Bacteria-triggered CD4+ T regulatory cells suppress Helicobacter hepaticus-induced colitis. J Exp Med 2002; 196(4):505-515.
104. Kullberg MC, Ward JM, Gorelick PL et al. Helicobacter hepaticus triggers colitis in specific-pathogen-free interleukin-10 (IL-10)-deficient mice through an IL-12 and gamma interferon-dependent mechanism. Infect Immun 1998; 66(11):5157-5166.
105. Lee EG, Boone DL, Chai S et al. Failure to regulate TNF-induced NF-kappaB and cell death responses in A20-deficient mice. Science 2000; 289(5488):2350-2354.
106. Lee YK, Puong KY, Ouwehand AC et al. Displacement of bacterial pathogens from mucus and Caco2 cell surface by lactobacilli. J Med Microbiol 2003; 52(Pt 10):925-930.
107. Letterio JJ, Roberts AB. Regulation of immune responses by TGF-β. Annu Rev Immunol 1998; 16:137-161.
108. Leveille-Webster CR, Arias IM. The biology of the P-glycoprotein. J Membr Biol 1995; 143:89-94.
109. Levings MK, Bacchetta R, Schultz U et al. The role of IL-10 and TGF-β in the differentiation and effector function of T regulatory cells. Int Arch Allergy Immunol 2002; 129(4):263-276.
110. Lichtman SN, Keku J, Clark RI et al. Biliary tract disease in rats with experimental small bowel bacterial overgrowth. Hepatology 1991; 13(4):766-772.

111. Liu H, Hu B, Xu D et al. CD4+CD25+ regulatory T cells cure murine colitis: The role of IL-10, TGF-β, and CTLA-4. J Immunol 2003; 171(10):5012-5017.
112. Liu Z, Geboes K, Colpaert S et al. Prevention of experimental colitis in SCID mice reconstituted with CD45RBhigh CD4+ T cells by blocking the CD40-CD154 interactions. J Immunol 2000; 164(11):6005-6014.
113. Liu Z, Geboes K, Heremans H et al. Role of interleukin-12 in the induction of mucosal inflammation and abrogation of regulatory T cell function in chronic experimental colitis. Eur J Immunol 2001; 31(5):1550-1560.
114. Lodes MJ, Cong Y, Elson CO et al. Bacterial flagellin is a dominant antigen in Crohn's disease. J Clin Invest 2004; 113(9):1296-1306.
115. Ludviksson BR, Ehrhardt RO, Strober W. TGF-β production regulates the development of the 2,4,6-trinitrophenol-conjugated keyhole limpet hemocyanin-induced colonic inflammation in IL-2-deficient mice. J Immunol 1997; 159(7):3622-3628.
116. Ludviksson BR, Strober W, Nishikomori R et al. Administration of mAb against alpha E β7 prevents and ameliorates immunization-induced colitis in IL-$2^{-/-}$ mice. J Immunol 1999; 162(8):4975-4982.
117. Mack DR, Michail S, Wei S et al. Probiotics inhibit enteropathogenic E. coli adherence in vitro by inducing intestinal mucin gene expression. Am J Physiol 1999; 276(4 Pt):G941-G950.
118. Mackay F, Browning JL, Lawton P et al. Both the lymphotoxin and tumor necrosis factor pathways are involved in experimental murine models of colitis. Gastroenterology 1998; 115(6):1464-1475.
119. Macpherson AJ, Uhr T. Induction of protective IgA by intestinal dendritic cells carrying commensal bacteria. Science 2004; 303(5664):1662-1665.
120. Madsen KL, Doyle JS, Jewell LD et al. Lactobacillus species prevents colitis in interleukin 10 gene-deficient mice. Gastroenterology 1999; 116(5):1107-1114.
121. Madsen K, Cornish A, Soper P et al. Probiotic bacteria enhance murine and human intestinal epithelial barrier function. Gastroenterology 2001; 121(3):580-591.
122. Madsen KL, Doyle JS, Tavernini MM et al. Antibiotic therapy attenuates colitis in interleukin 10 gene-deficient mice. Gastroenterology 2000; 118(6):1097-1105.
123. Mahler M, Bristol IJ, Leiter EH et al. Differential susceptibility of inbred mouse strains to dextran sulfate sodium-induced colitis. Am J Physiol 1998; 274(3 Pt 1):G544-G551.
124. Mahler M, Bristol IJ, Sundberg JP et al. Genetic analysis of susceptibility to dextran sulfate sodium-induced colitis in mice. Genomics 1999; 55(2):147-156.
125. Mahler M, Most C, Schmidtke S et al. Genetics of colitis susceptibility in IL-10-deficient mice: Backcross versus F2 results contrasted by principal component analysis. Genomics 2002; 80(3):274-282.
126. Malmstrom V, Shipton D, Singh B et al. CD134L expression on dendritic cells in the mesenteric lymph nodes drives colitis in T cell-restored SCID mice. J Immunol 2001; 166(11):6972-6981.
127. Malloy KJ, Salaun L, Cahill R et al. CD4+CD25+ T(R) cells suppress innate immune pathology through cytokine-dependent mechanisms. J Exp Med 2003; 197(1):111-119.
128. Mangell P, Nejdfors P, Wang M et al. Lactobacillus plantarum 299v inhibits Escherichia coli-induced intestinal permeability. Dig Dis Sci 2002; 47(3):511-516.
129. Mannon P, Fuss IJ, Mayer L et al. Anti-interleukin-12 treats active Crohn's disease. Gastroenterology 2004; 126(4):A22, abstract.
130. Mao Y, Noback S, Kasravi B et al. The effects of Lactobacillus strains and oat fiber on methotrexate-induced enterocolitis in rats. Gastroenterology 1996; 111(2):334-344.
131. Matsumoto S, Okabe Y, Setoyama H et al. Inflammatory bowel disease-like enteritis and caecitis in a senescence accelerated mouse P1/Yit strain. Gut 1998; 43(1):71-78.
132. Mashimo H, Wu DC, Podolsky DK et al. Impaired defense of intestinal mucosa in mice lacking intestinal trefoil factor. Science 1996; 274:204.
133. Mayer L, Eisenhard D. Lack of induction of suppressor T cells by intestinal epithelial cells from patients with inflammatory bowel disease. J Clin Invest 1990; 86(4):1255-1260.
134. Mizoguchi A, Mizoguchi E, Chiba C et al. Role of appendix in the development of inflammatory bowel disease in TCR-α mutant mice. J Exp Med 1996; 184(2):707-715.
135. Mizoguchi A, Mizoguchi E, Saubermann LJ et al. Limited CD4 T-cell diversity associated with colitis in T-cell receptor alpha mutant mice requires a T helper 2 environment. Gastroenterology 2000; 119(4):983-995.
136. Mizoguchi A, Mizoguchi E, Smith RN et al. Suppressive role of B cells in chronic colitis of T cell receptor alpha mutant mice. J Exp Med 1997; 186(10):1749-1756.
137. Mizoguchi A, Mizoguchi E, Takedatsu H et al. Chronic intestinal inflammatory condition generates IL-10-producing regulatory B cell subset characterized by CD1d upregulation. Immunity 2002; 16(2):219-230.
138. Monteleone G, Biancone L, Marasco R et al. Interleukin 12 is expressed and actively released by Crohn's disease intestinal lamina propria mononuclear cells. Gastroenterology 1997; 112(4):1169-1178.

139. Morris GP, Beck PL, Herridge MS et al. Hapten-induced model of chronic inflammation and ulceration in the rat colon. Gastroenterology 1989; 96(3):795-803.
140. Mottet C, Uhlig HH, Powrie F. Cutting edge: Cure of colitis by CD4+CD25+ regulatory T cells. J Immunol 2003; 170(8):3939-3943.
141. Nakamura K, Kitani A, Fuss I et al. TGF-β 1 plays an important role in the mechanism of CD4+CD25+ regulatory T cell activity in both humans and mice. J Immunol 2004; 172(2):834-842.
142. Nakamura K, Kitani A, Strober W. Cell contact-dependent immunosuppression by CD4+CD25+ regulatory T cells is mediated by cell surface-bound transforming growth factor beta. J Exp Med 2001; 194(5):629-644.
143. Nelson PA, Akselband Y, Dearborn SM et al. Effect of oral beta interferon on subsequent immune responsiveness. Ann NY Acad Sci 1996; 778:145-155.
144. Neurath MF, Fuss I, Kelsall BL et al. Experimental granulomatous colitis in mice is abrogated by induction of TGF-β-mediated oral tolerance. J Exp Med 1996; 183(6):2605-2616.
145. Neurath MF, Fuss I, Kelsall BL et al. Antibodies to interleukin 12 abrogate established experimental colitis in mice. J Exp Med 1995; 182(5):1281-1290.
146. Neurath MF, Fuss I, Pasparakis M et al. Predominant pathogenic role of tumor necrosis factor in experimental colitis in mice. Eur J Immunol 1997; 27(7):1743-1750.
147. Neurath MF, Weigmann B, Finotto S et al. The transcription factor T-bet regulates mucosal T cell activation in experimental colitis and Crohn's disease. J Exp Med 2002; 195(9):1129-1143.
148. Nieuwenhuis EE, Neurath MF, Corazza N et al. Disruption of T helper 2-immune responses in Epstein-Barr virus induced gene 3-deficient mice. Proc Natl Acad Sci USA 2002; 99(26):16951-16956.
149. Obermeier F, Dunger N, Strauch UG et al. Contrasting activity of cytosine-guanosin dinucleotide oligonucleotides in mice with experimental colitis. Clin Exp Immunol 2003; 134(2):217-224.
150. Ohkusa T. Production of experimental ulcerative colitis in hamsters by dextran sulfate sodium and change in intestinal microflora. Jpn J Gastroenterol 1985; 82:1327-1336.
151. Onderdonk AB, Franklin ML, Cisneros RL. Production of experimental ulcerative colitis in gnotobiotic guinea pigs with simplified microflora. Infect Immun 1981; 32(1):225-231.
152. Onderdonk AB, Hermos JA, Dzink JL et al. Protective effect of metronidazole in experimental ulcerative colitis. Gastroenterology 1978; 74:521-526.
153. Oppmann B, Lesley R, Blom B et al. Novel p19 protein engages IL-12p40 to form a cytokine, IL-23, with biological activities similar as well as distinct from IL-12. Immunity 2000; 13(5):715-725.
154. Orchard TR, Dhar A, Simmons JD et al. MHC class I chain-like gene A (MICA) and its associations with inflammatory bowel disease and peripheral arthropathy. Clin Exp Immunol 2001; 126(3):437-440.
155. Panwala CM, Jones JC, Viney JL. A novel model of inflammatory bowel disease: Mice deficient fro the multiple drug resistance gene, mdr1a, spontaneously develop colitis. J Immunol 1998; 161:5733-5744.
156. Peng Y, Laouar Y, Li MO et al. TGF-beta regulates in vivo expansion of Foxp3-expressing CD4+CD25+ regulatory T cells responsible for protection against diabetes. Proc Natl Acad Sci USA 2004; 101(13):4572-4577.
157. Pfeiffer C, Stein J, Southwood S et al. Altered peptide ligands can control CD4 T lymphocyte differentiation in vivo. J Exp Med 1995; 181(4):1569-1574.
158. Podolsky DK. Mucosal immunity and inflammation. V. Innate mechanisms of mucosal defense and repair: The best offense is a good defense. Am J Physiol 1999; 273(3 Pt 1):G495-G499.
159. Poussier P, Ning T, Chen J et al. Intestinal inflammation observed in IL-2R/IL-2 mutant mice is associated with impaired intestinal T lymphopoiesis. Gastroenterology 2000; 118(5):880-891.
160. Powrie F, Carlino J, Leach MW et al. A critical role for transforming growth factor-beta but not interleukin 4 in the suppression of T helper type 1-mediated colitis by $CD45RB^{low}$ CD4+ T cells. J Exp Med 1996; 183(6):2669-2774.
161. Powrie F, Correa-Oliveira R, Mauze S et al. Regulatory interactions between CD45RBhigh and CD45RBlow CD4+ T cells are important for the balance between protective and pathogenic cell-mediated immunity. J Exp Med 1994; 179(2):589-600.
162. Powrie F, Leach MW. Genetic and spontaneous models of inflammatory bowel disease in rodents: Evidence for abnormalities in mucosal immune regulation. Ther Immunol 1995; 2(2): 115-123.
163. Powrie F, Leach MW, Mauze S et al. Inhibition of Th1 responses prevents inflammatory bowel disease in scid mice reconstituted with $CD45RB^{hi}$ CD4+ T cells. Immunity 1994; 1(7):553-562.
164. Proberts CS, Chott A, Turner JR et al. Persistent clonal expansion of peripheral blood CD4+ lymphocytes in chronic inflammatory bowel disease. J Immunol 1996; 157:3183-3191.
165. Rachmilewitz D, Katakura K, Karmeli F et al. Toll-like receptor 9 signaling mediates the anti-inflammatory effects of probiotics in murine experimental colitis. Gastroenterology 2004; 126(2):520-528.
166. Rachmilewitz D, Karmeli F, Takabayashi K et al. Immunostimulatory DNA ameliorates experimental and spontaneous murine colitis. Gastroenterology 2002; 122(5):1428-41.

167. Rath HC, Schultz M, Freitag R et al. Different subsets of enteric bacteria induce and perpetuate experimental colitis in rats and mice. Infect Immun 2001; 69(4):2277-2285.
168. Rath HC, Wilson KH, Sartor RB. Differential induction of colitis and gastritis in HLA-B27 transgenic rats selectively colonized with Bacterioides vulgatus or Escherichia coli. Infect Immunol 1999; 67(6):2969-2974.
169. Rayes N, Seehofer D, Muller AR et al. Influence of probiotics and fibre on the incidence of bacterial infections following major abdominal surgery – results of a prospective trial. Z Gastroenterol 2002; 40(10):869-876.
170. Read S, Malmstrom V, Powrie F. Cytotoxic T lymphocyte-associated antigen 4 plays an essential role in the function of CD25+CD4+ regulatory cells that control intestinal inflammation. J Exp Med 2000; 192(2):295-302.
171. Rivera-Nieves J, Bamias G, Vidrich A et al. Emergence of perianal fistulizing disease in the SAMP1/YitFc mouse, a spontaneous model of chronic ileitis. Gastroenterology 2003; 124(4):972-982.
172. Rodloff AC, Widera P, Ehlers S et al. Suppression of blastogenic transformation of lymphocytes by Bacterioides fragilis in vitro and in vivo. Int J Med Microbiol 274:406-416.
173. Roncarolo MG, Bacchetta R, Bordignon C et al. Type 1 T regulatory cells. Immunol Rev 2001; 182:68-79.
174. Rudolph U, Finegold MJ, Rich SS et al. Ulcerative colitis and adenocarcinoma of the colon in G alpha i2-deficient mice. Nat Genet 1995; 10(2):143-150.
175. Sadlack B, Merz H, Schorle H et al. Ulcerative colitis-like disease in mice with a disrupted interleukin-2 gene. Cell 75:203-205.
176. Sadlack B, Merz H, Schorle H et al. Ulcerative colitis-like disease in mice with a disrupted interleukin-2 gene. Cell 1993; 75(2):253-261.
177. Sakaguchi S, Sakaguchi N, Shimizu J et al. Immunologic tolerance maintained by CD25+CD4+ regulatory T cells: Their common role in controlling autoimmunity, tumor immunity, and transplantation tolerance. Immunol Rev 2001; 182:18-32.
178. Sakaguchi S, Takahashi T, Yamazaki S et al. Immunologic self tolerance maintained by T-cell-mediated control of self-reactive T cells: Implications for autoimmunity and tumor immunity. Microbes Infect 2001; 3(11):911-918.
179. Sartor RB. Colitis in HLA-B27/β2 microglobulin transgenic rats. Int Rev Immunol 2000; 19(1):39-50.
180. Sartor RB. Therapeutic manipulation of the enteric microflora in inflammatory bowel diseases: Antibiotics, probiotics, and prebiotics. Gastroenterology 2004; 126(6):1620-1633.
181. Sartor RB, Veltkamp C. Interactions between enteric bacteria and the immune system which determine mucosal homeostasis vs. chronic intestinal inflammation: Lessons from rodent models. In: Rogler G, Kullman F, eds. IBD at the End of Its First Century. Dordrecht: Kluwer Academic, 2000:30-41.
182. Satsangi J, Parkes M, Louis E et al. Two stage genome-wide search inflammatory bowel disease provides evidence for susceptibility loci on chromosomes 3, 7, and 12. Nat Genet 1996; 14:199-202.
183. Schultz M, Tonkonogy SL, Sellon RK et al. IL-2-deficient mice raised under germfree conditions develop delayed mild focal intestinal inflammation. Am J Physiol 1999; 276(6 Pt 1):G1461-G1472.
184. Schultz M, Veltkamp C, Dieleman LA et al. Lactobacillus plantarum 299V in the treatment and prevention of spontaneous colitis in interleukin-10-deficient mice. Inflamm Bowel Dis 2002; 8(2):71-80.
185. Shaikh R, Santee S, Granger SW et al. Constitutive expression of LIGHT on T cells leads to lymphocyte activation, inflammation, and tissue destruction. J Immunol 2001; 167:6330-6337.
186. Sheil B, McCarthy J, O'Mahony L et al. Is the mucosal route of administration essential for probiotic function? Subcutaneous administration is associated with attenuation of murine colitis and arthritis. Gut 2004; 53(5):694-700.
187. Simpson SJ, Shah S, Comskey M et al. T cell-mediated pathology in two models of experimental colitis depends predominantly on the interleukin 12/Signal transducer and activator of transcription (Stat)-4 pathway, but is not conditional on interferon gamma expression by T cells. J Exp Med 1998; 187(8):1225-1234.
188. Snapper SB, Rosen FS, Mizoguchi E et al. Wiskott-Aldrich syndrome protein-deficient mice reveal a role for WASP in T but not B cell activation. Immunity 1998; 9(1):81-91.
189. Spencer SD, Di Marco F, Hooley J et al. The orphan receptor CRF2-4 is an essential subunit of the interleukin 10 receptor. J Exp Med 1998; 187(4):571-578.
190. Spencer DM, Veldman GM, Banerjee S et al. Distinct inflammatory mechanisms mediate early versus late colitis in mice. Gastroenterology 2002; 122(1):94-105.
191. Strober W, Fuss IJ, Blumberg RS. The immunology of mucosal models of inflammation. Annu Rev Immunol 2002; 20:495-549.

192. Strober W, Kelsall B, Marth T. Oral tolerance. J Clin Immunol 1998; 18(1):1-30.
193. Stuber E, Strober W, Neurath M. Blocking the CD40L-CD40 interaction in vivo specifically prevents the priming of T helper 1 cells through the inhibition of interleukin 12 secretion. J Exp Med 1996; 183(2):693-698.
194. Sundberg JP, Elson CO, Bedigian H et al. Spontaneous heritable colitis in a new substrain of C3H/HeJ mice. Gastroenterology 1994; 107(6):1726-1735.
195. Suzuki A, Hanada T, Mitsuyama K et al. CIS3/SOC3/SS13 plays a negative regulatory role in STAT3 activation and intestinal inflammation. J Exp Med 2001; 193(4):471-481.
196. Takahashi T, Sakaguchi S. Naturally arising CD25+CD4+ regulatory T cells in maintaining immunologic self-tolerance and preventing autoimmune disease. Curr Mol Med 2003; 3(8):693-706.
197. Takahashi T, Sakaguchi S. [T cell mediated immunoregulation and autoimmune disease]. Tanpakushitsu Kakusan Kos 2002; 47(16 Suppl):2331-2335.
198. Takahashi T, Sakaguchi S. The role of regulatory T cells in controlling immunologic self-tolerance. Int Rev Cytol 2003; 225:1-32.
199. Takeda K, Clausen BE, Kaisho T et al. Enhanced Th1 activity and development of chronic enterocolitis in mice devoid of Stat3 in macrophages and neutrophils. Immunity 1999; 10(1):39-49.
200. Takeda K, Kaisho T, Akira S. Toll-like receptors. Annu Rev Immunol 2003; 21:335-376.
201. Ten Hove T, Corbaz A, Amitai H et al. Blockade of endogenous IL-18 ameliorates TNBS-induced colitis by decreasing local TNF-α production in mice. Gastroenterology 2001; 121(6):1372-1379.
202. Tlaskalova H, Stepankova R, Hudcovic T et al. The role of bacterial microflora in development of dextran sodium sulfate (DSS) induced colitis in Immunocompetent and immunodeficient mice. Microb Ecol Health Dis 1999; 11:115-16, abstract.
203. Tomczak MF, Erdman SE, Poutahidis T et al. NF-kappa B is required within the innate immune system to inhibit microflora-induced colitis and expression of IL-12 p40. J Immunol 2003; 171(3):1484-1492.
204. Toy LS, Yio XY, Lin A et al. Defective expression of gp180, a novel CD8 ligand on intestinal epithelial cells, in inflammatory bowel disease. J Clin Invest 1997; 100(8):2062-2071.
205. Toyabe S, Seki S, Iiai T et al. Requirement of IL-4 and liver NK1+ T cells for concanavalin A-induced hepatic injury in mice. J Immunol 1997; 159(3):1537-1542.
206. Van de Wal Y, Corazza N, Allez M et al. Delineation of CD1d-restricted antigen presentation pathway associated with human and mouse intestinal epithelial cells. Gastroenterology 2003; 124(5):1420-1431.
207. Vezys V, Lefrancois L. Cutting edge: Inflammatory signals drive organ-specific autoimmunity to normally cross-tolerizing endogenous antigen. J Immunol 2002; 169(12):6677-6680.
208. Vezys V, Olson S, Lefrancois L. Expression of intestine-specific antigen reveals novel pathways of CD8 T cell tolerance induction. Immunity 2000; 12(5):505-514.
209. Wahl SM, Chen W. TGF-β: How tolerant can it be? Immunol Res 2003; 28(3):167-179.
210. Watanabe M, Ueno Y, Yajima T et al. Interleukin 7 transgenic mice develop chronic colitis with decreased interleukin 7 protein accumulation in the colonic mucosa. J Exp Med 1998; 187(3):389-402.
211. Watanabe T, Kitani A, Murrary P et al. NOD2 is a negative regulator of Toll-like receptor 2 mediated Th1 response. Natu Immunol manuscript in press, 2004.
212. Weiner HL. Oral tolerance: Immune mechanisms and the generation of Th3-type TGF-β-secreting regulatory cells. Microbes Infect 2001; 3(11):947-954.
213. Weiner HL. Oral tolerance: Immune mechanisms and treatment of autoimmune diseases. Immunol Today 1997; 18(7):335-343.
214. Wirtz S, Finotto S, Kanzler S et al. Cutting edge: Chronic intestinal inflammation in STAT-4 transgenic mice: Characterization of disease and adoptive transfer by TNF-plus IFN-γ-producing CD4+ T cells that respond to bacterial antigens. J Immunol 1999; 162(4):1884-1888.
215. Xiao S, Sung SS, Fu SM et al. Combining Fas mutation with interleukin-2 deficiency prevents Colitis and Lupus: Implicating interleukin-2 for auto-reactive T cell expansion and Fas ligand for colon epithelial cell death. J Biol Chem 2003; 278(52):52730-52738.
216. Yacyshyn B, Maksymowych W, Bowen-Yacyshyn MB. Differences in P-glycoprotein-170 expression and activity between Crohn's disease and ulcerative colitis. Hum Immunol 1999; 60(8):677-687.
217. Yamada Y, Marshall S, Specian RD et al. A comparative analysis of two models of colitis in rats. A comparative analysis of two models of colitis in rats. Gastroenterology 1992; 102(5):1524-1534.
218. Yamazaki M, Yajima T, Tanabe M et al. Mucosal T cells expressing high levels of IL-7 receptor are potential targets for treatment of chronic colitis. J Immunol 2003; 171(3):156-163.
219. Yoshida M, Shirai Y, Watanabe T et al. Differential localization of colitogenic Th1 and Th2 cells monospecific to a microflora-associated antigen in mice. Gastroenterology 2002; 123(6):1949-1961.
220. Yoshida M, Watanabe T, Usui T et al. CD4 T cells monospecific to ovalbumin produced by Escherichia coli can induce colitis upon transfer to BALB/c and SCID mice. Int Immunol 2001; 13(12):1561-1570.

Chapter 6

Overview of Role of the Immune System in the Pathogenesis of Inflammatory Bowel Disease

Thomas T. MacDonald and Giovanni Monteleone

Abstract

Crohn's disease (CD) and ulcerative colitis (UC) are caused by the over-activity of the immune system. Current and novel therapies are designed to dampen these over-active responses. Analysis of the types of immune responses ongoing in diseased mucosa of inflammatory bowel disease patients has revealed that CD and UC are fundamentally different diseases. The former has the molecular imprints of a Th1 dominant cell-mediated hypersensitivity response whereas the latter appears to involve antibody-mediated hypersensitivity.

Introduction

Over the years many notions have been put forward for the cause of IBD. These have ranged from the fanciful, such as toothpaste and cornflakes, to an infectious etiology, such as *Mycobacterium paratuberculosis*, filterable viruses, *Mycobacteria spp*, measles virus or even milk. Research in these areas however has been quite unproductive and there is growing realisation that IBD is (are) an extremely complex disorder(s) and the idea for example that most of the cases of Crohn's disease have a unifying underlying cause which if revealed would lead to a "cure", is now extremely improbable. As discussed in the next chapter, many experimental manipulations in animal models can produce an identical lesion in the colon and it is almost certainly best to think of CD and UC as complex diseases with common histological manifestations. One of the best analogies is asthma, attacks of which can be precipitated by exercise, air-borne allergens, foods or even cold air. No-one would try to find the cause of asthma.

Whatever the complex genetic and environmental etiology of CD and UC, the unifying theme is that these are diseases caused by excessive immune reactivity in the gut wall. Apart from enteral nutrition, the therapeutic basis for which is still unknown, all treatment strategies are based on suppressing or modulating the immune system. The challenge is to accomplish this in a targeted and specific way which is effective, but avoids side-effects such as infection. A major part of this effort involves clarification of the immune/inflammatory pathways in CD and UC to design disease-specific therapy. In the course of these studies it has become very clear that CD and UC are immunologically different diseases, although they share end-stage effector pathways of tissue damage. In this chapter we will review the role of the different components of the immune system in CD and UC.

Immune Mechanisms in Inflammatory Bowel Disease, edited by Richard S. Blumberg and Markus F. Neurath. ©2006 Eurekah.com and Springer Science+Business Media.

Crohn's Disease, but Not Ulcerative Colitis Bears the Stigmata of a Th1 Mediated Immune Response

Case reports of prolonged remission coinciding with HIV-induced CD4 T cell loss suggested a primary role for CD4 T cells in CD.[1] The first molecular differences however were described by Mullin and colleagues who showed that transcripts for IL-2 were high in CD but in not UC.[2] Using assays to measure the number of cytokine secreting cells, Breese et al then showed that CD but not UC lesions contained abundant cells which secreted IL-2 and interferon-γ and that transcripts for the latter were obvious in CD resection specimens, but not UC resection specimens.[3] Finally, using isolated lamina propria CD4 cells, Fuss and colleagues clearly showed that samples from CD patients made markedly more interferon-γ than either controls or UC patients when activated with anti-CD3/CD28 antibodies.[4] One interesting observation was that UC patients did appear to make more IL-5 than CD patients. This result has been mis-interpreted as showing that UC may be a Th2 disease, when in fact, the same samples made 10-20 fold as much interferon-γ as IL-5.

The realisation that Th1 differentiation depended largely on IL-12 signaling through the IL-12β2 chain and activation of the STAT4 pathway led to an investigation of IL-12 in IBD. IL-12 is a heterodimer of a p35 and p40 chains, with the p35 chain ubiquitously expressed and the p40 chain needed to make bioactive IL-12. The IL-12R is made of a β1 and β2 chains. The β1 chain is present on Th1 and Th2 cells whereas the β2 chain is expressed on Th1 cells. In CD but not UC, bioactive IL-12 is present in macrophages in the mucosa and submucosa.[5,6] In addition in CD but not UC, there is high expression of IL-12Rβ2 chain and activated STAT4 is present in CD lamina propria mononuclear cells.[7]

IL-18 is a cytokine which synergises with IL-12 to drive Th1 responses. It is also abundant in lamina propria macrophages and epithelial cells of CD patients but not UC patients and importantly, CD tissue contains the active p18 form of IL-18 whereas both normal and UC bowel only contain the p24 IL-18 precursor which is constitutively expressed in epithelial cells.[8,9]

Th1 commitment seems to be stable after polarisation and there is a great deal of interest in the changes in chromatin remodeling which allow a polarised Th1 cells to remember to switch on Th1 cytokines when activated.[10] An analysis of transcription factors expressed in Th1 vs Th2 cells allowed Glimcher and colleagues to discover T-bet, a novel member of the T-box family of transcription factors.[11] T-bet drives chromatin remodelling of the interferon-γ locus and upregulates IL-12Rβ2 chain.[12,13] Importantly, T-bet is also increased by interferon-γ, providing a mechanism by which Th1 responses can be self-propagating.[14] In patients with CD but not UC, T-bet is markedly over-expressed,[15] and interestingly, TGFβ inhibits T-bet expression, providing a pathway by which Th1 responses can be switched off in inflamed intestine. Figure 1 illustrates how T-bet and IL-12 interact to polarise and activate Th1 cells.

While the Th1 nature of the cell-mediated immune response in the mucosa of Crohn's disease is now well established, the nature of the antigens driving this response are still unknown. Clinical observations suggest that the indigenous bacterial flora drives the Th1 response,[16] and patients with IBD lose tolerance to their own flora.[17] Animal models of IBD clearly show the crucial role of the flora in causing colitis, but they also show that it is not likely to be a single member of the flora. The gut microbiota is highly complex with 400 different bacterial species in the colon and ileum. There are also probably millions of candidate antigens made by the flora. Perhaps the best support for a pathogenic role of the flora comes from studies of children with phagocytic defects, such as glycogen storage disease type 1b or chronic granulomatous disease. A subgroup of these children develop a terminal ileitis and colitis which is essentially indistinguishable from Crohn's disease.[18] If phagocyte function is boosted, the Crohn's-like lesion ameliorates.[19] This is presumably because the indigenous microflora constantly translocates across the epithelium and is killed by subepithelial macrophages. However if macrophage killing function is impaired, the bacteria will persist and

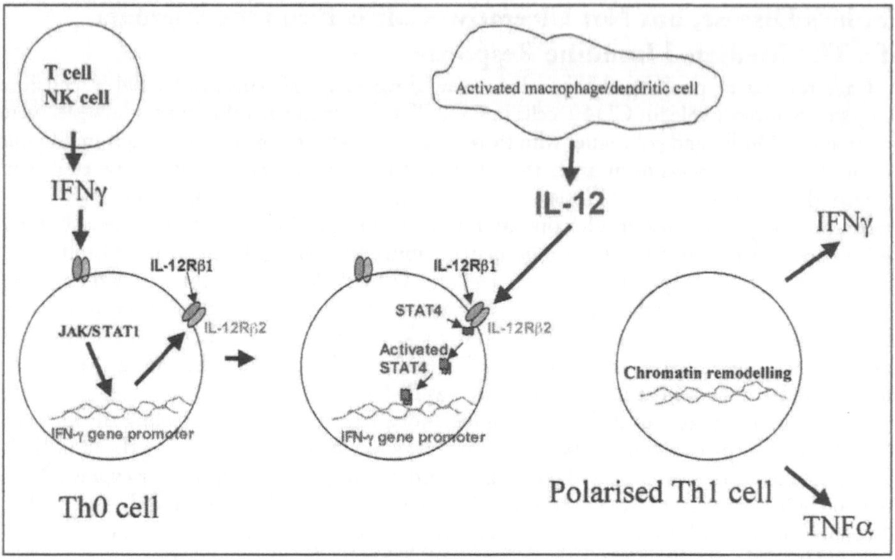

Figure 1. Two factors, IL-12 and T-bet synergise to drive Th1 immune responses in Crohn's disease. T-bet is induced by interferon-γ through STAT1 and can remodel the interferon-γ locus to induce Th1 polarisation. It also increases expression of the IL-12Rβ2 chain. IL-12 p70 binds to the IL-12R and signals through the β2 chain to activate STAT4 which goes to the nucleus where it also activates the interferon-γ promoter. In Crohn's disease but not UC, there are high local levels of IL-12, interferon-γ and T-bet.

initiate a T cell response. This is remarkably similar to the postulated function of NOD2/CARD15, the gene on chromosome 16 which is mutated in a subgroup of Crohn's patients and whose function is probably to recognise bacteria and switch on anti-bacterial killing mechanisms.[20,21]

The Role of T Cells in Ulcerative Colitis

There is a massive infiltrate of T cells into the inflamed mucosa of UC patients, but their role in damage is not understood. IL-12 is produced in large amounts by macrophages following activation with bacterial components and it is counter-intuitive that the infiltrating macrophages in UC mucosa would not be activated by the flora and make IL-12. However that is clearly the case and so perhaps IL-12 production is being actively down-regulated in UC. There is no global defect in IL-12 production in UC. IL-27 is a new cytokine made by dendritic cells and macrophages which is made up of a heterodimer of an IL-12p40 related protein, Epstein-Barr virus induced gene 3 (EBI3) and p28, a newly discovered IL-12p35 like molecule.[22] IL-27 activates naïve CD4 T cells and synergises with IL-12 to drive interferon-γ production. EBI3 is markedly increased in ulcerative colitis compared to Crohn's disease and normal mucosa.[23,24] Again however, IL-27 polarises cells along the Th1 pathway and this does not happen in UC. At the moment therefore the role of T cells in tissue damage in UC mucosa is still poorly understood.

The Role of Antibody in Ulcerative Colitis

One of the most striking features of ulcerative colitis is the massive and uniform presence of IgG plasma cells along diseased mucosa.[25] This contrasts markedly with Crohn's disease where IgG plasma cells are principally increased around ulcers.[26] Antibody-mediated hypersensitivity has long been a strong candidate as a principal mediator of tissue damage in UC. If rabbits are given a colonic irritant and injected intravenously with preformed immune complexes, these deposit in the colon wall and set off a local Arthus reaction which is virtually histologically identical to active UC.[27] The lesion in UC is also predominantly neutrophilic with severe epithelial damage and neutrophilic crypt abscesses. Taken together, these results suggest that the IgG plasma cells in CD might be secondary to ulceration and penetration of gut antigens into the mucosa, but that in UC they may be of primary pathogenic significance, ie, UC is an organ-specific, antibody-mediated, autoimmune disease.

Using immunohistochemistry, Halstensen and colleagues showed that there was colocalisation of IgG1 antibody and C1q, C4c, C3b and the terminal attack complex of complement on the apical aspect of colonic epithelial cellc in UC but not CD.[28,29] This often colocalises with an antibody which recognised tropomysin, a putative autoantigen in UC, which is recognised by tissue IgG from patients with UC.[30] Older data suggests that about half of patients with UC have autoantibodies to epithelial cells and goblet cells.[31,32] It is obvious however that there is no autoantigen which universally is present in UC patients and this has often been used to argue against autoantibody as being of pathogenic significance. However since UC may be a heterogeneous condition, there is the possibility that autoimmune responses against a constellation of autoantigens between individuals, some shared and some not, may activate complement and produce the same end-stage lesion.

Strong circumstantial evidence for the role of the B cell comes from the striking effects of appendectomy on UC. Prior appendectomy protects against the development of UC and even in affected cases, it ameliorates the natural history.[33-35] Appendectomy after diagnosis also modifies the natural history to a much milder course.[36] There is a substantial literature from the rabbit to suggest that the appendix plays an important role in generating the primary B cell repertoire and in sheep, Peyer's patches play the same role.[37,38] Removing the appendix may therefore deplete the repertoire of B cells with colon auto-reactive or cross-reactive specificities (Fig. 2).

More recent molecular analysis of the IgG and IgA response in UC has also supported a role for IgG antibodies.[39] Using PCR to analyse the third complementarity determing region (CDR3) of the immunoglobulin heavy chain, Thoree and colleagues demonstrated clonal relatedness of IgG and IgA1 in UC mucosa and these clones were spread along diseased bowel. These clones

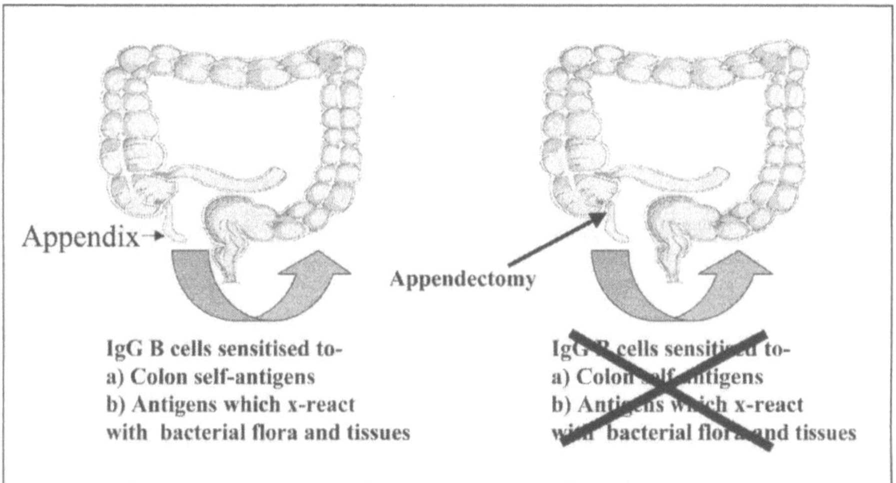

Figure 2. One of the most striking features of ulcerative colitis is that prior appendectomy or even appendectomy after diagnosis can alter the natural history of the disease. In other species the appendix is an important site of primary B cell development and it may be that in man, B cells with colon-self reactivity, or cross-reactive with antigens of the flora and colon, are generated in the appendix in man and migrate to the mucosa causing disease.

were also demonstrable in blood. Since IgA1 and IgG are characteristic of systemic and not mucosal antibody responses, the authors suggested that UC represents a peripheral type of antibody response which is expressed and expanded in the mucosa.

Nonspecific Effector Mechanisms Down-Stream of the Immune System Cause Tissue Injury in Crohn's Disease and Ulcerative Colitis

Regardless of the immunological differences between UC and CD, a common feature is the markedly elevated local concentrations of eucosanoid, leukotriene, free radical and cytokine mediators of inflammation. Inhibition of all of these pathways is effective in animal models of IBD, but it is only with antagonism of cytokines that therapeutic efficacy has been achieved in patients.[40] Nonetheless, there is virtually no evidence that cytokines such as IL-1β or TNFα damage the gut. Instead they prolong inflammation by activating NFκB dependent pathways, such as upregulation of adhesion molecules on vascular endothelium and increasing chemokine expression, which together continue to attract blood-borne inflammatory cells into the tissue. The ulceration and degradation of the mucosa in IBD however appears to be due to the cytokine-induced production of neutral endopeptidases, the matrix metalloproteinases (MMPs) in the tissue itself (Fig. 3).[41] By and large the MMPs which cause damage are not produced by inflammatory cells, with the exception of MMP2 and 9 (the gelatinases) produced by neutrophils and MMP12 (metalloelastase) produced by macrophages. Instead the collagenases (MMP1 and MMP13) and stromelysins (MMP3 and MMP10) are secreted by cytokine activated mucosal fibroblasts and MMP10 is made by epithelial cells.[42-45] In situ hybridisation and westen blotting has shown very high expression of these molecules in diseased tissues and around ulcers. MMP1,3,10 and 13 are secreted as zymogens and activated in the extracellular space by plasmin, free radicals and activated MMP's themselves. They have the capacity to proteolytically digest interstitial collagen in the lamina propria and basement membrane collagen IV. The stromelysins also degrade proteglycans in the lamina propria. We have shown that matrix metalloproteinases can rapidly destroy the mucosa in explants of human fetal gut where the resident T cells have been activated with anti-CD3 and exogenous IL-12.[46,47]

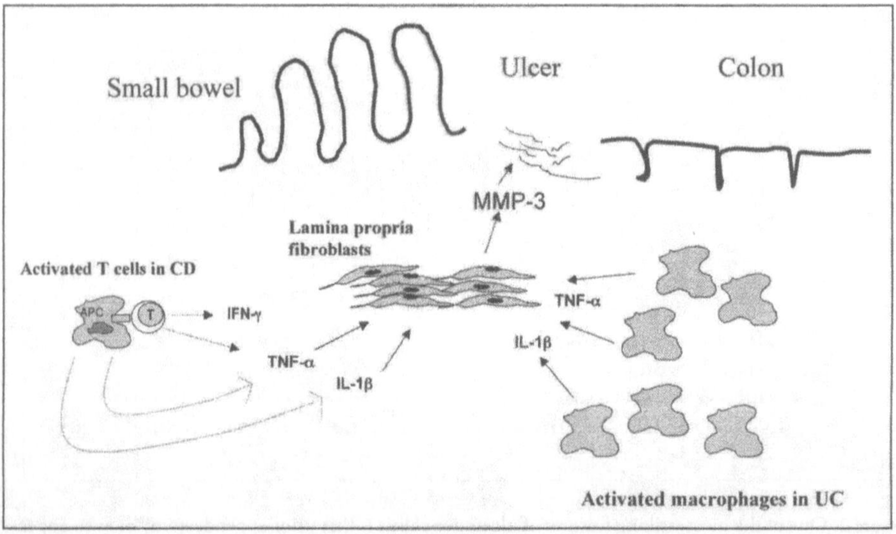

Figure 3. Although T cells cause Crohn's disease and antibodies probably cause UC, the down-stream effector pathways are the same. In both diseases there are high local concentrations of TNFα and IL-1β and these activate micosal fibroblasts to produce matrix metalloenzymes which degrade the mucosa and cause ulcers.

MMP's play a much wider role than merely degradation of extracellular matrix. The matrix plays a fundamental role in controlling cell survival, cell shape, growth and differentiation by sequestering growth factors and providing ligands for cell surface receptors.[48] Since MMP's alter the matrix, they must also alter all of these signals within the microenvironment. For example, latent TGFβ is sequestered in the matrix by decorin, a proteoglycan which is associated with collagen.[49] Degradation of decorin by MMP's produced by TNFα activated lamina propria myofibroblasts makes TGFβ bioavailable to drive collagen synthesis in fibroblasts. But TGFβ is also a potent inhibitor of MMP-1 and MMP3,[50] so release of TGFβ1 feeds back and inhibits MMP production by the myofibroblasts. However TGFβ itself activates a negative loop by inducing the expression of the inhibitory signalling molecule Smad7, which prevents signal transduction to the nucleus following TGFβ binding.[51] There is therefore a dampening of the TGFβ1 inhibitory loop, allowing TNFα to continue to drive MMP production. MMP3 and MMP7 can cleave the epithelial adherens junction protein E-cadherin and the fragments produced disrupt other cells by acting as a competitive inhibitor of E-cadherin homotypic binding between cells.[52,53] Since E-cadherin is important not only in maintaining epithelial integrity in the gut but also acts as a tumor suppressor,[54] MMP3 in the gut wall may cause loss of barrier function and may even promote carcinogenesis. MMP's can cleave pro-TNFα from the cell membrane increasing local concentrations in the gut wall,[55] but also degrade IL-1β which will decrease inflammation.[56]

An obvious therapeutic strategy for IBD is therefore to use MMP inhibitors, many of which have been produced. Unfortunately their translation into clinical use has been problematic because of side-effects.

The Immune System Fails to Negatively Regulate in CD and UC

Undoubtedly the chronic inflammation in the mucosa causes the long term problems in IBD. Paradoxically however the immune system has developed many strategies to down-regulate chronic inflammation and yet these do not appear to be operative in IBD. TGFβ is the most potent endogenous down-regulator of inflammation. It is made by virtually all cell types and although there has been interest in regulatory T cells which either secrete TGFβ1 or express it as a membrane form, large amounts are also made by activated macrophages. Mice in whom the TGFβ1 gene has been deleted die of generalised inflammation early in life, a component of which is a severe colitis.[57,58] TGFβ is also markedly increased in IBD tissues.[59]

TGF-β1 initiates signaling through the ligand-dependent activation of a complex of heterodimeric transmembrane serine/threonine kinases, consisting of type I (TGF-β1 RI) and type II (TGF-β1 RII) receptors.[60,61] Upon TGF-β1 binding, the receptors rotate relatively within the complex, resulting in phosphorylation and activation of TGF-β1RI by the constitutively active and auto-phosphorylated TGF-β1RII. TGF-β1 signals from the receptor to the nucleus using a set of proteins, termed Smads, based on their high homology to the Drosophila Mad and the *Caenorrhabditis elegans* Sma proteins. To date, nine different Smad genes which fall into three distinct functional sets have been identified: signal-transducing receptor-activated Smads, which include Smad1, 2, 3, 5, 8 and 9; a single common mediator, Smad4, and inhibitory Smad 6 and 7. Activated TGF-β1RI directly phosphorylates Smad2 and Smad3 at serine residues in the carboxy-terminal SSXS sequence. Once activated, Smad2 and Smad3 associate with Smad4 and translocate to the nucleus where Smad protein complexes participate in transcriptional control of target genes by binding to TGFβ regulatory elements in the promoters of many genes (Fig. 4).

We have recently shown that the inhibitory Smad, Smad7, is over-expressed in inflammatory bowel disease mucosa and that when lamina propria mononuclear cells (a mixture of T cells, plasma cells, stromal cells and macrophages) from Crohn's patients are stimulated with TGFβ1, there is no phosphorylation of Smad 2/3.[51] In addition, whereas TGFβ1 potently down-regulates pro-inflammatory cytokine production by normal lamina propria mononuclear cells (LPMC), it has no effect on cytokine production by cells from IBD patients. However if Smad7 is inhibited

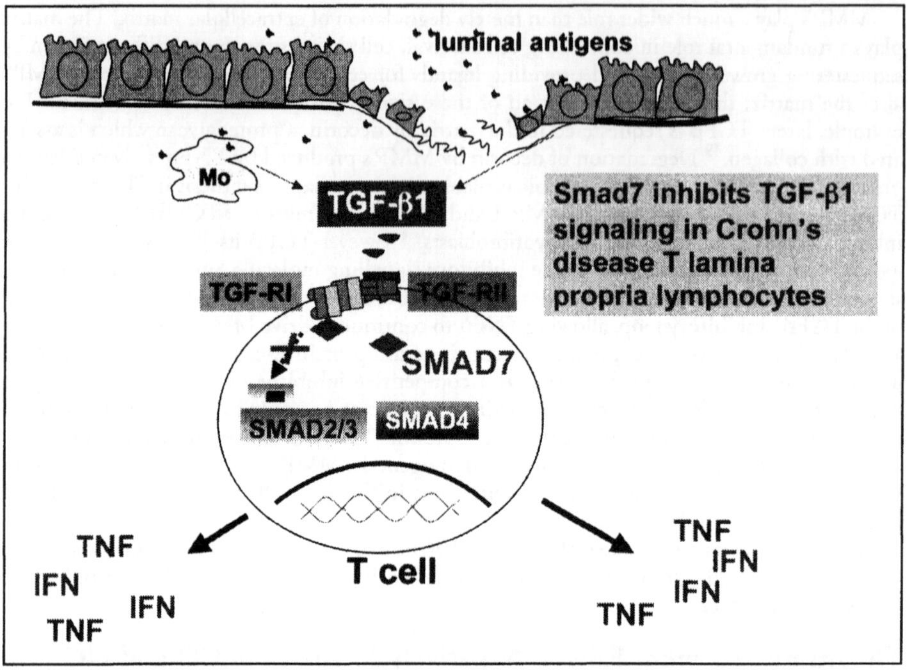

Figure 4. TGFβ is a potent negative regulator of pro-inflammatory cytokine and MMP production. In order to deliver this negative signal to the inflammatory cells, TGFβ binds to specific receptors and activates the SMAD signalling pathway. In IBD however, mononuclear cells contain a negative regulator of SMAD signaling, Smad 7, which prevents phosphorylation of Smad2/3. Thus despite high local concentrations of TGFβ1 in IBD, this failure in signalling negates its bioactivity and allows the continued production of pro-inflammatory cytokines by T cells

in LPMC from IBD patients by using a specific anti-sense oligonucleotide, TGFβ1 is then able to inhibit cytokine production. Furthermore, anti-sense to Smad 7 is also functional in explant cultures of Crohn's diseases where with Smad7 eliminated, endogenous TGFβ1 in the tissue can signal to inflammatory cells and reduce cytokine production.

The mechanism by which TGFβ1 inhibits inflammation in the gut when Smad 7 is abrogated has also been investigated.[62] The addition of TNFα to normal LPMC leads to the activation of NFκB within minutes. This is completely abrogated by pretreatment with TGFβ1. The mechanism appears to be due to the fact that TGFβ1 increases transcription of IκB α, so that NFκB remains sequestered in the cytoplasm. Indeed in reporter gene assays, TGFβ1 activates the IKBα promoter. In IBD, NFκB is activated because of excess local TNFα and this cannot be inhibited because of Smad 7. However if smad 7 is inhibited using anti-sense, endogenous TGFβ1 upregulates IKBα and rapidly prevents NFκB activation and migration of the active subunits to the nucleus. A crucial question therefore is what controls the expression of Smad7 in the gut and at the moment this is not known.

Microscopic Colitis (Collagenous Colitis and Lymphocytic Colitis)

These are rare diseases (annual incidence 1 per 100,000) which were only identified in the last 20-30 years. Patients present with chronic watery diarrhea, abdominal cramping, anorexia, nausea, and incontinence. At colonoscopy, the mucosa usually appears normal although some cases do show reddening of the mucosa, edema and some erosions. However the disease is much less obvious than classical IBD.

Histological analysis of the biopsies reveals that microscopic colitis consists of 2 diseases.[63] The first is called " lymphocytic colitis". The characteristic feature of this condition is an increase in the number of intraepithelial lymphocytes (IEL) from less than 5 per 100 epithelial cells to about 25 per 100 epithelial cells.[64] Epithelial cells may be flattened and some might show detachment from basement membrane. There is also an increase in lymphocytes and macrophages in the lamina propria. Neutrophils are uncommon. The increased epithelial lymphocytes are similar in phenotype to normal bowel, in that they are mostly CD8+ and express the αβ T cell receptor. Likewise, the increased lamina propria T cells are mostly CD4+, as in normal bowel.[65]

The second form of microscopic colitis is called "collagenous colitis". It derives its name from the thick band of collagen which is seen below the epithelium in these patients and which is very obvious by standard H&E staining or by routine stains for collagen. Detailed studies have shown that there is increased production of the messenger RNA for collagen in the mucosa of collagenous colitis patients.[66] There is also a mononuclear cell infiltrate, but there is much less increase in IEL than lymphocytic colitis. In collagenous colitis there is greater epithelial detachment than in lymphocytic colitis. There is no standard treatment for microscopic colitis. Very recently however promising results have been obtained in collagenous colitis using budenoside.[67]

Because of their rarity, there has been no functional studies of the types of immune responses ongoing in the mucosa of microscopic colitis patients. The role of IEL in lymphocytic colitis is not known, nor is it known if the collagen band in collagenous colitis is secondary to the inflammation or is a primary defect.

References

1. James SP. Remission of Crohn's disease after human immunodeficiency virus infection. Gastroenterology 1988; 95:1667-1669.
2. Mullin GE, Lazenby AJ, Harris ML et al. Increased interleukin-2 messenger RNA in the intestinal mucosal lesions of Crohn's disease but not ulcerative colitis. Gastroenterology 1992; 102:1620-1627.
3. Breese E, Braegger CP, Corrigan CJ et al. Interleukin-2- and interferon-gamma-secreting T cells in normal and diseased human intestinal mucosa. Immunology 1993; 78:127-131.
4. Fuss IJ, Neurath M, Boirivant M et al. Disparate CD4+ lamina propria (LP) lymphokine secretion profiles in inflammatory bowel disease. Crohn's disease LP cells manifest increased secretion of IFN-gamma, whereas ulcerative colitis LP cells manifest increased secretion of IL-5. J Immunol 1996; 157:1261-1270.
5. Monteleone G, Biancone L, Marasco R et al. Interleukin 12 is expressed and actively released by Crohn's disease intestinal lamina propria mononuclear cells. Gastroenterology 1997; 112:1169-1178.
6. Parronchi P, Romagnani P, Annunziato F et al. Type 1 T-helper cell predominance and interleukin-12 expression in the gut of patients with Crohn's disease. Am J Pathol 1997; 150:823-832.
7. Parrello T, Monteleone G, Cucchiara S et al. Up-regulation of the IL-12 receptor beta 2 chain in Crohn's disease. J Immunol 2000; 165:7234-7239.
8. Monteleone G, Trapasso F, Parrello T et al. Bioactive IL-18 expression is up-regulated in Crohn's disease. J Immunol 1999; 163:143-147.
9. Pizarro TT, Michie MH, Bentz M et al. IL-18, a novel immunoregulatory cytokine, is up-regulated in Crohn's disease: Expression and localization in intestinal mucosal cells. J Immunol 1999; 162:6829-6835.
10. Avni O, Lee D, Macian F et al. T(H) cell differentiation is accompanied by dynamic changes in histone acetylation of cytokine genes. Nat Immunol 2002; 3:643-651.
11. Szabo SJ, Kim ST, Costa GL et al. A novel transcription factor, T-bet, directs Th1 lineage commitment. Cell 2000; 100:655-669.
12. Mullen AC, Hutchins AS, High FA et al. Hlx is induced by and genetically interacts with T-bet to promote heritable T(H)1 gene induction. Nat Immunol 2002; 3:652-658.
13. Afkarian M, Sedy JR, Yang J et al. T-bet is a STAT1-induced regulator of IL-12R expression in naive CD4+ T cells. Nat Immunol 2002; 3:549-557.
14. Lighvani AA, Frucht DM, Jankovic D et al. T-bet is rapidly induced by interferon-gamma in lymphoid and myeloid cells. Proc Natl Acad Sci USA 2001; 98:15137-15142.

15. Neurath MF, Weigmann B, Finotto S et al. The transcription factor T-bet regulates mucosal T cell activation in experimental colitis and Crohn's disease. J Exp Med 2002; 195:1129-1143.
16. Shanahan F. Crohn's disease. Lancet 2002; 359:62-69.
17. Duchmann R, Kaiser I, Hermann E et al. Tolerance exists towards resident intestinal flora but is broken in active inflammatory bowel disease (IBD). Clin Exp Immunol 1995; 102:448-455.
18. Winkelstein JA, Marino MC, Johnston Jr RB et al. Chronic granulomatous disease. Report on a national registry of 368 patients. Medicine (Baltimore) 2000; 79:155-169.
19. Myrup B, Valerius NH, Mortensen PB. Treatment of enteritis in chronic granulomatous disease with granulocyte colony stimulating factor. Gut 1998; 42:127-130.
20. Ogura Y, Bonen DK, Inohara N et al. A frameshift mutation in NOD2 associated with susceptibility to Crohn's disease. Nature 2001; 411:603-606.
21. Hugot JP, Chamaillard M, Zouali H et al. Association of NOD2 leucine-rich repeat variants with susceptibility to Crohn's disease. Nature 2001; 411:599-603.
22. Pflanz S, Timans JC, Cheung J et al. IL-27, a heterodimeric cytokine composed of EBI3 and p28 protein, induces proliferation of naive CD4(+) T cells. Immunity 2002; 16:779-790.
23. Christ AD, Stevens AC, Koeppen H et al. An interleukin 12-related cytokine is up-regulated in ulcerative colitis but not in Crohn's disease. Gastroenterology 1998; 115:307-313.
24. Omata F, Birkenbach M, Matsuzaki S et al. The expression of IL-12 p40 and its homologue, Epstein-Barr virus-induced gene 3, in inflammatory bowel disease. Inflamm Bowel Dis 2001; 7:215-220.
25. Kett K, Rognum TO, Brandtzaeg P. Mucosal subclass distribution of immunoglobulin G-producing cells is different in ulcerative colitis and Crohn's disease of the colon. Gastroenterology 1987; 93:919-924.
26. Brandtzaeg P. The B cell system. Food Allergy & Intolerance. London: Bailliere Tindall, 1987; 118-155.
27. Mee AS, McLaughlin JE, Hodgson HJ et al. Chronic immune colitis in rabbits. Gut 1979; 20:1-5.
28. Halstensen TS, Mollnes TE, Garred P et al. Surface epithelium related activation of complement differs in Crohn's disease and ulcerative colitis. Gut 1992; 33:902-908.
29. Halstensen TS, Mollnes TE, Garred P et al. Epithelial deposition of immunoglobulin G1 and activated complement (C3b and terminal complement complex) in ulcerative colitis. Gastroenterology 1990; 98:1264-1271.
30. Halstensen TS, Das KM, Brandtzaeg P. Epithelial deposits of immunoglobulin G1 and activated complement colocalise with the M(r) 40 kD putative autoantigen in ulcerative colitis. Gut 1993; 34:650-657.
31. McGiven AR, Ghose T, Nairn RC. Autoantibodies in ulcerative colitis. Br Med J 1967; 2:19-23.
32. Marcussen H, Nerup J. Fluorescent anti-colon and organ-specific antibodies in ulcerative colitis. Scand J Gastroenterol 1973; 8:9-15.
33. Rutgeerts P, D'Haens G, Hiele M et al. Appendectomy protects against ulcerative colitis. Gastroenterology 1994; 106:1251-1253.
34. Shanahan F, O'Sullivan GC. Appendectomy, mucosal immunity, and colitis. Gastroenterology 1997; 112:1766-1767.
35. Okazaki K, Onodera H, Watanabe N et al. A patient with improvement of ulcerative colitis after appendectomy. Gastroenterology 2000; 119:502-506.
36. Cosnes J, Carbonnel F, Beaugerie L et al. Effects of appendectomy on the course of ulcerative colitis. Gut 2002; 51:803-807.
37. Pospisil R, Mage RG. Rabbit appendix: A site of development and selection of the B cell repertoire. Curr Top Microbiol Immunol 1998; 229:59-70.
38. Reynolds JD, Kennedy L, Peppard J et al. Ileal Peyer's patch emigrants are predominantly B cells and travel to all lymphoid tissues in sheep. Eur J Immunol 1991; 21:283-289.
39. Thoree VC, Golby SJ, Boursier L et al. Related IgA1 and IgG producing cells in blood and diseased mucosa in ulcerative colitis. Gut 2002; 51:44-50.
40. van Deventer SJ. Small therapeutic molecules for the treatment of inflammatory bowel disease. Gut 2002; 50(Suppl 3):III47-III53.
41. MacDonald TT, Pender SL. Proteolytic enzymes in inflammatory bowel disease. Inflamm Bowel Dis 1998; 4:157-164.
42. Saarialho-Kere UK, Vaalamo M, Puolakkainen P et al. Enhanced expression of matrilysin, collagenase, and stromelysin-1 in gastrointestinal ulcers. Am J Pathol 1996; 148:519-526.
43. Vaalamo M, Karjalainen-Lindsberg ML, Puolakkainen P et al. Distinct expression profiles of stromelysin-2 (MMP-10), collagenase-3 (MMP-13), macrophage metalloelastase (MMP-12), and tissue inhibitor of metalloproteinases-3 (TIMP-3) in intestinal ulcerations. Am J Pathol 1998; 152:1005-1014.

44. Pender SL, Breese EJ, Gunther U et al. Suppression of T cell-mediated injury in human gut by interleukin 10: Role of matrix metalloproteinases. Gastroenterology 1998; 115:573-583.
45. Heuschkel RB, MacDonald TT, Monteleone G et al. Imbalance of stromelysin-1 and TIMP-1 in the mucosal lesions of children with inflammatory bowel disease. Gut 2000; 47:57-62.
46. Pender SL, Tickle SP, Docherty AJ et al. A major role for matrix metalloproteinases in T cell injury in the gut. J Immunol 1997; 158:1582-1590.
47. Monteleone G, MacDonald TT, Wathen NC et al. Enhancing lamina propria Th1 cell responses with interleukin 12 produces severe tissue injury. Gastroenterology 1999; 117:1069-1077.
48. Lukashev ME, Werb Z. ECM signalling: Orchestrating cell behaviour and misbehaviour. Trends Cell Biol 1998; 437-41.
49. Imai K, Hiramatsu A, Fukushima D et al. Degradation of decorin by matrix metalloproteinases: Identification of the cleavage sites, kinetic analyses and transforming growth factor-beta1 release. Biochem J 1997; 322:809-14.
50. Uria JA, Jimenez MG, Balbin M et al. Differential effects of transforming growth factor-beta on the expression of collagenase-1 and collagenase-3 in human fibroblasts. J Biol Chem 1998; 273:9769-9777.
51. Monteleone G, Kumberova A, Croft NM et al. Blocking Smad7 restores TGFβ1 signalling in chronic inflammatory bowel disease. J Clin Invest 2001; 108:601-608.
52. Lochter A, Galosy S, Muschler J et al. Matrix metalloproteinase stromelysin-1 triggers a cascade of molecular alterations that leads to stable epithelial-to-mesenchymal conversion and a premalignant phenotype in mammary epithelial cells. J Cell Biol 1997; 139:1861-72.
53. Noe V, Fingleton B, Jacobs K et al. Release of an invasion promoter E-cadherin fragment by matrilysin and stromelysin-1. J Cell Sci 2001; 114:111-118.
54. Berx G, Cleton-Jansen AM, Nollet F et al. E-cadherin is a tumour/invasion suppressor gene mutated in human lobular breast cancers. EMBO J 1995; 14:6107-15.
55. Roghani M, Becherer JD, Moss ML et al. Metalloprotease-disintegrin MDC9: Intracellular maturation and catalytic activity. J Biol Chem 1999; 274:3531-40.
56. Ito A, Mukaiyama A, Itoh Y et al. Degradation of interleukin 1β by matrix metalloproteinases. J Biol Chem 1996; 271:14657-60.
57. Shull MM, Ormsby I, Kier AB et al. Targeted disruption of the mouse transforming growth factor-beta 1 gene results in multifocal inflammatory disease. Nature 1992; 359:693-699.
58. Kulkarni AB, Huh CG, Becker D et al. Transforming growth factor beta 1 null mutation in mice causes excessive inflammatory response and early death. Proc Natl Acad Sci USA 1993; 90:770-774.
59. Babyatsky MW, Rossiter G, Podolsky DK. Expression of transforming growth factors alpha and beta in colonic mucosa in inflammatory bowel disease. Gastroenterology 1996; 110:975-984.
60. Attisano L, Wrana JL. Signal transduction by the TGF-beta superfamily. Science 2002; 296:1646-1647.
61. Moustakas A, Souchelnytskyi S, Heldin CH. Smad regulation in TGF-beta signal transduction. J Cell Sci 2001; 114:4359-4369.
62. Monteleone G, Mann J, Monteleone I et al. A failure of transforming growth factor-beta1 negative regulation maintains sustained NF-kappa B activation in gut inflammation. J Biol Chem. 2004; 279:3925-3932.
63. Tahkalidis P et al. Microscopic colitis. J Gastro Hepatol 17:236-248.
64. Lazenby AJ, Yardley JH, Giardiello FM et al. Lymphocytic ("microscopic") colitis: A comparative histopathologic study with particular reference to collagenous colitis. Hum Pathol 1989; 20:18-28.
65. Mosnier JF, Larvol L, Barge J et al. Lymphocytic and collagenous colitis: An immunohistochemical study. Am J Gastroenterol 1996; 91:709-713.
66. Gunther U, Schuppan D, Bauer M et al. Fibrogenesis and fibrolysis in collagenous colitis. Patterns of procollagen types I and IV, matrix-metalloproteinase-1 and -13, and TIMP-1 gene expression. Am J Pathol 1999; 155:493-503.
67. Miehlke S, Heymer P, Bethke B et al. Budesonide treatment for collagenous colitis: A randomized, double-blind, placebo-controlled, multicenter trial. Gastroenterology 2002; 123:978-984.

Chapter 7

The Role of the Epithelial Barrier in Inflammatory Bowel Disease

Edward E.S. Nieuwenhuis and Richard S. Blumberg

Introduction

Inflammatory Bowel Diseases (IBD) represent a heterogeneous group of diseases with two distinct disorders of unknown etiology, Crohn's disease (CD) and ulcerative colitis (UC). Both diseases are characterized by chronic intestinal inflammation. Genetic susceptibility, environmental triggers and immune dysregulation have been described as the main factors involved in the establishment and development of IBD.[1] One hypothesis on the etiology of IBD is that these diseases represent an aberrant immune response by the mucosal immune system, to either pathogenic or resident luminal bacteria.[2] According to this hypothesis, resident bacteria can persistently stimulate the mucosal and systemic immune system, thereby perpetuating the inflammatory response. This hypothesis is supported by the recent evidence of a mutation of the bacterial sensing gene *NOD2* being strongly associated with susceptibility to CD.[3,4] The importance of luminal bacteria in the pathogenesis of IBD draws particular attention to the role played by the intestinal epithelial cell. In addition to their important barrier, absorption, and transport function, intestinal epithelial cells (IEC) play an important role in innate and adaptive immune functions including management of the luminal microbial ecology. The inducible secretion of regulatory molecules by IEC is consistent with the notion that these cells may play a role in initiating and regulating physiologic and pathological mucosal responses, associated with the development of IBD. Moreover, the IEC may be an important target of the dysregulated immune response, due to the effects of either barrier disruptive cytokines (IFN-γ, IL-14, IL-13, TNF-α) and cytotoxic T cells, which may further enhance inflammation and prevent appropriate repair processes.

Bacteria and the Epithelium

Epithelial cells are often the first host cells to detect and subsequently interact with luminal antigens and may, therefore, be regarded as crucial cells in discriminating between friend and foe, self and nonself or pathogen and commensal. Through differential expression of inflammatory and anti-inflammatory signals such as cytokines, chemokines and adhesion molecules, epithelial cells can initiate and orchestrate the local immune response to microorganisms. Pattern recognition molecules, which are mostly membrane bound receptors, can recognize characteristic structural features of bacteria, such as lipopolysaccharides, peptidoglycans or other specific bacterial motifs. Many of these molecules are functionally present on the epithelial cells and are, as such, involved in the earliest, innate, host response to microbes.[5] A genetic defect in this crucial compartment of the innate immune system may result in the development of IBD. Indeed, NOD-2 potentially expressed by IECs, may be aberrant in a subset of CD subjects.[3,4]

Immune Mechanisms in Inflammatory Bowel Disease, edited by Richard S. Blumberg and Markus F. Neurath.

Bacteria and IBD

The human intestinal tract contains 10^{14} bacteria—more than 400 different species—which is 10 times more than the total number of cells in the human body. The gut associated lymphoid tissue (GALT), which comprises the largest complement of lymphocytes in the human body, is separated from the microorganisms by a single cell layer of polarized epithelium. Studies both in human patients as well as in animal models have established a critical role for the interaction between luminal bacteria and the intestinal epithelial and immune cells in the pathogenesis of IBD.[6,7] In this respect, microorganisms such as *paramyxovirus* (measles), *Mycobacterium paratuberculosis* and *Listeria monocytogenes* have been implicated in the pathogenesis of IBD.[8,9,10]

Several studies report beneficial effects of the use of antibiotics in the treatment of IBD, hypothetically through eradication of harmful bacteria.[11] Specifically pathogenic microorganisms or classes of bacteria or particular bacterial products that are immunostimulatory however have not yet been determined. Although various clinical trials in CD indicate that broad-spectrum antibiotics, such as tetracycline, ampicillin, trimethoprim-sulfa and ciprofloxacin, decrease disease activity, only metronidazole has become an established antimicrobial drug in the treatment of CD.[12] Antibiotics are not routinely used in UC. In UC, metronidazole is not effective, though it may have a role in the treatment of pouchitis. A recent study in UC patients showed however, that ciprofloxacin may be beneficial when combined with 5-ASA agents and steroids.[13] In addition, important data have also been generated from experimental models that are associated with the occurrence of 'spontaneous' colitis such as the IL-10 or IL-2 knock-out mouse models. In most of these mouse models inflammation is absent whenever the animals are raised in a germ free or specific pathogen free environment, thereby linking the severity of colitis to the grade of microbial colonization.[14,15] In analogy with the distinct association between *Helicobacter pylori* and peptic ulcer disease, many attempts have been made to determine whether the presence of specific luminal bacteria is associated with the occurrence of IBD. For several reasons, this has not been an easy task. Analysis of the microbiota in IBD patients is extremely complicated, due to the completely inhomogeneous and asymmetric distribution of the many different bacterial strains along the cephalocaudal axis of the intestine as well as along the crypt-villus units. Furthermore, only a minimal proportion of the flora is readily detectable by conventional culture techniques. Recent molecular methods such as 16S rRNA sequence analysis of the intestinal flora have not been perfected and are at best semi-quantitative. Next, it is not just the number of bacteria present in the intestine but also specific bacterial-epithelial adhesion that may account for a local inflammatory response. Novel, extremely adhesive bacteria have occasionally been reported in association with IBD. More generally, a phenomenon of enhanced bacterial adhesion in IBD that may be the result of alterations that occur in the epithelial cells during IBD such as aberrant epithelial glycan expression has been reported. Once identified conclusively, antibiotic treatment aiming specifically at eradication or prevention of overgrowth of harmful organisms may become a novel modality in the treatment of IBD.

As a final point, the emerging and ever expanding field of probiotics as IBD treatment is based on the concept that this treatment modality may re-establish or even enhance the proposed beneficial anti-inflammatory properties of the healthy enteric flora. The administration of probiotic bacteria such as *Lactobacilus gg* was associated with clear anti-inflammatory effects in various animal models such as the IL-10 ko mouse model.[16] In humans, at least two studies provide compelling evidence for beneficial effects of probiotics in IBD. One study has reported maintenance of remission in patients with pouchitis, by using a cocktail of 8 probiotic bacterial strains.[17] Other groups have shown that the use of a nonpathogenic *E. coli* strain in UC patients equals the effect of the widely used drug mesalazine.[18]

The working mechanisms responsible for the 'probiotic effects' have not been elucidated yet, but may involve interference in the signal transduction pathway for cytokine production by the IECs. It has been shown that certain components of nonpathogenic bacteria may attenuate pro-inflammatory responses by blocking degradation of IkB, which is the counter regulatory factor of NFkB, a key factor in the signal transduction pathway that leads to inflammation.[19] As such, the IEC may in essence integrate the presence of the overall lumenal micro ecology in the form of a physiologic phenotype that is both represented by IEC structure and function but also as paracrine effect on subjacent cells within the epithelium and lamina propria.

In the remainder of this chapter we will discuss the role of the epithelial cell as an effector cell in IBD in mediating these interactions between the luminal microbiota and subjacent cells within the lamina propria with a specific focus on pattern recognition molecules, chemokines, cytokines, mucins, defensins and cryptidins.

The Epithelial Cell as Effector Cell in IBD

The chronic inflammatory intestinal response that characterizes IBD may be the result of, or contributed by an exaggerated host defense reaction of the intestinal epithelium to non-pathogenic and possibly commensal flora. Pattern recognition molecules such as Toll-like receptors (TLRs), appear to be the primary microbial sensors of microorganisms and are expressed on intestinal epithelial cells. It has been suggested therefore, that altered expression of these molecules may play a crucial role in the dysregulated innate immune response that characterizes IBD. From in vitro experiments it is known that LPS is able to activate signaling pathways in intestinal epithelial cell lines through TLRs that are involved in the secretion of various mediators such as chemokines and cytokines.[5] Primary IECs of normal mucosa constitutively express TLR3 and TLR5 at high levels, while TLR2 and TLR4 are only detectable at low levels. In active CD (but not in UC), TLR3 is significantly downregulated in IECs in active CD. In contrast, TLR4 is strongly upregulated in both UC and CD, while TLR2 and TLR5 expression remains unchanged in IBD. Taken together, these data associate specific changes in selective TLR expression on the intestinal epithelium during active IBD[5,20] (Table 1). How and if bacterial components such as the gram-negative bacterial cell wall component LPS stimulate these types of receptors in vivo in order to initiate, maintain or limit disease remains to be elucidated. It is unknown, whether damage of the mucosal barrier, which is a common feature of IBD, may directly lead to increased sensitivity of the intestinal epithelium to LPS.

Recently, the first susceptibility gene associated with CD (NOD-2) has been identified on chromosome 16. Mutations of the NOD-2 gene occur twice as frequently in CD patients in comparison to the general population.[3,4] NOD-2 is involved in the activation of nuclear factor NF-κB in monocytes and acts as an intracellular receptor for components of microbes, specifically peptidoglycan.[21] Recent evidence that NOD-2 may be expressed in IECs raises the possibility that this genetic defect may phenotypically manifest itself as altered IEC function.[22-24]

Table 1. Pattern recognition molecule expression in human epithelia

Pattern Recognition Molecule	Expression in CD	Expression in in UC	Reference
TLR 2	-	-	5
TLR 3	↓	-	5
TLR 4	↑	↑	5
TLR 5	-	-	5
NOD-2	polymorphism	-	3, 4
CD14	polymorphism		55

↑ (upregulated); ↓ (downregulated)

Chemokines are chemoattractant cytokines known to play a significant role in the trafficking and activation of leukocytes. Intestinal epithelial cells can play a role in initiating and regulating mucosal innate and acquired immune responses through the secretion of these molecules. In addition to this, chemokine receptor expression has also been identified on IECs. Therefore, it has been suggested that these cells may take part in a functional chemokine-network, involving immune cells as well as epithelial cells within the mucosa. In patients with UC, enterocytes derived from colonic specimens are the major source of neutrophil directing chemokines such as ENA-78 and MCP-1.[25-28] Chemokines involved in the recruitment of $CD4^+$ T cells, such as IP-10, MIG and MIP-3 are also expressed in colonic epithelial cells. Immunohistochemical analysis of colonic tissue from healthy subjects and patients with UC or CD revealed that MIP-3 immunoreactivity is primarily localized within the crypt epithelial cells while its production is significantly elevated in CD compared with UC or normal colon.[29,30] Taken together, these findings indicate that IECs are an important source of various chemokines and may therefore play a role in the recruitment of neutrophils and T lymphocytes to the epithelial layer in IBD, which may initiate and/or promote intestinal inflammation.

Both CD and UC are characterized by a breakdown of mucosal tolerance, which is associated with the development of an exaggerated immune response against components of the resident intestinal microbial flora. Each type of IBD appears to be associated with a specific mucosal cytokine secretion pattern. In this regard, several studies indicate that in CD, T-helper-1 (Th1) cytokines such as IL-2, IFN-γ and cytokines that are involved in directing T cells toward a Th1-type profile such as IL-12 and IL-18 predominate. Proinflammatory cytokines such as IL-1, IL-6, TNF-α and IL-8 have been associated with both UC and CD. In contrast, ulcerative colitis is thought to be mediated by T cells producing IL-5 rather than IFN-γ.[31,32] Intestinal epithelial cells are capable of producing a wide range of pro-inflammatory cytokines such as IL-1β and TNF-α, both implicated in the pathogenesis of IBD.[33,34] On the other hand, epithelial cells are also capable of expressing TGF-β and IL-10, cytokines that are involved in suppressing these typical pro-inflammatory (Th1-type) responses.[33,35,36,37] As such, it has been shown that signaling through CD1d at the epithelial surface can induce IL-10 production by this cell and may therefore dampen epithelial proinflammatory signals.[38] The importance of TNF-α (tumor necrosis factor-α) in the pathogenesis of CD has been confirmed by several clinical trials involving a chimeric anti-TNF-α antibody, as a successful treatment of CD.[11,39] In preliminary studies, treatment with recombinant human IL-10 has also been shown to be beneficial in CD. These findings set the stage for the epithelial cell to be a central player in the complex network of pro and counter regulatory cytokines in IBD (Table 2).

Table 2. Chemokine expression in human epithelia

Chemokines	Epithelium Expression in CD	Epithelium Expression in in UC	Reference
ENA-78	↑	↑	25
MCP-1,2,3	↑	↑	26, 27
IL-8	↑	↑	26
MIP1a and b	↑	↑	26, 56
IP-10, MIG	Constitutive	Constitutive	29
MIP-3α	↑	–	30

↑ (upregulated); ↓ (downregulated)

Mucus forms a gel-layer covering the gastrointestinal tract, acting as a semi-permeable barrier between lumen and epithelium. Mucins, the building blocks of the mucus-gel, determine the thickness and properties of mucus. In patients with IBD, alterations in both membrane-bound and secretory mucins have been described involving genetic mutations in mucin genes, changes in mucin mRNA and protein levels, degree of glycosylation, sulfation, and degradation of mucins.[40,41] As mucins are strategically positioned between the vulnerable mucosa and the bacterial environment, changes in mucin structure and/or quantity likely influence their protective functions and therefore constitute possible etiological factors in the pathogenesis of IBD. Animal models for IBD have demonstrated how changes in mucins, in particular in MUC2, imposed by immunological or microbial factors, may contribute to the development and/or perpetuation of chronic IBD.

Intestinal epithelial cells express various antimicrobial peptides such as defensins and cryptidins, which are part of innate immunity and contribute to the local intestinal host defense of the intestinal barrier. This system may be defective in IBD. In humans, defensins are produced throughout the gastrointestinal tract, including the mouth, esophagus, stomach, small intestine and colon. In the human gut, there are at least 9 different defensins (HNP 1- 4 in the neutrophils, HD5 and 6 in the Paneth cells, HBD-1 in the small intestine and colon, HBD-2 in the gastric mucosa and colon, and HBD-3 in colon). The β-defensin HBD-1 is constitutively expressed in colonic tissue while the β-defensin HBD-2 is induced in IBD, albeit more in ulcerative colitis then CD.[42-44] This difference indicates that it is the type of IBD that dictates this component of the mucosal innate defense. Therefore, defensins may control bacterial invasion in IBD, although the functional significance remains to be established. Recently, a novel intestinal protein named angiogenin 4 (Ang4) was identified, belonging to a class of proteins originally believed to be involved in the development of blood vessels. Ang4, which is released from Paneth cells, appears to be a potent killer of certain kinds of gut microbes and may therefore be regarded as part of the arsenal of microbicidal proteins deployed by Paneth cells.[45] Interestingly, the production of Ang4 appears to be under the control of a commensal intestinal bacterium called *Bacteroides thetaiotaomicron*, which may lead to new insights in the working mechanisms of probiotic bacteria.

Luminal Antigens, Fate and Processing

One of the primary sites for priming the intestinal adaptive immune system is located within the Peyers patches (PP). A wide range of luminal antigens gain access to the GALT through the M cells, typically located in the epithelium of PP. As M cells do not seem to express the classical antigen-presenting molecule MHC-II it is generally believed that M cells deliver antigens to professional antigen presenting cells such as the dendritic cells (DC), abundantly present within the PPs. This pathway is specifically important in the uptake and presentation of nonsoluble (particulate) antigens. Recent data describe a new mechanism for bacterial uptake in the mucosa tissues that is mediated by dendritic cells (DCs). DCs open the tight junctions between epithelial cells, send dendrites outside the epithelium and directly sample bacteria.[46] Interestingly, this process does not seem to disrupt the barrier integrity because DCs express typical tight-junction proteins such as occludin, claudin-1 and zonula occludens-1. Finally, a third pathway for transfer of luminal antigens potentially from bacteria may exist that involves apical to basal transcytosis of antigen-antibody complexes by the neonatal Fc-Receptor for IgG.[47] Such a pathway may also serve to deliver antigens to potential antigen presenting cells such as the DC.

Another important route by which antigen can be taken up and subsequently become presented is through the intestinal epithelial cell (IEC). It has been shown that IEC express both MHC-II molecules as well as the nonclassical antigen-presenting molecule CD1. As the IEC is the first host cell to come in contact with dietary or microbial antigen, this cell can be regarded as a key cell in signaling to the immune cells of the GALT. Therefore, antigen presentation by IEC to MHC-II restricted and/or CD1 restricted immune cells that are located adjacent to and in the underlying mucosa may function as the earliest activation pathway of the adaptive immune system in IBD.[48,49] Recently, IL-13 production was demonstrated to be a significant pathologic factor in an animal model for UC, since its neutralization by IL-13 fusion antibodies prevented colitis. Subsequently, CD1-restricted NK-T cells were shown to be the source of the IL-13. These data thus describe a cellular mechanism involving CD1 and CD1-restricted cells that may explain the pathogenesis of ulcerative colitis and involve a pathway by which CD1d-restricted cells may both secrete a cytokine that is deleterious to the IEC barrier (IL-13) and promote destruction through NK-T cell mediated cytolysis.[50]

IEC and Mucosal Barrier Function/Permeability

One of the hallmarks of IBD is activation of cells belonging to the GALT. Upon activation, these cells secrete soluble products that are associated with tissue injury. IFN-γ, for example, increases mucosal permeability by injuring epithelial cell tight junctions and by inducing intracellular adhesion molecules, which enhance neutrophil transmigration. Activated T-cells also appear to be an important mediator of villous atrophy and crypt hyperplasia and induce epithelial cell malabsorption through downregulation of Na^+-K^+ ATP-ase activity.[51] In the case of destruction, the integrity of the intestinal mucosal surface barrier is rapidly re-established through a 'resealing-process' that includes epithelial cell migration, epithelial cell proliferation, and differentiation. A complex network of regulatory peptides, such as growth factors and cytokines, regulates this process. These factors, often released by injured epithelial cells, may play an essential role in the regulation of differential epithelial cell functions to preserve normal homeostasis and integrity of the intestinal mucosa. Whether epithelial repair and restoration pathways are aberrant in IBD is unknown.

In IBD, epithelial integrity may become disrupted as a result of ongoing intestinal inflammation leading to chronically defective intestinal barrier function. In turn, this altered state may allow the uncontrolled uptake of antigens and in particular pro-inflammatory molecules, including luminal bacteria and bacterial products. Results from IL-2 deficient mice have also shown that the IEC is more avid in its uptake of luminal antigens, which creates a certain type of transcellular leakiness.[52] It is however not clear whether the IBD-associated defects in epithelial barrier function may be regarded as a primary etiological factor or just as a result of ongoing inflammation. The former hypothesis may be supported by the interesting finding of a familial tendency to enhanced mucosal permeability in relatives of Crohn's patients.

Other specific abnormalities in mucosal barrier function have been found in patients with UC. These include reduction in colonic mucin species IV and alterations in mucin structure and lectin binding. Epithelial cell starvation via the hydrogen sulfide pathway has also been mentioned as a factor that may lead to deterioration of the mucosal barrier function.[53] Finally, the enhanced susceptibility to colonic injury of mice deficient in intestinal trefoil factor demonstrates the importance of locally produced growth factors to mucosal homeostasis, healing and/or protection from injury.[54]

Conclusion

The epithelial barrier protects the intestine from harmful antigens. The intestinal epithelial cell produces a wide array of molecules that play a role in initiating, maintaining, but also preventing and controlling mucosal inflammation. A defect in epithelial cell regulated barrier function may lead to continued exposure of intestinal immune cells to bacterially derived molecules leading to destructive intestinal inflammation that characterizes IBD.

References

1. Podolsky DK. Inflammatory bowel disease. N Engl J Med 2002; 347(6):417-29.
2. Sartor RB. Innate immunity in the pathogenesis and therapy of IBD. J Gastroenterol 2003; 38(Suppl 15):43-7.
3. Ogura Y, Bonen DK, Inohara N et al. A frameshift mutation in NOD2 associated with susceptibility to Crohn's disease. Nature 2001; 411(6837):603-6.
4. Hugot JP, Chamaillard M, Zouali H et al. Association of NOD2 leucine-rich repeat variants with susceptibility to Crohn's disease. Nature 2001; 411(6837):599-603.
5. Cario E, Podolsky DK. Differential alteration in intestinal epithelial cell expression of toll-like receptor 3 (TLR3) and TLR4 in inflammatory bowel disease. Infect Immun 2000; 68(12):7010-7.
6. Blumberg RS, Saubermann LJ, Strober W. Animal models of mucosal inflammation and their relation to human inflammatory bowel disease. Curr Opin Immunol 1999; 11(6):648-56.
7. Swidsinski A, Ladhoff A, Pernthaler A et al. Mucosal flora in inflammatory bowel disease. Gastroenterology 2002; 122(1):44-54.
8. Montgomery SM, Morris DL, Pounder RE et al. Paramyxovirus infections in childhood and subsequent inflammatory bowel disease. Gastroenterology 1999; 116(4):796-803.
9. Greenstein RJ. Is Crohn's disease caused by a mycobacterium? Comparisons with leprosy, tuberculosis, and Johne's disease. Lancet Infect Dis 2003; 3(8):507-14.
10. Liu Y, van Kruiningen HJ, West AB et al. Immunocytochemical evidence of Listeria, Escherichia coli, and Streptococcus antigens in Crohn's disease. Gastroenterology 1995; 108(5):1396-404.
11. Podolsky DK. The future of IBD treatment. J Gastroenterol 2003; 38(Suppl 15):63-6.
12. Colombel JF, Cortot A, van Kruiningen HJ. Antibiotics in Crohn's disease. Gut 2001; 48(5):647.
13. Mimura T, Rizzello F, Helwig U et al. Four-week open-label trial of metronidazole and ciprofloxacin for the treatment of recurrent or refractory pouchitis. Aliment Pharmacol Ther 2002; 16(5):909-17.
14. Madsen KL. Inflammatory bowel disease: Lessons from the IL-10 gene-deficient mouse. Clin Invest Med 2001; 24(5):250-7.
15. Waidmann M, Bechtold O, Frick JS et al. Bacteroides vulgatus protects against Escherichia coli-induced colitis in gnotobiotic interleukin-2-deficient mice. Gastroenterology 2003; 125(1):162-77.
16. Madsen KL, Malfair D, Gray D et al. Interleukin-10 gene-deficient mice develop a primary intestinal permeability defect in response to enteric microflora. Inflamm Bowel Dis 1999; 5(4):262-70.
17. Gionchetti P, Rizzello F, Venturi A et al. Oral bacteriotherapy as maintenance treatment in patients with chronic pouchitis: A double-blind, placebo-controlled trial. Gastroenterology 2000; 119(2):305-9.
18. Rembacken BJ, Snelling AM, Hawkey PM et al. Nonpathogenic Escherichia coli versus mesalazine for the treatment of ulcerative colitis: A randomised trial. Lancet 1999; 354(9179):635-9.
19. Haller D, Russo MP, Sartor RB et al. IKK beta and phosphatidylinositol 3-kinase/Akt participate in nonpathogenic Gram-negative enteric bacteria-induced RelA phosphorylation and NF-kappa B activation in both primary and intestinal epithelial cell lines. J Biol Chem 2002; 277(41):38168-78.
20. Cario E, Rosenberg IM, Brandwein SL et al. Lipopolysaccharide activates distinct signaling pathways in intestinal epithelial cell lines expressing Toll-like receptors. J Immunol 2000; 164(2):966-72.
21. Inohara N, Ogura Y, Fontalba A et al. Host recognition of bacterial muramyl dipeptide mediated through NOD2. Implications for Crohn's disease. J Biol Chem 2003; 278(8):5509-12.
22. Lala S, Ogura Y, Osborne C et al. Crohn's disease and the NOD2 gene: A role for paneth cells. Gastroenterology 2003; 125(1):47-57.
23. Rosenstiel P, Fantini M, Brautigam K et al. TNF-alpha and IFN-gamma regulate the expression of the NOD2 (CARD15) gene in human intestinal epithelial cells. Gastroenterology 2003; 124(4):1001-9.
24. Gutierrez O, Pipaon C, Inohara N et al. Induction of Nod2 in myelomonocytic and intestinal epithelial cells via nuclear factor-kappa B activation. J Biol Chem 2002; 277(44):41701-5.
25. Z'Graggen K, Walz A, Mazzucchelli L et al. The C-X-C chemokine ENA-78 is preferentially expressed in intestinal epithelium in inflammatory bowel disease. Gastroenterology 1997; 113(3):808-16.
26. Banks C, Bateman A, Payne R et al. Chemokine expression in IBD. Mucosal chemokine expression is unselectively increased in both ulcerative colitis and Crohn's disease. J Pathol 2003; 199(1):28-35.
27. Reinecker HC, Loh EY, Ringler DJ et al. Monocyte-chemoattractant protein 1 gene expression in intestinal epithelial cells and inflammatory bowel disease mucosa. Gastroenterology 1995; 108(1):40-50.
28. Vainer B, Sorensen S, Nielsen OH et al. Subcellular localization of intercellular adhesion molecule-1 in colonic mucosa in ulcerative colitis. Ultrastruct Pathol 2002; 26(2):113-21.

29. Dwinell MB, Lugering N, Eckmann L et al. Regulated production of interferon-inducible T-cell chemoattractants by human intestinal epithelial cells. Gastroenterology 2001; 120(1):49-59.
30. Kwon JH, Keates S, Bassani L et al. Colonic epithelial cells are a major site of macrophage inflammatory protein 3alpha (MIP-3alpha) production in normal colon and inflammatory bowel disease. Gut 2002; 51(6):818-26.
31. Fuss IJ, Neurath M, Boirivant M et al. Disparate CD4+ lamina propria (LP) lymphokine secretion profiles in inflammatory bowel disease. Crohn's disease LP cells manifest increased secretion of IFN-gamma, whereas ulcerative colitis LP cells manifest increased secretion of IL-5. J Immunol 1996; 157(3):1261-70.
32. Bouma G, Strober W. The immunological and genetic basis of inflammatory bowel disease. Nat Rev Immunol 2003; 3(7):521-33.
33. Eckmann L, Jung HC, Schurer-Maly C et al. Differential cytokine expression by human intestinal epithelial cell lines: Regulated expression of interleukin 8. Gastroenterology 1993; 105(6):1689-97.
34. Taylor CT, Dzus AL, Colgan SP. Autocrine regulation of epithelial permeability by hypoxia: Role for polarized release of tumor necrosis factor alpha. Gastroenterology 1998; 114(4):657-68·
35. Ishiguro Y. Mucosal proinflammatory cytokine production correlates with endoscopic activity of ulcerative colitis. J Gastroenterol 1999; 34(1):66-74.
36. ering N, Kucharzik T, Gockel H et al. Human intestinal epithelial cells down-regulate IL-8 expression in human intestinal microvascular endothelial cells; Role of transforming growth factor-beta 1 (TGF-beta1). Clin Exp Immunol 1998; 114(3):377-84.
37. Autschbach F, Braunstein J, Helmke B et al. In situ expression of interleukin-10 in noninflamed human gut and in inflammatory bowel disease. Am J Pathol 1998; 153(1):121-30.
38. Colgan SP, Hershberg RM, Furuta GT et al. Ligation of intestinal epithelial CD1d induces bioactive IL-10: Critical role of the cytoplasmic tail in autocrine signaling. Proc Natl Acad Sci USA 1999; 96(24):13938-43.
39. van Dullemen HM, van Deventer SJ, Hommes DW et al. Treatment of Crohn's disease with anti-tumor necrosis factor chimeric monoclonal antibody (cA2). Gastroenterology 1995; 109(1):129-35.
40. Einerhand AW, Renes IB, Makkink MK et al. Role of mucins in inflammatory bowel disease: Important lessons from experimental models. Eur J Gastroenterol Hepatol 2002; 14(7):757-65.
41. Buisine MP, Desreumaux P, Leteurtre E et al. Mucin gene expression in intestinal epithelial cells in Crohn's disease. Gut 2001; 49(4):544-51.
42. Wehkamp J, Schwind B, Herrlinger KR et al. Innate immunity and colonic inflammation: Enhanced expression of epithelial alpha-defensins. Dig Dis Sci 2002; 47(6):1349-55.
43. Wehkamp J, Fellermann K, Herrlinger KR et al. Human beta-defensin 2 but not beta-defensin 1 is expressed preferentially in colonic mucosa of inflammatory bowel disease. Eur J Gastroenterol Hepatol 2002; 14(7):745-52.
44. Fahlgren A, Hammarstrom S, Danielsson A et al. Increased expression of antimicrobial peptides and lysozyme in colonic epithelial cells of patients with ulcerative colitis. Clin Exp Immunol 2003; 131(1):90-101.
45. Hooper LV, Stappenbeck TS, Hong CV et al. Angiogenins: A new class of microbicidal proteins involved in innate immunity. Nat Immunol 2003; 4(3):269-73.
46. Rescigno M, Urbano M, Valzasina B et al. Dendritic cells express tight junction proteins and penetrate gut epithelial monolayers to sample bacteria. Nat Immunol 2001; 2(4):361-7.
47. Dickinson BL, Badizadegan K, Wu Z et al. Bidirectional FcRn-dependent IgG transport in a polarized human intestinal epithelial cell line. J Clin Invest 1999; 104(7):903-11.
48. van de Wal Y, Corazza N, Allez M et al. Delineation of a CD1d-restricted antigen presentation pathway associated with human and mouse intestinal epithelial cells. Gastroenterology 2003; 124(5):1420-31.
49. Hershberg RM, Mayer LF. Antigen processing and presentation by intestinal epithelial cells - polarity and complexity. Immunol Today 2000; 21(3):123-8.
50. Heller F, Fuss IJ, Nieuwenhuis EE et al. Oxazolone colitis, a Th2 colitis model resembling ulcerative colitis, is mediated by IL-13-producing NK-T cells. Immunity 2002; 17(5):629-38.
51. MacDonald TT, Spencer J. Evidence that activated mucosal T cells play a role in the pathogenesis of enteropathy in human small intestine. J Exp Med 1988; 167(4):1341-9.
52. Telega GW, Baumgart DC, Carding SR. Uptake and presentation of antigen to T cells by primary colonic epithelial cells in normal and diseased states. Gastroenterology 2000; 119(6):1548-59.
53. Roediger WE, Duncan A, Kapaniris O et al. Reducing sulfur compounds of the colon impair colonocyte nutrition: Implications for ulcerative colitis. Gastroenterology 1993; 104(3):802-9.
54. Mashimo H, Wu DC, Podolsky DK et al. Impaired defense of intestinal mucosa in mice lacking intestinal trefoil factor. Science 1996; 274(5285):262-5.

55. Klein W, Tromm A, Griga T et al. A polymorphism in the IL11 gene is associated with ulcerative colitis. Genes Immun 2002; 3(8):494-6.
56. Vainer B, Nielsen OH, Horn T. Expression of E-selectin, sialyl Lewis X, and macrophage inflammatory protein-1alpha by colonic epithelial cells in ulcerative colitis. Dig Dis Sci 1998; 43(3):596-608.
57. Jung HC, Eckmann L, Yang SK et al. A distinct array of proinflammatory cytokines is expressed in human colon epithelial cells in response to bacterial invasion. J Clin Invest 1995; 95(1):55-65.
58. McGee DW, Bamberg T, Vitkus SJ et al. A synergistic relationship between TNF-alpha, IL-1 beta, and TGF-beta 1 on IL-6 secretion by the IEC-6 intestinal epithelial cell line. Immunology 1995; 86(1):6-11.
59. Casola A, Estes MK, Crawford SE et al. Rotavirus infection of cultured intestinal epithelial cells induces secretion of CXC and CC chemokines. Gastroenterology 1998; 114(5):947-55.
60. Schuerer-Maly CC, Eckmann L, Kagnoff MF et al. Colonic epithelial cell lines as a source of interleukin-8: Stimulation by inflammatory cytokines and bacterial lipopolysaccharide. Immunology 1994; 81(1):85-91.

CHAPTER 8

Involvement of Dendritic Cells in the Pathogenesis of Inflammatory Bowel Disease

Francisco Leon, Lesley E. Smythies, Phillip D. Smith and Brian L. Kelsall

Introduction

Dendritic cells (DC) are antigen-presenting cells (APCs) with unique properties that allow them to efficiently process antigens and activate naïve T cells. Mucosal DCs have a particular capacity to induce regulatory T cell differentiation in the steady state, but they are also potent inducers of effector T cell responses, indeed constituting the main APC in the gut. In addition, mucosal DCs may contribute to innate defense by the production of cytokines following direct exposure to pathogens. DCs may also play a role in the maintenance of secondary T cell responses within inflamed mucosa, such as in inflammatory bowel disease (IBD). In contrast, resident intestinal macrophages do not appear to act as APC, or produce inflammatory cytokines, but are specialized in mucosal defense against pathogens by their ability to scavenge, phagocytose, and kill microorganisms. However, during active mucosal inflammation like IBD, fully inflammatory monocytes are recruited from the blood, which can contribute to tissue injury and damage. The present chapter will focus on the role of mucosal APC, mainly dendritic cells in the pathogenesis of IBD. A summary of current concepts of DC biology will be followed by a review of the growing knowledge regarding DC and macrophages in IBD.

Current Views on Dendritic Cell Biology

Sensors of Infection and Induction of Immunity

Dendritic cells constitute a family of bone-marrow-derived APCs that are uniquely capable of inducing the differentiation of naïve T cells. The most commonly held paradigm of DC function is that DCs are present in a less than fully mature (or activated) state within all vascularized tissues, where they efficiently capture or sample foreign and self-antigens. Upon exposure to microbial products or inflammatory signals, DCs mature and migrate to T cell areas of organized lymphoid tissues, where they interact with and activate migrating naïve T cells. This maturation process can occur in response to pathogens via interactions between pattern recognition receptors (PRR), such as toll-like receptors (TLRs) on DCs, and "pathogen-associated molecular patterns" (PAMPs) expressed by organisms,[1] or in response to infection or inflammation, via recognition of inflammatory cytokines, e.g., TNF-α, IL-1, or type I IFNs,[2] intracellular compounds, e.g., uric acid,[3] and specific immune responses, e.g., via CD40-CD40L interactions (see refs. 4-7). The DC maturation process is associated with the loss of endocytic and phagocytic receptors, the up-regulation of costimulatory molecules, the extension of dendrites that increase the surface area available for interaction with T cells and a shift in chemokine receptor expression (e.g., from CCR6 to CCR7) to promote DC migration to lymphoid tissues. These mature or activated cells are well suited to binding and activating

Immune Mechanisms in Inflammatory Bowel Disease, edited by Richard S. Blumberg and Markus F. Neurath.

naïve T cells. As a consequence of this unique biology, a primary function of DCs is to detect and induce immune responses to invading pathogens.

"Steady-State" Dendritic Cells and the Maintenance of Self-Tolerance

Recent findings complicate the relatively simplistic view of DC maturation and function described above. First, immature DC populations reside not only within peripheral tissues, but also within discrete regions of organized secondary lymphoid organs. These lymphoid tissue-resident DC can process systemically administered or blood-borne soluble antigens,[8] as well as cell-associated foreign or self antigens, for presentation to T cells, the later by a process termed "cross-presentation".[9,10] Second, DCs are not simply stationary cells that are waiting to be activated to mature and migrate. In fact, the turnover of most DC populations within lymphoid tissues is very rapid (on the order of days) and DCs within all tissues migrate continuously in the absence of inflammation or infection from nonlymphoid and peripheral sites, like the intestine, to organized lymphoid tissues via the lymph. The DCs that migrate in the "steady-state" are thought to be in an immature, or altered state of activation and continuously present self-antigens, most likely acquired from dying apoptotic cells, for the induction and maintenance of self-tolerance (see ref. 11). In addition, the turnover rate of DCs in tissues and draining lymph nodes is increased substantially by inflammatory cytokines, such as TNF-α, that are induced during systemic or local inflammation or infection. Therefore, a more reasonable view of DC biology is that of continuously migrating immature or partially mature tolerogenic cells, which upon contact with pathogens or inflammatory stimuli become fully activated, resulting in more rapid turnover, migration to draining lymphoid tissues, or to T cell zones within lymphoid tissues, and the induction of effector, rather than tolerogenic T cell responses.

Dendritic Cell Subsets vs. the Plasticity of Dendritic Cell Function

A third complicating factor in understanding DC function is the fact that multiple DC sub-populations have now been identified in mice and humans (see refs. 12,13). In the mouse, at least 8 different DC and preDC populations can be defined.[13-15] Three main DC subsets exist in the spleen ($CD11c^+/CD11b^+$, subdivided into $CD4^+$ or $CD4^-$, and $CD11c^+/CD11b^-/CD8\alpha^+$). In addition to the $CD11c^+/CD11b^-/CD8\alpha^+$ and the $CD11c^+/CD11b^+/CD4^-/CD8\alpha^-$ ("double negative") subsets, mesenteric lymph nodes and Peyer's patches contain another predominant population, $CD11c^+/CD11b^-/CD4^-/CD8\alpha^-$ ("triple negative") DC.[15,16] Additionally, three subpopulations of the recently characterized plasmacytoid preDCs (pDCs) that express intermediate levels of CD11c have been described in lymphoid tissues and blood.[14,17-22]

In humans, most studies of DCs have focused on circulating precursors in peripheral blood. Human DCs are characterized as $HLA\text{-}DR^{hi}$ cells that do not express typical markers of T cells, B cells, monocytes, and NK cells ["lineage negative" (lin^-)]. $HLA\text{-}DR^{hi}/lin^-$ cells include three non-overlapping subsets of classical $CD11c^{hi}$ precursor DCs that express CD1b/CD1c, CD16, or BDCA3, and $CD11c^-$ precursor plasmacytoid DCs that express CD123, BDCA2, and BDCA4 (see refs. 4,13).

There is considerable functional specialization of murine DC subpopulations, although this is not absolute. The adoptive transfer of antigen-pulsed $CD8^+$ DCs to naïve recipient mice preferentially induced Th1 responses, which was dependent on DC production of IL-12p40 (thus either IL-12-p70 or IL-23), while the transfer of $CD8^-$ DCs induced Th2 responses,[23,24] which was dependent to some extent on DC production of IL-10.[25] These and other studies resulted in the original designation of $CD8^+$ DCs at "DC1" cells and $CD8^-$ (or $CD11b^+$) DCs as "DC2". In addition, $CD8^+$, but not $CD8^-$, DCs can internalize apoptotic cells,[26] and are the major cells responsible for T cell tolerance following intravenous injection of antigen-loaded apoptotic cells.[26,27] $CD8^+$ DCs are also responsible for the cross-presentation of self-antigens to $CD8^+$ T cells in vivo.[10,28,29]

Studies of human DCs also provide some support for the concept of functional specialization and separate "DC1" and "DC2" lineages. CD154-activated mature DCs derived from monocytes ("myeloid DC") in culture with GM-CSF and IL-4 were initially found to preferentially induce Th1 responses in vitro, while mature plasmacytoid DCs derived upon culture of CD123+ blood precursors with IL-3 and CD154-activated preferentially induced Th2 responses.[13] Functional specialization is also suggested by the expression of different TLR receptors on different subpopulations of DC.[1,30]

Despite evidence for functional specialization, significant "plasticity" also characterizes the function of DC subsets (reviewed in refs. 5,6,13,31) Recent studies argue that the induction of T cell differentiation by DC is directly influenced by the signals DCs receive during antigen uptake, activation, and T cell priming. Indeed, the ability of DCs to mediate either tolerance or immunization likely depends on their state of activation, which, in turn, depends on the form of antigen and the presence or absence of activating tissue factors. Furthermore, exposure of DCs to certain cytokines or growth factors, such as IL-10 or TGF-β, or to certain pathogens or their products, such as *B. pertussis* fimbrial hemagluttinin (FHA),[32] can promote tolerogenic responses via active engagement of regulatory receptors expressed by DCs. In addition, in vitro studies have shown that different microbial products can direct DCs to drive either Th1, Th2 or regulatory T cell responses, which are at least partially dependent on the differential production of IL-10 and IL-12 by the DCs (see ref. 5). This flexibility of DC function may be particularly relevant to the intestine, where DCs are continuously exposed to a large variety of commensal bacteria, which may promote a regulatory DC phenotype.

Intestinal Dendritic Cells

The importance of DCs in the induction of mucosal immune responses was initially suggested by studies performed prior to the availability of specific reagents to identify DCs. These studies focused on MHC class II positive cells with the morphology of DCs in Peyer's patches (PPs),[33-45] the lamina propria of the intestine,[33,34,43,46-48] and in the lymph draining the intestine.[48-53] These studies indicated the presence and location of DCs in intestinal tissues, and the ability of DCs to act as APCs, primarily through their ability to stimulate a mixed lymphocyte reaction (MLR).

Populations of Intestinal Dendritic Cells

Peyer's Patches and Isolated Lymphoid Follicles

Early studies suggested that PP DCs can act as APCs for T cell responses in the mouse[54] and in the human,[42,45,55] as well as to induce T cell help for immunoglobulin class switching of B cells to IgA.[56,57] More recent studies have used DC-reactive monoclonal antibodies[58-60] and have described the cells' immaturity[61] and their capacity to process oral antigen in vivo.[60,62] Isolated lymphoid follicles in the colon (cILF) are similar in structure to single follicles of PPs, and have been shown to contain DCs in the sub-epithelial dome (SED) and para-follicular regions, similar to the localization of DCs in PPs. No data has directly addressed DCs in the recently described ILF in the small intestine.[63]

PP DCs have special characteristics. PP and MLN, but not spleen or LN DCs can induce the mucosal homing receptor α4β7 as well as the chemokine receptor CCR9 on T cells in vitro.[64] In addition, PP DCs bias towards a Th2/TR1 response: PP DCs induce the differentiation of T cells that upon in vitro restimulation produce high levels of IL-4 and IL-10 and lower levels of IFN-γ compared to T cells primed with spleen DCs, which produced predominantly IFN-γ.[65-67] PP DCs produce IL-10 in response to activation via the costimulatory molecules CD40L,[65] and receptor activator of NFkB (RANK),[68] whereas similarly stimulated spleen DCs produce predominantly IL-12. $CD11b^+$ DCs produce high levels of IL-10 and low levels of bioactive IL-12p70[15] and induce IL-10-producing T cells. In contrast, $CD8\alpha^+$ and $CD8\alpha^-$ $CD4^-CD11b^{lo}$ PP DCs produce IL-12 and little or no IL-10 and induce the differentiation of

IFN-γ-producing T cells. Murine PP plasmacytoid pDCs[18] may also be particularly capable of inducing IL-10-producing regulatory T cells.[22]

DCs have also been demonstrated in the human and macaque PPs, where some of the cells appear to coexpress CD4, CCR5, and DC-SIGN, suggesting they may be potential targets of HIV-1 infection.[69] Interestingly, M-DC8+ cells, which likely represent DCs and produce high levels of TNF-α, were also detected in PPs in inflamed tissues.[70] and may have a role in driving T cell responses during inflammation.

Mesenteric Lymph Nodes

Intestinal "veiled cells" are a heterogeneous population of cells that migrate from all regions of the intestinal tract via the lymph to draining MLNs,[51] where they become trapped.[49,50] These veiled cells process fed antigen[53] or antigen delivered by direct intestinal injection,[52,53] and are capable of stimulating primed T cells in vitro and priming naive T cells in vivo.[53] Under steady-state conditions, DCs constitutively migrate from the intestine to the MLNs; pathogen products enhance this migration.[51] Recently, a population of rat MLN DCs (OX41-,CD4-) were shown to contain apoptotic bodies from intestinal epithelial cells in the steady state, suggesting that DCs in the intestine continuously sample apoptotic epithelial cells for cross-presentation to T cells in the MLN.[71]

Many of the phenotypic and functional features of MLN DCs are similar to those of PP DCs, except that MLN DCs appear to represent more "mature" DCs that have migrated from the PP or possibly the intestinal lamina propria. MLN DCs from antigen-fed mice produce IL-10, and possibly TGF-β, and preferentially stimulate antigen-specific CD4+ T cells to produce IL-4, IL-10 and TGF-β.[72,73] Additionally, MLN DCs have the capacity to induce the α4β7 integrin on T cells.[74,75]

In humans, DCs in the MLN are HLA-DR^{hi}, large dendriform cells in the T cell areas that express costimulatory molecules. The attachment of these cells to numerous CD4+ T cells and IgD+ naive B cells in vivo, suggests that human MLN DCs may be involved in the induction of T-cell help for primary B cell responses.

Intestinal Lamina Propria

Dendritic cells are a major antigen presenting cell population in the intestinal lamina propria (LP) of the mouse,[43] rat[33,47,48] and human.[46] Irregularly shaped, strongly MHC class II-positive LP DCs cells have been identified just below the basement membrane and, in the rat. DCs have also been detected within the epithelium of the intestine.[47] Murine small bowel DCs may also be induced to migrate into the epithelium and extend their processes into the intestinal lumen by a well orchestrated process,[76] suggesting that luminal sampling of bacteria and dietary products may be immunologically relevant. The phenotype of LP DCs is not yet well defined, even in animal models, but the same three CD11c^{hi} DC subsets that had been defined in the PP, as well as CD11c^{lo} pDCs, were detected in the murine LP.[77] Others found a predominance of CD8α$^-$ /CD11b^{lo} DCs in the LP, especially in the murine terminal ileum.[78] Interestingly, the number of DCs in the LP of the uninflamed colon is markedly less than that of the uninflamed small bowel, and colonic DCs are concentrated in the cILF mentioned above.

Few studies have focused on LP DCs in humans. DCs seem to be present in the LP and in poorly defined lymphoid aggregates in the small bowel.[79] Immature CD11c$^+$HLA$^-$DR$^+$lin- DCs have been detected in colonic and rectal biopsies.[80] CD83$^+$ DCs and DC-SIGN$^+$ cells have been identified in the colonic LP.[81]

Functional studies of lamina propria DCs from all species have confirmed that these cells are potent inducers of the MLR.[33,43,46,48] The aforementioned CD8α$^-$ /CD11b^{lo} DCs in the terminal ileum (but not of the more proximal small intestine) of mice constitutively express IL-23[78] and are associated with bacteria in vivo, suggesting that DCs in the terminal ileum normally process endogenous commensal bacterial. CD83$^+$ DCs and DC-SIGN$^+$ cells may produce IL-12 and IL-18 during intestinal inflammation in Crohn's disease.[81]

DC Trafficking to the Intestine

Chemokines and chemokine receptors that govern the localization and migration of DC populations in the intestine are just beginning to be unraveled. In murine PPs in the steady state, CD11b$^+$ DCs appear to be recruited to SED by a number of chemokines expressed constitutively by epithelial cells in the follicle associated epithelium (FAE), including CCL9[82] and possibly CCL20;[16] the chemokines that attract CD8a$^-$CD4$^-$CD11b$^-$ DCs to this site have not been identified. Upon activation in vivo with a soluble antigen preparation from *T. gondii* tachyzoites (STAg) the SED CD11b$^+$, and possibly some CD8α$^-$ CD4$^-$ CD11b$^-$, DCs migrated to the PP interfollicular regions (IFR).[16] This correlated with an up-regulation of CCR7, the receptor for CCL7 and CCL8, which are constitutively expressed in the IFR. These studies support the hypothesis that activation of DCs in the SED, as would occur following exposure to organisms entering via M-cells, results in DC migration to the IFR, where T cell priming could occur. In contrast, CD8α$^+$ DCs constitutively expressed CCR7 and not CCR6 and migrated towards CCL7, but not CCL20, in vitro, suggesting that CD8α$^+$ DCs are resident in the IFR throughout much of their life cycle.

Studies of chemokines responsible for the migration of DCs to the LP in the steady state are lacking, although epithelial cell-expressed CCL25, the ligand for CCR9 and CCR10, is a possible candidate chemokine in the small bowel, and CCL28, the ligand for CCR3 and CCR10, may be an important chemokine in the colon (see refs. 82-84). During inflammation, however, a large number of inflammatory chemokines, including CCL20/CCR6, CCL2/CCR2, CCL5/CCR5 or CCR1, and CXCL-12/CXCR4, are produced by epithelial cells that could attract DCs (see refs. 83,85,86). CCL20 mRNA expression and protein production are up-regulated in intestinal epithelial cells in vitro by stimulation with LPS, TNF-α or IL-1α, or infection with enteric bacterial pathogens.[87] In addition, increased expression of CCL20 has been detected in the epithelium of inflamed human colon.[87] CCR6 is expressed by more immature DCs and is down-regulated following activation, coincident with an up-regulation of CCR7[88] (see ref. 83). Accordingly, immature DCs may be attracted to mucosal tissues as immature cells and after activation by inflammatory signals, migrate to the PP or to MLN for priming naïve or activation of central memory T cells.

Intestinal Dendritic Cells and the Induction of Tolerance and Immunity to Pathogens

Currently, the most accepted model holds that under-steady state conditions precursor DCs continuously enter the mucosal LP, PPs, and cILFs, develop into immature DCs, and localize to different regions due to the differential constitutive expression of specific chemokines, such as CCL9 and CCL20 in the FAE of the murine PP.[16,82] After transport of antigens across M cells or epithelial cells, or possibly via the uptake of apoptotic bodies from epithelial cells, DCs migrate from the LP to the MLN or from the SED of the PP and cILF to the respective IFRs in the steady state. This migration is accompanied by an up-regulation of chemokine receptors for T cell zone chemokines, such as CCR7, but low levels of costimulatory molecules and cytokines.

Following migration of antigen loaded, "quiescent" DCs to T cell zones in the PP, cILF, or MLN, the DCs stimulate T cells to differentiate into regulatory T cells that can mediate bystander tolerance following subsequent antigen encounter. Intestinal stromal cells may promote a suppressive environment that conditions DCs to induce regulatory T cells. While the precise nature of these regulatory cells in relation to the CD25$^+$ regulatory T cells originally described by Sakaguchi and colleagues[89] is not clear, the cells seem to suppress primarily via a cytokine-dependent mechanism in vitro and in vivo, involving TGFβ, IL-10, and possibly IL-4 (see refs. 90,91). The subset of DC that may induce regulatory T cells is not yet clear, although CD11b$^+$ DCs or pDCs (either CD11c^{int}/B220$^+$, or CD45RB^{hi}/ B220$^+$) in the mouse are potential candidates.[14,15,22]

In contrast to tolerogenic responses to innocuous antigens, mucosal pathogens induce active local and systemic immunity. Initial encounter of pathogens with the cellular components of the innate mucosal barrier, epithelial cells and underlying DCs and macrophages, involves the recognition of microbial PAMPs by PRRs, such as TLR receptors.[1,92-94] TLR signaling of epithelial cells results in the production of proinflammatory cytokines and chemokines, such as IL-1, IL-8, IL-6, TNFα, IFN-α, CCL5, and CCL20, which can recruit and activate neutrophils, macrophages and DCs.[95] In addition to this indirect mechanism of DC activation by epithelial cell derived cytokines, DCs may be directly activated by invading pathogens via TLRs and other surface receptors. Such activation results in fully immunogenic DCs with high levels of MHC, costimulatory, and adhesion molecules, as well as cytokines. The phenotype of the ensuing T cell response is influenced by direct and indirect effects of the particular pathogen encountered. This will include the particular subset of DC engaged (which will differ because of the expression of different PRRs), and the compilation of signals that the DC receives from direct interaction with the pathogen, as well as from tissue factors present at the site of encounter, and in the lymph node during T cell priming (see refs. 5,6,13,31).

Dendritic Cells in Inflammatory Bowel Disease

DCs appear to play an important role in the pathogenesis of IBD. Current information suggests that DCs have two major functions. First, studies in mice indicate that DCs are involved in priming abnormal T cell immune responses to endogenous flora in organized lymphoid tissues, such as PP in the terminal ileum, ILF in the colon, or MLN. Such DCs may also help maintain T cell reactivity by the restimulation of central memory T cells within these lymphoid tissues. Second, DCs may promote the persistence of inflammatory T cell responses by direct interactions with LP T cells within the inflamed tissue. This interaction likely involves DCs that are induced to mature within the tissues, but that do not migrate following activation. Locally activated DCs could enhance ongoing T cell responses by direct cognate interaction and by the production of cytokines such as IL-12, IL-23 or possibly IL-27, which could expand effector T cell populations and/or prolong their survival (see Fig. 1).

DCs in Animal Models of IBD

Rodent models of colitis have provided strong evidence that the interaction between commensal bacterial flora and the mucosal immune system plays an essential role in the pathogenesis of IBD (reviewed in refs. 96,97). Four different models of experimental IBD suggest that DCs are instrumental in the establishment and maintenance of inflammation. In the SCID/CD45RB^{hi} transfer model of colitis, in which disease is induced by adoptive transfer of wild-type CD45RB^{hi} T cells into a SCID recipient,[96] CD11c$^+$ DCs in the MLNs of colitic mice increased in number and 20-30% expressed high levels of OX40L (CD134L), a marker of DC activation.[98] Furthermore, blocking CD134-CD134L interactions prevented the induction of colitis and decreased the proliferation of T cells, as well as reduced the number of T cells that expressed α4β7 in the MLN. In contrast, CD134L$^+$ DCs were not found in the inflamed LP, suggesting that activated MLN and not LP DCs were critical for the onset of colitis through interactions between DC-expressed CD134L and CD134 on T cells. Because the accumulation of CD134L$^+$ DCs was prevented by the cotransfer of CD4$^+$ CD45RB^{low} T cells, it was suggested that regulatory T cells may function in part to prevent DC activation.[98] How and where CD4$^+$ CD45RB^{lo} T cells may prevent DC maturation is not yet clear; however, one can predict that cytokines such IL-10 and TGFβ, which are important for the suppressive function of CD4$^+$ CD45RB^{low} cells in vivo and which can prevent DC maturation, are likely involved.

In a similar model of colitis in which total CD4$^+$ CD45RB^{hi} T cells are transferred into RAG-deficient mice, colitic mice can be treated with CD4$^+$ CD25$^+$ T cells. Following transfer after the onset of colitis, CD4$^+$ CD25$^+$ T cells expand not only in the MLN, but also in the inflamed colon in close proximity to CD11c$^+$ DCs.[99] These data suggest that DCs are important for the activation and expansion of CD25$^+$ T cells in both draining lymph nodes and at sites of inflammation, and are consistent with a requirement for highly activated DCs for $CD4^+CD25^+$ T cell expansion.[100] In addition, the data suggest that $CD4^+$ $CD25^+$ T cells may act in the lymph node or locally within inflamed tissue to suppress T cell effector function. A slight variation of this model is one in which total $CD4^+$ T cells are transferred to RAG-1-deficient mice resulting in colitis in some animal facilities. In studies of this model, nonreconstituted, noncolitic RAG-1-deficient mice have clusters of subepithelial $CD11c^+$ DC in the colon, which likely represent poorly formed cILF. Upon reconstitution, $CD4^+$ T cells appear in the $CD11c^+$ DC clusters where they expand in number during the early or "emerging" stages of the disease.[101] The degree of expansion within the clusters was proportional to the severity of the resultant intestinal inflammation, suggesting a role in for DCs within cILF in the initiation (or maintenance) of inflammation.[101] Coincident with disease development in this model, DC in the LP increase in number, display a mature/activated phenotype, and likely produce IL-23,[102] consistent with a role for local activated DCs in the induction or expansion of pathogenic (as well as regulatory) T cells directly within inflamed tissues. Also consistent with these data, in the $Gi2\alpha^{-/-}$ model of colitis, we have observed a major increase in the number of activated $CD11b^+$ DCs within inflamed LP (Jianping He, et al, manuscript in preparation).

Finally, in $IL\text{-}2^{-/-}$ mice, the inhibition of RANKL-RANK interaction with osteoprotegerin (OPG), a decoy receptor for T cell produced RANKL, reduces the spontaneous colitis.[103] This effect of OPG may be due to inhibition of RANK-mediated DC survival. Consistent with a role for RANK-RANKL interactions in DC function in the intestine, PP DCs express RANK and when stimulated with soluble RANKL produce IL-10, whereas spleen DCs produced predominantly IL-12. In addition, treatment of mice with soluble RANKL enhanced the sensitivity to oral tolerance induced to soluble ovalbumin. These studies suggest that RANK stimulation can enhance either the tolerogenic capacity of DCs under steady state conditions, or the immunogenic capacity of DCs in inflamed tissue, possibly through effects on DC survival.

Taken together, studies in mice have provided important experimental support for a central role for DCs in regulating intestinal inflammation in IBD. DCs are likely involved in the induction and expansion of both pathogenic and regulatory T cells within organized lymphoid tissues and at sites of inflammation. What is not at all clear from these studies, however, is how DCs, and in particular different DC sub-populations, are involved in the overall balance of pathogenic vs. regulatory cell activation in the pathogenesis of disease in these models. For example, it would be logical to predict that intestinal $CD11b^+$ DCs are more involved in regulatory T cell activation and expansion, given the in vitro studies showing the propensity of these DC populations to induce IL-10-producing T cells. However, the finding that $CD11b^+$ DCs that express OX-40L may be directly involved in pathogenic T cell expansion,[98] and are preferentially increased in number within the inflamed colon (Jianping He, et al, manuscript in preparation) is counter to this hypothesis.

DC in Human IBD

Studies of DCs in human IBD have focused on cells in the normal and inflamed colonic lamina propria. Similar to murine models, early studies pointed to a role for activated DCs at sites of inflammation[104-107] However, the lack of DC-specific markers makes the interpretation of those studies complicated (reviewed in refs. 108,109). More recently, DCs have been identified with more precision in human gut, and the results of different studies are somewhat controversial.[80,81,109-114] While some authors have not observed a significant increase in the numbers or maturation status of LP DCs in IBD,[80] others have provided evidence for an increase in both immature and mature DCs within inflamed tissues. On the one hand, CCL20 is over-expressed in the intestinal epithelium associated with human IBD lesions, and this over-expression was reported to correlate with an increase in the number of immature $Langerin^+$ DC in situ.[110] Furthermore, TNF-α increased the expression of CCL-20 in colonic explant cultures from normal patients, suggesting that immature DCs (as well as T cells) expressing CCR6 are attracted to enhanced expression of CCL20 by epithelial cells

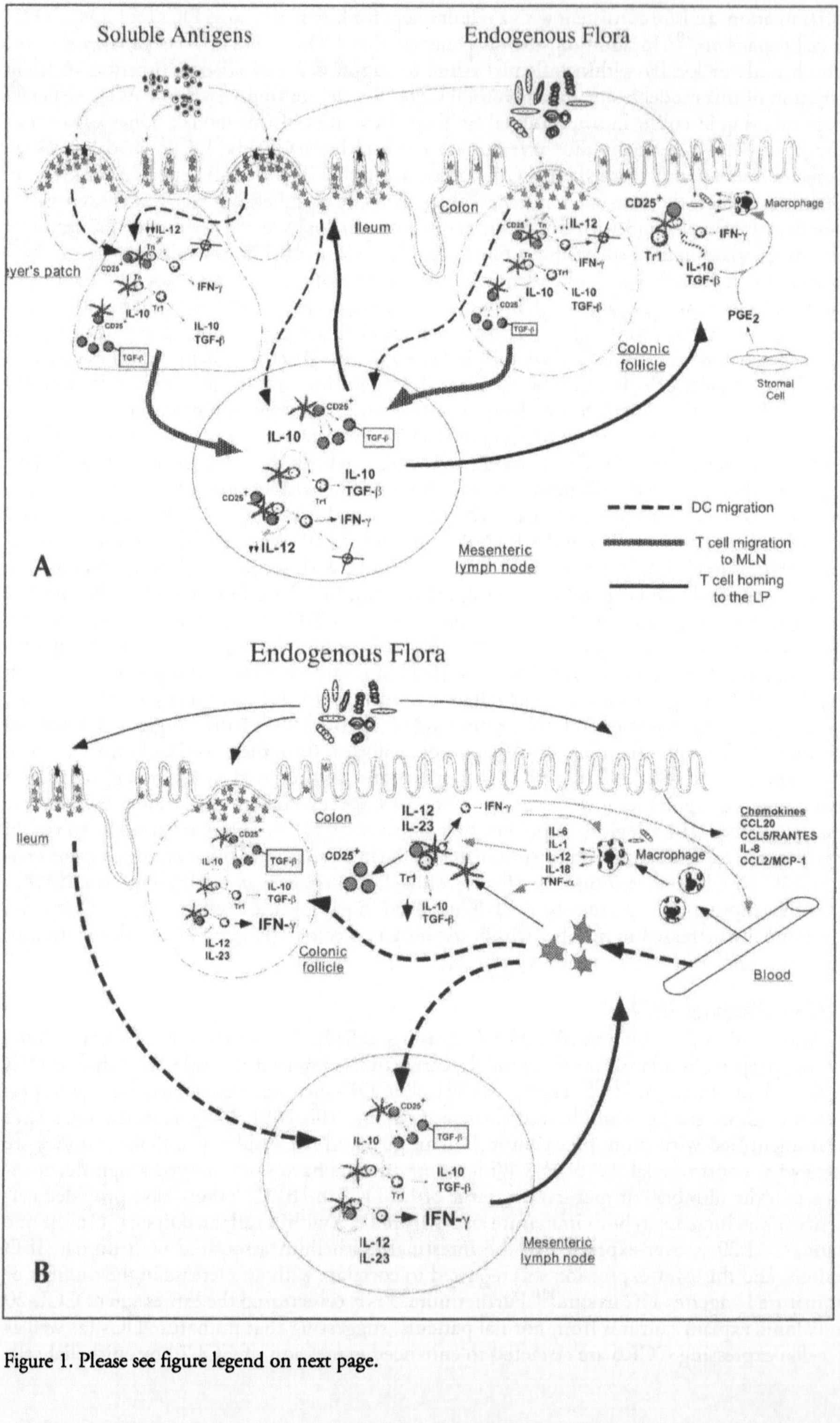

Figure 1. Please see figure legend on next page.

Figure 1. Dendritic cell function in the uninflamed mucosa and Th1-mediated inflammatory bowel disease. A) Model for mucosal DC induction of tolerance to soluble antigens and commensal bacteria in uninflamed mucosa. Soluble antigens are taken-up by DCs in the small intestinal LP or organized lymphoid tissues, such as PPs following transport across intestinal epithelial cells or M cells or, or are taken up directly by DC stellate processes extending into the epithelium and intestinal lumen. Alternatively, DCs may capture antigens associated with apoptotic bodies from dying epithelial cells. Commensal bacteria are more likely taken up by M cells overlying cILF, or by DCs extending into the epithelium and lumen of the terminal ileum. Uptake by DCs in the colonic lamina propria is possible, but less likely, since few DCs are present in this site. Antigen loaded DCs traffic to T cell regions of organized lymphoid tissues, resulting in the induction of regulatory T cells, either by the induction of IL-10-producing regulatory T cells (Tr1 cells), or the expansion of $CD25^+$ regulatory T cells. Under steady-state conditions, the generation of Th1 cells is limited either by the lack of DC-activating signals (in the case of soluble proteins), or in the presence of DC-"inactivating" signals (PAMPs) that inhibit IL-12, and enhance IL-10 production (in the case of endogenous flora). The lack of Th1 cells in these conditions may be due to abortive T cell activation and expansion, resulting in initial production of IFN-γ followed by T cell death in the absence of IL-12 or high levels of T cell costimulation. Regulatory T cells are therefore preferentially induced to both soluble antigens and endogenous flora, which migrate to the intestinal LP, where they act to prevent untoward immune reactivity to soluble proteins and commensal microorganisms that gains access to the LP via incidental breaks in the mucosal barrier. This suppression may occur via contact dependent or cytokine-mediated mechanisms. In uninflamed mucosa, macrophages in the LP phagocytose and kill microorganisms, but do not produce inflammatory cytokines or present antigens to T cells, likely due to conditioning by locally produced factors, such as TGF-β and PGE_2. B) Model for the role of DCs and macrophages in the pathogenesis of Th1-mediated intestinal inflammation. Enhanced DC production of cytokines such as IL-12 and/or IL-23, or under-production of IL-10, may result in abnormal Th1-differentiation (or the lack of regulatory T cell differentiation and expansion) in response to commensal bacteria. Chemokines produced locally in established inflammation attract both DC precursors and monocytes into inflamed sites. Monocytes recruited from the blood develop into inflammatory macrophages that produce pro-inflammatory chemokines and cytokines that can activate the endothelium, induce further chemokine production from surrounding cells, such as epithelial cells, and contribute directly to tissue damage. DC precursors may have one of several fates. First, they may be activated locally by factors, such as TNF-α and present endogenous antigens to T cells within tissues. This could result in the enhanced expansion or survival of effector T cells, a function normally attributed only to macrophages. Second, immature DCs may process antigens within inflamed tissues and then migrate to T cell zones in MLNs or cILF, where the DC could prime naïve T cells and expand memory T cells. In addition, $CD25^+$ regulatory T cells likely expand locally and in draining lymphoid tissues in response to activated DCs. However, these $CD25^+$ regulatory T cells may be less effective in their suppressive function under conditions of inflammation, since pro-inflammatory cytokines, such as IL-6, can act to overcome any suppression of T cell activation by these regulatory T cells. IL-10-producing regulatory T cells may also be generated during inflammatory responses, but they may be less effective against DCs fully activated in an intensely proinflammatory local microenvironment.

in the inflamed intestine.[110] On the other hand, others have shown an increase in numbers and maturation of DCs within inflamed tissue.[81,109,111]

In Crohn's disease (CD), an increase in $CD83^+$ LP DC has been recently documented.[81] These cells were found in aggregates in the LP, and their increased presence was accompanied by an increase in IL-12- and IL-18-producing $CD83^-$ $CD80^+$ DC-$SIGN^+$ cells that may represent a second population of DCs.[81] Another study showed increased numbers of lin^- $CD86^+$ $CD40^+$ DC in the peripheral blood and in the LP of patients with both CD and ulcerative colitis (UC), consistent with the presence of activated DCs in both blood and tissues.[111] The significance of the enhanced expression of CD86 and CD40 on blood DCs is not clear, since DC normally migrate in immature form from blood to tissues and after maturation to T cell zones of lymphoid tissues via the lymph. Thus, while possibly reflecting the cytokine environment in the blood that could result in early maturation, the effect on migration to tissue is unknown. Regarding DC subsets in IBD, our preliminary studies of CD patients demonstrate increases in the number of $CD86^+$ $CD11c^+$ myeloid DCs in peripheral blood and mature $fascin^{hi}$ myeloid DCs in the colonic LP [Leon, et al, manuscript in preparation]. DCs from IBD tissues also have been shown to express elevated levels of CD40 that are reduced after treatment with anti-TNF-α.[109] Finally, M-DC8⁺ monocytes, which may be precursors of DC with a high potential to secrete TNF-α,[112] are also elevated in the colonic LP of CD,[70] suggesting a novel pathogenic mechanism in CD in addition to the potential initiation and maintenance of the Th1 response.

The role of DCs in the pathogenesis of UC has received less investigative attention, but recent evidence indicates an increase in the number of $CD83^+$ and $CD86^+$ LP cells which are most likely DCs, and which produce macrophage inhibitory factor (MIF). MIF in turn can induce IL-1 and IL-8 production by monocytes and DCs,[113] which may contribute to neutrophil recruitment and activation. Also in UC, colonic LP contains numerous basal aggregates composed of lymphocytes and $CD80^+$ dendritiform cells that most likely represent activated DCs.[114] Since DCs generated in vitro from peripheral monocytes from patients with UC showed an increased immunostimulatory capacity,[115] it is possible that immature DCs from UC patients are in general more sensitive to activation signals or have an accelerated differentiation program in response to factors such as GM-CSF. Finally, UC is characterized by an increase within LP tissues of EBI-3, one subunit of the IL-12-related cytokine IL-27,[116] which is produced by macrophages and DCs. In addition, $EBI\text{-}3^{-/-}$ mice have few NKT cells, and are resistant to oxazolone colitis,[117] a murine model of UC that is dependent on the presence NKT cells.[118] These data suggest that DC (or macrophage)-derived IL-27 may potentially be important in the pathogenesis of UC through an effect on NKT cell activation.[119]

Intestinal Macrophages

Macrophages are well represented in the intestinal tract LP in both normal and inflamed tissues. However, the phenotype and function of LP macrophages differs significantly in these two conditions. In normal small bowel and colon, tissue macrophages act primarily as phagocytic cells that clear pathogens, macromolecules, and dead and dying cells. However, the cells are down-regulated phenotypically and functionally for responses to activating signals from cytokines or pathogens that typically induce the production of pro-inflammatory mediators and cytokines and enhance the antigen-presenting capacity of monocytes or tissue macrophages from nonmucosal sites. Intestinal macrophages are likely derived directly from circulating monocytes, which upon exposure to signals, including TGF-β in the local LP tissue microenvironment, results in the down-regulation of their innate receptors, pro-inflammatory activities, and antigen presentation capabilities, but the retention of avid host defense and scavenger functions through strong phagocytic and cytotoxic activities.[120-122]

Therefore, in contrast to blood monocytes, which express an extensive array of innate receptors as well as integrins, intestinal macrophages lack or are markedly down-regulated for most innate response receptors, notably CD14,[123,124] Fcγ and Fcα receptors[121,122] and TLR2 and TLR4.[125] Consistent with these findings, intestinal macrophages do not respond to LPS or other stimuli of bacterial and nonbacterial origin with the production of IL-1, IL-6, IL-8, TNF-α, IL-10, IL-12, RANTES and TGF-β indicating a global suppression of inflammatory cytokine and chemokine production. Compared to blood monocytes, intestinal macrophages lack constitutive and inducible expression of CD80, CD86, and CD40, and are poor at presenting antigens to T cells. In contrast, human intestinal macrophages avidly phagocytose particulate antigens and microorganisms, such as *C. albicans, S. typhimurium,* and *E. coli,* and despite their lack of respiratory burst capacity, are fully capable of killing these organisms. Thus, intestinal macrophages in the steady-state are noninflammatory, nonantigen-presenting cells that retain the capacity to phagocytose and kill invading microbes.

Intestinal Macrophages in IBD

During active IBD, blood monocytes are recruited to the inflamed tissue, and the phenotype of intestinal macrophages more closely resembles that of blood monocytes.[124,126-129] Consequently, more $CD14^+$ mononuclear phagocytes are present in IBD lesions than in normal, un-inflamed intestinal mucosa.[123] The preferential localization of experimentally inoculated, radiolabeled

$CD14^+$ monocytes to regions of intestinal inflammation[130] supports the concept that blood monocytes populate the intestinal mucosa, particularly inflammatory lesions. The factor(s) that initiate monocyte recruitment to inflamed mucosa likely include inflammatory CC chemokines (MIP-1α,β, RANTES, MCP-1, 2, 3, 4, and 5), noncysteine-containing chemotactic ligands (C5a, C3a and TGF-β), and pathogen-derived peptides (*Helicobacter pylori* urease and f-met-leu-phe).

Once in the tissue, blood-derived macrophages release a variety of proinflammatory cytokines, such as TNF-α, macrophage infiltrating factor (MIF), IL-1, IL-6, IL-12 and IL-18, which are critically involved in the onset and the development of Crohn's disease.[131] In particular, TNF-α has been implicated in the disruption of the epithelial barrier, induction of apoptosis of villous epithelial cells, and the secretion of chemokines from intestinal epithelial cells that result in recruitment of T cells and polymorphonuclear leukocytes. In addition, TNF-α can directly activate B cells, augment IFN-γ production by T cells, and induce matrix metalloproteinases from intestinal stromal cells that could directly lead to tissue injury. Macrophages producing TNF-α may also induce the maturation of DCs within tissues and lead to their migration to T cell zones in PPs, cILF, or MLNs. IL-6 is likely involved in the induction of adhesion molecule expression by vascular endothelium, in the direct or indirect augmentation of TNF-α, IL-1 and IFN-γ, and in the survival on lamina propria T-cells. IL-6 may be important in overcoming suppression of effector T cells by $CD4^+$ $CD25^+$ regulatory T cells by acting to augment activation of effector cells. Finally macrophage-derived IL-12 and IL-18 can act to induce and augment IFN-γ-producing T cells, through the cytokines' effects on T cell differentiation, expansion and survival, as well as directly on IFN-γ production. As a result of this involvement, IL-12 and IL-18 have been targeted in the therapy of murine models of IBD[132] and, along with TNF-α and IL-6, are being targeted for the treatment of human CD[133].

Summary

In conclusion, during inflammation, DCs are likely activated by inflammatory signals and induced to migrate to T cell zones of organized lymphoid tissues where the cells induce T cell responses. In addition to their established role in T cell priming and the induction of tolerance, DCs may act to enhance (or possibly suppress) T cell responses at sites of mucosal inflammation. Determining the importance of DCs in this regard, as well as establishing a potential role for DCs in continuous activation of naïve or central memory cells in lymph nodes draining inflammatory sites, will elucidate the role of DCs as a potential therapeutic target for chronic inflammatory diseases, like IBD. Resident intestinal macrophages are noninflammatory and do not efficiently present antigens to intestinal T cells, yet are avidly phagocytic and able to kill internalized organisms. During intestinal inflammation, monocytes are recruited from the blood, become inflammatory macrophages in the inflamed tissue, and are major contributors to tissue destruction and perpetuation of inflammation via their production of chemokines and pro-inflammatory cytokines. Macrophages may also contribute directly to DC activation and maturation, which would drive DCs to present antigens from the bacterial flora to T cells locally within tissue or to more efficiently traffic to T cell zones of lymphoid tissue.

Thus, DCs and macrophages have evolved functional niches that promote cooperation in the prevention of untoward intestinal inflammation in the steady state and in the eradication of invasive microorganisms during infection. The balance between suppressing inflammation and promoting host defense is altered in humans with IBD allowing a persistent inflammatory response to commensal bacteria. Based on studies from animal models, the pathogenesis of IBD likely involves either the lack of appropriate regulation from T cells, or an over-production of effector T cells. The end result of these potential mechanisms is the abnormal induction and/or survival of effector T cells and the production of factors such as cytokines by inflammatory macrophages and neutrophils that result in tissue destruction. The destructive process likely involves normally tolerizing DCs, which in the microenvironment of the inflamed mucosa activate T cell responses to normal flora in both draining lymphoid tissues and at sites of inflammation, with macrophages and neutrophils contributing the bulk of inflammatory and destructive cytokines.

Acknowledgements

L.S. and P.D.S. was supported by NIH grants DK-47322, DK-54495, HD-41361, DK-64400 and the Research Service of the Veterans Affairs

References

1. Kaisho T, Akira S. Regulation of dendritic cell function through toll-like receptors. Curr Mol Med 2003; 3(8):759-71.
2. Honda K, Sakaguchi S, Nakajima C et al. Selective contribution of IFN-α/β signaling to the maturation of dendritic cells induced by double-stranded RNA or viral infection. Proc Natl Acad Sci USA 2003; 100(19):10872-7.
3. Shi Y, Evans JE, Rock KL. Molecular identification of a danger signal that alerts the immune system to dying cells. Nature 2003; 425(6957):516-21.
4. Banchereau J, Briere F, Caux C et al. Immunobiology of dendritic cells. Annu Rev Immunol 2000; 18:767-811.
5. Kapsenberg ML. Dendritic-cell control of pathogen-driven T-cell polarization. Nat Rev Immunol 2003; 3(12):984-93.
6. Reis e Sousa C, Diebold SD, Edwards AD et al. Regulation of dendritic cell function by microbial stimuli. Pathol Biol (Paris) 2003; 51(2):67-8.
7. Rescigno M. Dendritic cells and the complexity of microbial infection. Trends Microbiol 2002; 10(9):425-61.
8. Wilson NS, El-Sukkari D, Villadangos JA. Dendritic cells constitutively present self antigens in their immature state in vivo, and regulate antigen presentation by controlling the rates of MHC class II synthesis and endocytosis. Blood 2003.
9. Larsson M, Fonteneau JF, Bhardwaj N. Cross-presentation of cell-associated antigens by dendritic cells. Curr Top Microbiol Immunol 2003; 276:261-75.
10. Belz GT, Carbone FR, Heath WR. Cross-presentation of antigens by dendritic cells. Crit Rev Immunol 2002; 22(5-6):439-48.
11. Steinman RM, Hawiger D, Nussenzweig MC. Tolerogenic dendritic cells. Annu Rev Immunol 2003; 21:685-711.
12. Ardavin C. Origin, precursors and differentiation of mouse dendritic cells. Nat Rev Immunol 2003; 3(7):582-90.
13. Shortman K, Liu YJ. Mouse and human dendritic cell subtypes. Nat Rev Immunol 2002; 2(3):151-61.
14. Wakkach A, Fournier N, Brun V et al. Characterization of dendritic cells that induce tolerance and T regulatory 1 cell differentiation in vivo. Immunity 2003; 18(5):605-17.
15. Iwasaki A, Kelsall BL. Unique functions of CD11b+, CD8 alpha+, and double-negative Peyer's patch dendritic cells. J Immunol 2001; 166(8):4884-90.
16. Iwasaki A, Kelsall BL. Localization of distinct Peyer's patch dendritic cell subsets and their recruitment by chemokines macrophage inflammatory protein (MIP)-3alpha, MIP-3beta, and secondary lymphoid organ chemokine. J Exp Med 2000; 191(8):1381-94.
17. Nakano H, Yanagita M, Gunn MD. CD11c(+)B220(+)Gr-1(+) cells in mouse lymph nodes and spleen display characteristics of plasmacytoid dendritic cells. J Exp Med 2001; 194(8):1171-8.
18. Asselin-Paturel C, Brizard G, Pin JJ et al. Mouse strain differences in plasmacytoid dendritic cell frequency and function revealed by a novel monoclonal antibody. J Immunol 2003; 171(12):6466-77.
19. Asselin-Paturel C, Boonstra A, Dalod M et al. Mouse type I IFN-producing cells are immature APCs with plasmacytoid morphology. Nat Immunol 2001; 2(12):1144-50.
20. O'Keeffe M, Hochrein H, Vremec D et al. Mouse plasmacytoid cells: Long-lived cells, heterogeneous in surface phenotype and function, that differentiate into CD8(+) dendritic cells only after microbial stimulus. J Exp Med 2002; 196(10):1307-19.
21. Martin P, Del Hoyo GM, Anjuere F et al. Characterization of a new subpopulation of mouse CD8alpha+ B220+ dendritic cells endowed with type 1 interferon production capacity and tolerogenic potential. Blood 2002; 100(2):383-90.
22. Bilsborough J, George TC, Norment A et al. Mucosal CD8alpha+ DC, with a plasmacytoid phenotype, induce differentiation and support function of T cells with regulatory properties. Immunology 2003; 108(4):481-92.
23. Maldonado-Lopez R, De Smedt T, Michel P et al. CD8alpha+ and CD8alpha- subclasses of dendritic cells direct the development of distinct T helper cells in vivo. J Exp Med 1999; 189(3):587-92.

24. Pulendran B, Smith JL, Caspary G et al. Distinct dendritic cell subsets differentially regulate the class of immune response in vivo. Proc Natl Acad Sci USA 1999; 96(3):1036-41.
25. Maldonado-Lopez R, Maliszewski C, Urbain J et al. Cytokines regulate the capacity of CD8alpha(+) and CD8alpha(-) dendritic cells to prime Th1/Th2 cells in vivo. J Immunol 2001; 167(8):4345-50.
26. Iyoda T, Shimoyama S, Liu K et al. The CD8+ dendritic cell subset selectively endocytoses dying cells in culture and in vivo. J Exp Med 2002; 195(10):1289-302.
27. Liu K, Iyoda T, Saternus M et al. Immune tolerance after delivery of dying cells to dendritic cells in situ. J Exp Med 2002; 196(8):1091-7.
28. den Haan JM, Lehar SM, Bevan MJ. CD8(+) but not CD8(-) dendritic cells cross-prime cytotoxic T cells in vivo. J Exp Med 2000; 192(12):1685-96.
29. Belz GT, Behrens GM, Smith CM et al. The CD8alpha(+) dendritic cell is responsible for inducing peripheral self-tolerance to tissue-associated antigens. J Exp Med 2002; 196(8):1099-104.
30. Boonstra A, Asselin-Paturel C, Gilliet M et al. Flexibility of mouse classical and plasmacytoid-derived dendritic cells in directing T helper type 1 and 2 cell development: Dependency on antigen dose and differential toll-like receptor ligation. J Exp Med 2003; 197(1):101-9.
31. Mazzoni A, Segal DM. Controlling the toll road to dendritic cell polarization. J Leukoc Biol 2004.
32. McGuirk P, McCann C, Mills KH. Pathogen-specific T regulatory 1 cells induced in the respiratory tract by a bacterial molecule that stimulates interleukin 10 production by dendritic cells: A novel strategy for evasion of protective T helper type 1 responses by Bordetella pertussis. J Exp Med 2002; 195(2):221-31.
33. Mayrhofer G, Pugh CW, Barclay AN. The distribution, ontogeny and origin in the rat of Ia-positive cells with dendritic morphology and of Ia antigen in epithelia, with special reference to the intestine. Eur J Immunol 1983; 13(2):112-22.
34. Bland PW, Whiting CV. Differential control of major histocompatibility complex class II I-Ek alpha protein expression in the epithelium and in subsets of lamina propria antigen-presenting cells of the gut. Immunology 1993; 79(1):107-11.
35. Sminia T, Janse EM, Wilders MM. Antigen-trapping cells in Peyer's patches of the rat. Scand J Immunol 1982; 16(6):481-5.
36. Wilders MM, Sminia T, Plesch BE et al. Large mononuclear Ia-positive veiled cells in Peyer's patches. II. Localization in rat Peyer's patches. Immunology 1983; 48(3):461-7.
37. Wilders MM, Drexhage HA, Sminia T et al. Veiled cells in the gastrointestinal tract. Acta Chir Scand Suppl 1985; 525:93-111.
38. Soesatyo M, Biewenga J, Kraal G et al. The localization of macrophage subsets and dendritic cells in the gastrointestinal tract of the mouse with special reference to the presence of high endothelial venules. An immuno- and enzyme-histochemical study. Cell Tissue Res 1990; 259(3):587-93.
39. Sminia T, van der Ende MB. Macrophage subsets in the rat gut: An immunohistochemical and enzyme-histochemical study. Acta Histochem 1991; 90(1):43-50.
40. Soesatyo M, van den Berg TK, van Rees EP et al. Ontogeny of reticulum cells in the rat intestine and their possible role in the development of the lymphoid microenvironment. Reg Immunol 1992; 4(1):46-52.
41. Mahida YR, Patel S, Jewell DP. Mononuclear phagocyte system of human Peyer's patches: An immunohistochemical study using monoclonal antibodies. Clin Exp Immunol 1989; 75(1):82-6.
42. Nagura H, Ohtani H, Masuda T et al. HLA-DR expression on M cells overlying Peyer's patches is a common feature of human small intestine. Acta Pathol Jpn 1991; 41(11):818-23.
43. Pavli P, Woodhams CE, Doe WF et al. Isolation and characterization of antigen-presenting dendritic cells from the mouse intestinal lamina propria. Immunology 1990; 70(1):40-7.
44. Brandtzaeg P, Bjerke K. Immunomorphological characteristics of human Peyer's patches. Digestion 1990; 46 (Suppl 2):262-73.
45. Bjerke K, Halstensen TS, Jahnsen F et al. Distribution of macrophages and granulocytes expressing L1 protein (calprotectin) in human Peyer's patches compared with normal ileal lamina propria and mesenteric lymph nodes. Gut 1993; 34(10):1357-63.
46. Pavli P, Hume DA, Van De Pol E et al. Dendritic cells, the major antigen-presenting cells of the human colonic lamina propria. Immunology 1993; 78(1):132-41.
47. Maric I, Holt PG, Perdue MH et al. Class II MHC antigen (Ia)-bearing dendritic cells in the epithelium of the rat intestine. J Immunol 1996; 156(4):1408-14.
48. Liu LM, MacPherson GG. Rat intestinal dendritic cells: Immunostimulatory potency and phenotypic characterization. Immunology 1995; 85(1):88-93.
49. Pugh CW, MacPherson GG, Steer HW. Characterization of nonlymphoid cells derived from rat peripheral lymph. J Exp Med 1983; 157(6):1758-79.
50. MacPherson GG. Properties of lymph-borne (veiled) dendritic cells in culture. I. Modulation of phenotype, survival and function: Partial dependence on GM-CSF. Immunology 1989; 68(1):102-7.

51. MacPherson GG, Jenkins CD, Stein MJ et al. Endotoxin-mediated dendritic cell release from the intestine. Characterization of released dendritic cells and TNF dependence. J Immunol 1995; 154(3):1317-22.
52. Liu LM, MacPherson GG. Lymph-borne (veiled) dendritic cells can acquire and present intestinally administered antigens. Immunology 1991; 73(3):281-6.
53. Liu LM, MacPherson GG. Antigen acquisition by dendritic cells: Intestinal dendritic cells acquire antigen administered orally and can prime naive T cells in vivo. J Exp Med 1993; 177(5):1299-307.
54. Spalding DM, Koopman WJ, Eldridge JH et al. Accessory cells in murine Peyer's patch. I Identification and enrichment of a functional dendritic cell. J Exp Med 1983; 157(5):1646-59.
55. Spencer J, Finn T, Isaacson PG. Human Peyer's patches: An immunohistochemical study. Gut 1986; 27(4):405-10.
56. Spalding DM, Williamson SI, Koopman WJ et al. Preferential induction of polyclonal IgA secretion by murine Peyer's patch dendritic cell-T cell mixtures. J Exp Med 1984; 160(3):941-6.
57. George A, Cebra JJ. Responses of single germinal-center B cells in T-cell-dependent microculture. Proc Natl Acad Sci USA 1991; 88(1):11-5.
58. Witmer MD, Steinman RM. The anatomy of peripheral lymphoid organs with emphasis on accessory cells: Light-microscopic immunocytochemical studies of mouse spleen, lymph node, and Peyer's patch. Am J Anat 1984; 170(3):465-81.
59. Witmer-Pack MD, Hughes DA, Schuler G et al. Identification of macrophages and dendritic cells in the osteopetrotic (op/op) mouse. J Cell Sci 1993; 104 (Pt 4):1021-9.
60. Kelsall BL, Strober W. Distinct populations of dendritic cells are present in the subepithelial dome and T cell regions of the murine Peyer's patch. J Exp Med 1996; 183(1):237-47.
61. Ruedl C, Hubele S. Maturation of Peyer's patch dendritic cells in vitro upon stimulation via cytokines or CD40 triggering. Eur J Immunol 1997; 27(6):1325-30.
62. Ruedl C, Rieser C, Bock G et al. Phenotypic and functional characterization of CD11c+ dendritic cell population in mouse Peyer's patches. Eur J Immunol 1996; 26(8):1801-6.
63. Hamada H, Hiroi T, Nishiyama Y et al. Identification of multiple isolated lymphoid follicles on the antimesenteric wall of the mouse small intestine. J Immunol 2002; 168:57-64.
64. Mora JR, Bono MR, Manjunath N et al. Selective imprinting of gut-homing T cells by Peyer's patch dendritic cells. Nature 2003; 424(6944):88-93.
65. Iwasaki A, Kelsall BL. Freshly isolated Peyer's patch, but not spleen, dendritic cells produce interleukin 10 and induce the differentiation of T helper type 2 cells. J Exp Med 1999; 190(2):229-39.
66. Everson MP, Lemak DG, McDuffie DS et al. Dendritic cells from Peyer's patch and spleen induce different T helper cell responses. J Interferon Cytokine Res 1998; 18(2):103-15.
67. Everson MP, McDuffie DS, Lemak DG et al. Dendritic cells from different tissues induce production of different T cell cytokine profiles. J Leukoc Biol 1996; 59(4):494-8.
68. Williamson E, Bilsborough JM, Viney JL. Regulation of mucosal dendritic cell function by receptor activator of NF-kappa B (RANK)/RANK ligand interactions: Impact on tolerance induction. J Immunol 2002; 169(7):3606-12.
69. Jameson B, Baribaud F, Pohlmann S et al. Expression of DC-SIGN by dendritic cells of intestinal and genital mucosae in humans and rhesus macaques. J Virol 2002; 76(4):1866-75.
70. de Baey A, Mende I, Baretton G et al. A subset of human dendritic cells in the T cell area of mucosa-associated lymphoid tissue with a high potential to produce TNF-alpha. J Immunol 2003; 170(10):5089-94.
71. Huang FP, Platt N, Wykes M et al. A discrete subpopulation of dendritic cells transports apoptotic intestinal epithelial cells to T cell areas of mesenteric lymph nodes. J Exp Med 2000; 191(3):435-44.
72. Alpan O, Rudomen G, Matzinger P. The role of dendritic cells, B cells, and M cells in gut-oriented immune responses. J Immunol 2001; 166(8):4843-52.
73. Akbari O, DeKruyff RH, Umetsu DT. Pulmonary dendritic cells producing IL-10 mediate tolerance induced by respiratory exposure to antigen. Nat Immunol 2001; 2(8):725-31.
74. Stagg AJ, Kamm MA, Knight SC. Intestinal dendritic cells increase T cell expression of alpha4beta7 integrin. Eur J Immunol 2002; 32(5):1445-54.
75. Johansson-Lindbom B, Svensson M, Wurbel MA et al. Selective generation of gut tropic T cells in gut-associated lymphoid tissue (GALT): Requirement for GALT dendritic cells and adjuvant. J Exp Med 2003; 198(6):963-9.
76. Rescigno M, Urbano M, Valzasina B et al. Dendritic cells express tight junction proteins and penetrate gut epithelial monolayers to sample bacteria. Nat Immunol 2001; 2(4):361-7.
77. Mowat AM. Anatomical basis of tolerance and immunity to intestinal antigens. Nat Rev Immunol 2003; 3(4):331-41.
78. Becker C, Wirtz S, Blessing M et al. Constitutive p40 promoter activation and IL-23 production in the terminal ileum mediated by dendritic cells. J Clin Invest 2003; 112(5):693-706.

79. Moghaddami M, Cummins A, Mayrhofer G. Lymphocyte-filled villi: Comparison with other lymphoid aggregations in the mucosa of the human small intestine. Gastroenterology 1998; 115(6):1414-25.
80. Bell SJ, Rigby R, English N et al. Migration and maturation of human colonic dendritic cells. J Immunol 2001; 166(8):4958-67.
81. te Velde AA, van Kooyk Y, Braat H et al. Increased expression of DC-SIGN+IL-12+IL-18+ and CD83+IL-12-IL-18- dendritic cell populations in the colonic mucosa of patients with Crohn's disease. Eur J Immunol 2003; 33(1):143-51.
82. Zhao X, Sato A, Dela Cruz CS et al. CCL9 is secreted by the follicle-associated epithelium and recruits dome region Peyer's patch CD11b+ dendritic cells. J Immunol 2003; 171(6):2797-803.
83. Caux C, Vanbervliet B, Massacrier C et al. Regulation of dendritic cell recruitment by chemokines. Transplantation 2002; 73(1 Suppl):S7-11.
84. Kunkel EJ, Campbell DJ, Butcher EC. Chemokines in lymphocyte trafficking and intestinal immunity. Microcirculation 2003; 10(3-4):313-23.
85. Papadakis KA. Chemokines in inflammatory bowel disease. Curr Allergy Asthma Rep 2004; 4(1):83-9.
86. Ajuebor MN, Swain MG. Role of chemokines and chemokine receptors in the gastrointestinal tract. Immunology 2002; 105(2):137-43.
87. Izadpanah A, Dwinell MB, Eckmann L et al. Regulated MIP-3alpha/CCL20 production by human intestinal epithelium: Mechanism for modulating mucosal immunity. Am J Physiol Gastrointest Liver Physiol 2001; 280(4):G710-9.
88. Dieu MC, Vanbervliet B, Vicari A et al. Selective recruitment of immature and mature dendritic cells by distinct chemokines expressed in different anatomic sites. J Exp Med 1998; 188(2):373-86.
89. Sakaguchi S, Fukuma K, Kuribayashi K et al. Organ-specific autoimmune diseases induced in mice by elimination of T cell subset. I evidence for the active participation of T cells in natural self-tolerance; Deficit of a T cell subset as a possible cause of autoimmune disease. J Exp Med 1985; 161(1):72-87.
90. Wu HY, Weiner HL. Oral tolerance. Immunol Res 2003; 28(3):265-84.
91. Weiner HL. Oral tolerance: Immune mechanisms and the generation of Th3-type TGF-beta-secreting regulatory cells. Microbes Infect 2001; 3(11):947-54.
92. Didierlaurent A, Sirard JC, Kraehenbuhl JP et al. How the gut senses its content. Cell Microbiol 2002; 4(2):61-72.
93. Barton GM, Medzhitov R. Toll-like receptors and their ligands. Curr Top Microbiol Immunol 2002; 270:81-92.
94. Gewirtz AT. Intestinal epithelial toll-like receptors: To protect. And serve? Curr Pharm Des 2003; 9(1):1-5.
95. Kagnoff MF, Eckmann L. Epithelial cells as sensors for microbial infection. J Clin Invest 1997; 100(1):6-10.
96. Strober W, Fuss IJ, Blumberg RS. The immunology of mucosal models of inflammation. Annu Rev Immunol 2002; 20:495-549.
97. Uhlig HH, Powrie F. Dendritic cells and the intestinal bacterial flora: A role for localized mucosal immune responses. J Clin Invest 2003; 112(5):648-51.
98. Malmstrom V, Shipton D, Singh B et al. CD134L expression on dendritic cells in the mesenteric lymph nodes drives colitis in T cell-restored SCID mice. J Immunol 2001; 166(11):6972-81.
99. Mottet C, Uhlig HH, Powrie F. Cure of colitis by CD4+CD25+ regulatory T cells. J Immunol 2003; 170:3939-43.
100. Yamazaki S, Iyoda T, Tarbell K et al. Direct expansion of functional CD25+ CD4+ regulatory T cells by antigen-processing dendritic cells. J Exp Med 2003; 198:235-47.
101. Leithauser F, Trobonjaca Z, Moller P et al. Clustering of colonic lamina propria CD4(+) T cells to subepithelial dendritic cell aggregates precedes the development of colitis in a murine adoptive transfer model. Lab Invest 2001; 81(10):1339-49.
102. Krajina T, Leithauser F, Moller P et al. Colonic lamina propria dendritic cells in mice with CD4+ T cell-induced colitis. Eur J Immunol 2003; 33(4):1073-83.
103. Ashcroft AJ, Cruickshank SM, Croucher PI et al. Colonic dendritic cells, intestinal inflammation, and T cell-mediated bone destruction are modulated by recombinant osteoprotegerin. Immunity 2003; 19:849-61.
104. Seldenrijk CA, Drexhage HA, Meuwissen SG et al. Dendritic cells and scavenger macrophages in chronic inflammatory bowel disease. Gut 1989; 30(4):486-91.
105. Oshitani N, Sawa Y, Hara J et al. Functional and phenotypical activation of leucocytes in inflamed human colonic mucosa. J Gastroenterol Hepatol 1997; 12(12):809-14.

106. Waraich T, Sarsfield P, Wright DH. The accessory cell populations in ulcerative colitis: A comparison between the colon and appendix in colitis and acute appendicitis. Hum Pathol 1997; 28(3):297-303.
107. Oshitani N, Kitano A, Kakazu T et al. Functional diversity of infiltrating macrophages in inflamed human colonic mucosa ulcerative colitis. Clin Exp Pharmacol Physiol 1998; 25(1):50-3.
108. MacDonald TT. The mucosal immune system. Parasite Immunol 2003; 25(5):235-46.
109. Stagg AJ, Hart AL, Knight SC et al. The dendritic cell: Its role in intestinal inflammation and relationship with gut bacteria. Gut 2003; 52(10):1522-9.
110. Kaser A, Ludwiczek O, Holzmann S et al. Increased expression of CCL20 in human inflammatory bowel disease. J Clin Immunol 2004; (24):74-85.
111. Vuckovic S, Florin TH, Khalil D et al. CD40 and CD86 upregulation with divergent CMRF44 expression on blood dendritic cells in inflammatory bowel diseases. Am J Gastroenterol 2001; 96(10):2946-56.
112. de Baey A, Mende I, Riethmueller G et al. Phenotype and function of human dendritic cells derived from M-DC8(+) monocytes. Eur J Immunol 2001; 31(6):1646-55.
113. Murakami H, Akbar SM, Matsui H et al. Macrophage migration inhibitory factor activates antigen-presenting dendritic cells and induces inflammatory cytokines in ulcerative colitis. Clin Exp Immunol 2002; 128(3):504-10.
114. Yeung MM, Melgar S, Baranov V et al. Characterisation of mucosal lymphoid aggregates in ulcerative colitis: Immune cell phenotype and TcR-gammadelta expression. Gut 2000; 47:215-27.
115. Ikeda Y, Akbar F, Matsui H et al. Characterization of antigen-presenting dendritic cells in the peripheral blood and colonic mucosa of patients with ulcerative colitis. Eur J Gastroenterol Hepatol 2001; 13(7):841-50.
116. Christ AD, Stevens AC, Koeppen H et al. An interleukin 12-related cytokine is up-regulated in ulcerative colitis but not in Crohn's disease. Gastroenterology 1998; 115:307-13.
117. Nieuwenhuis EE, Neurath MF, Corazza N et al. Disruption of T helper 2-immune responses in Epstein-Barr virus-induced gene 3-deficient mice. Proc Natl Acad Sci USA 2002; 99:16951-6.
118. Heller F, Fuss IJ, Nieuwenhuis EE et al. Oxazolone colitis, a Th2 colitis model resembling ulcerative colitis, is mediated by IL-13-producing NK-T cells. Immunity 2002; 17:629-38.
119. Fuss IJ, Heller F, Boirivant M et al. Nonclassical CD1d-restricted NK T cells that produce IL-13 characterize an atypical Th2 response in ulcerative colitis. J Clin Invest 2004; 113:1490-7.
120. Smith PD, Janoff EN, Mosteller-Barnum M et al. Isolation and purification of CD14-negative mucosal macrophages from normal human small intestine. J Immunoll Methods 1997; 202(1):1-11.
121. Smith PD, Smythies LE, Mosteller-Barnum M et al. Intestinal macrophages lack CD14 and CD89 and consequently are down-regulated for LPS- and IgA-mediated activities. J Immunol 2001; 167(5):2651-2656.
122. Smythies LE, Sellers M, Clements RH et al. Human intestinal macrophages display profound inflammatory energy despite avid phagocytic and bacteriocidal activity. J Clin Invest 2005; In Press.
123. Grimm MC, Pavli P, Doe WF. Evidence for a CD14$^+$ population of monocytes in inflammatory bowel disease mucosa implications for pathogenesis. Clin Expl Immunol 1995; 100(2):291-297.
124. Rogler G, Hausmann M, Vogl D et al. Isolation and phenotypic characterization of colonic macrophages. Clin Exp Immunol 1998; 112(2):205-215.
125. Hausmann M, Kiessling S, Mestermann S et al. Toll-like receptors 2 and 4 are up-regulated during intestinal inflammation. Gastroenterology 2002; 122(7):1987-2000.
126. Allison MC, Poulter LW. Changes in phenotypically distinct mucosal macrophage populations may be a prerequisite for the development of inflammatory bowel disease. Clin Exp Immunol 1991; 85(3):504-509.
127. Burgio VT, Fais S, Boirivant M et al. Peripheral monocyte and naive T-cell recruitment and activation in Crohn's disease. Gastroenterology 1995; 109(4):1029-1038.
128. Mahida YR, Patel S, Gionchetti P et al. Macrophage subpopulations in lamina propria of normal and inflamed colon and terminal ileum. Gut 1989; 30(6):826-834.
129. Rugtveit J, Brandtzaeg P, Halstensen TS et al. Increased macrophage subset in inflammatory bowel disease: Apparent recruitment from peripheral blood monocytes. Gut 1994; 35(5):669-674.
130. Grimm MC, Pullman WE, Benner GM et al. Direct evidence of monocyte recruitment to inflammatory bowel disease mucosa. J Gastroenterol Hepatol 1995; 10(4):387-395.
131. Grip O, Janciauskiene S, Lindgren S. Macrophages in inflammatory bowel disease. Curr Drug Targets Inflamm Allergy 2003; 2(2):155-60.
132. Lochner M, Forster I. Anti-interleukin-18 therapy in murine models of inflammatory bowel disease. Pathobiology 2002-2003; 70:164-9.
133. Kanai T, Uraushihara K, Totsuka T et al. Macrophage-derived IL-18 targeting for the treatment of Crohn's disease. Curr Drug Targets Inflamm Allergy 2003; 2:131-6.

Chapter 9

Alterations of T Lymphocytes in Inflammatory Bowel Diseases

Charles O. Elson, Yingzi Cong and Casey T. Weaver

The intestine contains an abundance of cells of the innate and acquired immune system, and among these T cells represent a major component. Intestinal T cells are distributed among the various compartments in the mucosal immune system, including GALT, the epithelial layer, and the lamina propria. Intestinal T cells are quite distinct from peripheral T cells in phenotype and function, however there have been remarkably few differences between T cells isolated from normal vs inflamed mucosa, perhaps because the 'normal' gut is in a state of low grade inflammation.[1,2] Most studies on intestinal T cells focus on human or mouse isolates and there can be some significant species differences between these two. Secondly, one should keep in mind that human "lamina propria" isolates are actually mixtures of true lamina propria lymphocytes plus cells from the abundant microscopic lymphoid follicles that are present throughout the human intestine. Lamina propria (LP) T cells have been most studied in humans with or without IBD (Table 1).[3] The ratio of $CD4^+$ to $CD8^+$ T cells is 2:1, similar to peripheral blood. Most lamina propria T cells bear $\alpha\beta$ T cell receptors, express activation markers, and are $CD45RO^+$ memory cells. Despite these features, LP T cells proliferate poorly to stimulation via the T cell receptor/CD3 complex. Rather, they respond preferentially to stimulation via TCR ligation plus costimulation via CD2 and CD28.[4] The reason for this shift in TCR-mediated pathways is unclear but may be due to the presence of a soluble factor in the LP microenvironment. LP T cells can produce cytokines when stimulated in vitro, but do not produce detectable cytokines in situ. Even in progressive, fatal experimental colitis mediated by $CD4^+$ T cells, only a few $CD4^+$ T cells per thousand are actively producing cytokines in situ at any one time.[5,6]

Intraepithelial (IEL) T cells also have an activated, memory phenotype and proliferate poorly to TCR/CD3 stimulation.[7] Most IEL are $CD8^+$ T cells rather than $CD4^+$, at least in the small intestine. Colon IEL T cells have not been extensively studied but in mice these are mainly $CD4^+$. Most IEL T cells in humans bear $\alpha\beta$ TCR but in mice about half are $\alpha\beta$ and the rest $\gamma\delta$ T cells. IEL $\alpha\beta$ T cells are thymic-dependent and are derived from the GALT. In mice there is also a thymic-independent subset of IEL T cells which are CD8 $\alpha\alpha^+CD5^-$. These originate in "cryptopatches", small clusters of precursor T lymphocytes that are scattered about the intestine. IEL T cells have a full set of cytotoxic functions when tested *ex* vivo. Their normal role in vivo remains unclear but they presumably play an important role in host defense. IEL T cells are not detectably altered during IBD in humans and there is no evidence at present for a pathogenic role of IEL in either human or experimental IBD.

LP $CD4^+$ T cells reacting to the enteric flora are the effector cells causing experimental IBD in many experimental models.[8] $CD4^+$ T cells make up the main cell populations that infiltrate mucosal tissues in all IBD models studied, and the inflammation is ameliorated if $CD4^+$ T cells are deleted.[9,10] $CD8^+$ T cells are present in the intestine but have not been shown to play a decisive pathogenic role in experimental IBD, e.g., no major effect on inflammation was obtained when $CD8^+$ T cells were deleted.[9] So far the data indicate that chronic intestinal

Immune Mechanisms in Inflammatory Bowel Disease, edited by Richard S. Blumberg and Markus F. Neurath.

Table 1. Properties of lamina propria T cells

- $CD4^+$ or $CD8^+$, mainly TCRαβ
- Express activation, memory markers
- Altered signaling pathways, favoring CD2 rather than CD3
- Produce cytokines, but proliferate poorly in vitro
- Functions: cytokines, helper activity for B cells, regulation

inflammation is mediated by either an excessive $CD4^+$ Th1 response with increased IL-12/IFNγ secretion or an excessive $CD4^+$ Th2 response with increased IL-4/IL-13 production.[11-16] Th1 mediated lesions are marked by transmural cellular infiltration sometimes associated with granuloma, and whereas epithelial cell layer changes are clearly present, they are not a dominant feature. A similar histopathology is observed in Crohn's disease which appears to Th1 mediated also.[17-23] In contrast, Th2-mediated inflammation is characterized by a more superficial colonic inflammation and epithelial hyperplasia, somewhat similar to ulcerative colitis.[16,19,24,25] However, ulcerative colitis has not been clearly shown to be a Th2-mediated inflammation. Because of the key role that $CD4^+$ T cells appear to play in IBD, the focus of this chapter will be on the various $CD4^+$ T cell subsets and how they might cause or prevent chronic intestinal inflammation.

$CD4^+$ T Cell Subsets in the Intestine

$CD4^+$ T cells display extensive diversity in terms of phenotype, function and anatomical distribution. After generation in the thymus and migration to the peripheral immune organs such as GALT, $CD4^+$ T cells are in naive status and lack most effector function. Naïve T cell activation and differentiation requires interaction between TCR/CD3 complex on $CD4^+$ T cells and antigen/MHC on APCs, plus a second signal provided via interactions between a number of costimulatory molecules on APCs and their ligands on $CD4^+$ T cells (e.g., B7/CD28, B7H/ICOS, CD40/CD40L, and OX40L/OX40).[26-30] Once activated $CD4^+$ T cells can differentiate into functionally distinct subsets. T-helper 1 (Th1) cells preferentially produce IFNγ and IL-2, Th2 cells preferentially secrete IL-4, IL-5, and IL-13.[31-35] In addition, regulatory $CD4^+$ T cells (Treg) are also induced by antigen and these frequently produce IL-10 and TGFβ.[35,142] Multiple factors determine $CD4^+$ T cell polarization into these subsets, but one key factor is cytokine signals, with IL-12 driving Th1 and IL-4 driving Th2, and in at least in some instances IL-10 and TGFβ driving Treg cell differentiation.[32,33,36-38] Thus mice whose IL-12 p40 or IL-12 receptor beta2 chain are deficient have impaired Th1 responses,[39,40] whereas the mice lack IL-4 or IL-4 receptor have defective Th2 responses.[41-43] In addition, IL-10 deficient or TGFβ deficient mice lack Treg cells, and these mice develop various inflammatory diseases.[44-46] IFNγ and IL-18 also modulates Th1 differentiation; mice deficient in these cytokines exhibit impaired Th1 response also. IL-18 acts by markedly enhancing IL-12 dependent Th1 cell differentiation.[47-50] In an analogous fashion, IL-13 appears to modulate Th2 cell differentiation in either an IL-4 dependent or IL-4 independent manner.[51-53]

In addition to these cytokines, transcription factors, such as signal transducer and activator of transcription (STAT) proteins, have been shown to be involved in $CD4^+$ T cell differentiation.[54,55] Activation and phosphorylation of STAT 4 and STAT 1 mediate the action of IL-12 in driving Th1 cell development and IFNγ regulation,[56,57] whereas STAT 6 is required for IL-4 induced Th2 cell differentiation.[58-61] That said, STAT proteins are expressed in both Th1 and Th2 cells and may not have a unique role in directly regulating the Th1 and Th2 cell development. Recently, two transcription factors, T-bet and GATA-3, have been described as unique regulators of Th1 and Th2 differentiation, respectively.[62,63] T-bet is a novel transcription factor of the T-box family, and has been found to be expressed by Th1 but not Th2 cells.[62] T-bet deficient mice have an impaired IFNγ production in $CD4^+$ but not $CD8^+$ T cells, and retroviral transduction of primary T cells or even fully polarized Th2 cells with T-bet induces high levels of IFNγ production and suppresses IL-4 and IL-5 production.[62,64] GATA-3 is a pleiotropic transcription factor of the C4 zinc-finger family that binds to a DNA sequence characterized by a 5'-GATA-3' core element. GATA-3 is selectively expressed in Th2 but not Th1 cells, and is important for IL-5 expression in T cells by transactivation of the IL-5 promoter.[59,65] Ectopic expression of GATA-3 is sufficient to modestly induce Th2-specific cytokine expression even in committed Th1 cells.[66] Repressor of GATA (ROG), a recently cloned specific repressor of GATA-3, has been shown to suppress the GATA-3 functions on Th2-cytokine gene promoters.[67] GATA-3 suppresses Th1 development by downregulation of STAT4 and not through effects on IL-12Rβ2 chain or T-bet.[68] Various other transcription factors such as C-maf and NFATc1 have also been shown to induce or enhance Th2 cytokine production.[69-71]

The molecular basis of the Treg cell development and function is not clear. Although CTLA-4, GITR, TGFβ, and IL-10 have been implicated in the function of T reg cells, none of these molecules are specific for this cell type.[72-75] Recently a key transcription factor, Foxp3, has been shown to be specifically expressed by and essential for $CD4^+CD25^+$ Treg development and function.[76] $CD4^+CD25^+$ cells in Foxp3-deficient mice do not have regulatory activity, and $CD4^+CD25$- cells ectopically expressing Foxp3 can block the induction of IBD and gastritis in lymphopenic mice and inhibit proliferation of naive T cells in vitro. Conversely, transgenic mice over- expressing Foxp3 contain more regulatory T cells than wild-type counterparts, and crossing the Foxp3 transgenics to CTLA-4 deficient mice delays the lethality of the latter.[77-79] However, how Foxp3 is induced and how it drives T cells to become regulatory remain open questions, and whether Foxp3 promotes IL-10 production by Treg cells is still controversial. $CD4^+$ T cell subset development is summarized in Figure 1. The reader is referred to recent reviews for details about the individual models.

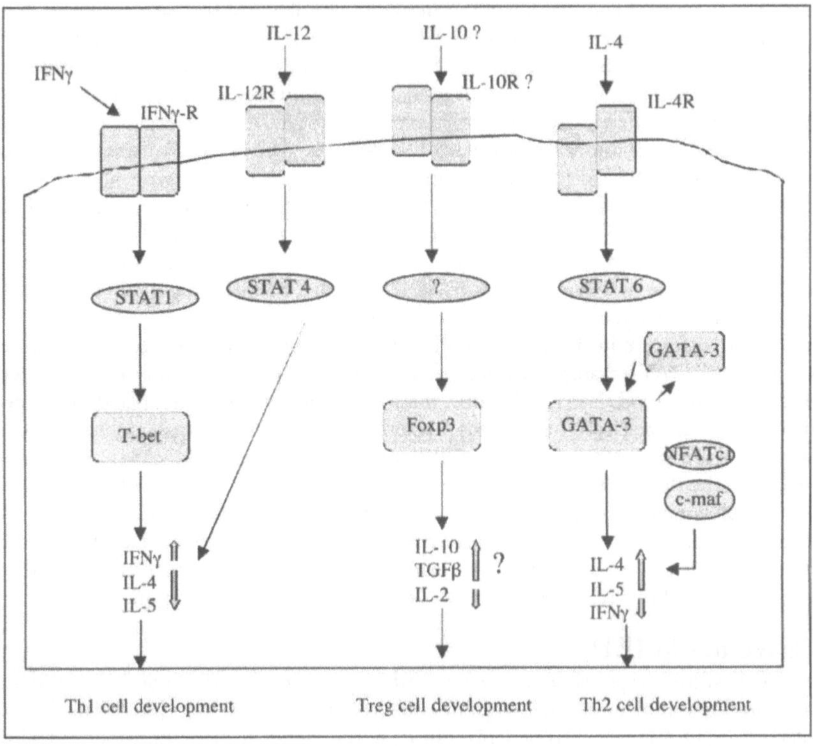

Figure 1. Check points in T cell subsets development.

Intestinal T Effector Cells

Th1 Effectors in IBD

In Crohn's disease, LP T cells exhibit a Th1 phenotype and thus produce large amounts of IFNγ and TNFα; in addition, IL-12 secretion, the driving force of Th1 differentiation, is increased.[80] Multiple animal models also indicate that most common immunologic mechanisms leading to mucosal inflammation are involved a dysregulated Th1 cell pathway that responds to enteric bacterial antigen stimulation. Multiple Th1 type models of mucosal inflammation have been developed in recent years, only a few of these models will be discussed briefly here.

$CD4^+CD45RB^{hi}$ adoptive transfer model. Normal $CD4^+$ T cells can be separated into two different subpopulations, one naïve cells expressing high levels of CD45RB ($CD4^+$ $CD45RB^{hi}$), and memory cells expressing low levels of CD45RB ($CD4^+$ CD45RBlo). Adoptive transfer of $CD4^+$ $CD45RB^{hi}$ cells into SCID or $RAG^{-/-}$ mice results in colitis and wasting over the following weeks to 2-3 months. Transfer of the reciprocal $CD4^+$ $CD45RB^{lo}$ subset does not result in disease nor does transfer of unseparated $CD4^+$ T cells. Whereas $CD45RB^{hi}$ cells populate the small intestine as well as the large intestine of the SCID recipients, inflammation is limited to the colon. $CD4^+$ T cells from colitic recipient mice proliferate and produce high levels of IFNγ and the other Th1 cytokines in response to APCs pulsed with enteric bacterial antigens.[81,82] Treatment with anti-IL-12, anti-IFNγ, or anti-TNF, but not administration of IL-4 prevents the disease development, indicating that the colitis is mediated by Th1 effector cells.[11,83,84] When $CD4^+$,$CD45RB^{hi}$ T cells are transferred to SCID recipients with a reduced flora, or to recipients that are treated with antibiotics the colitis is ameliorated.[85] These results indicate that the bacterial flora is driving the colitis and indeed the T cells became oligoclonal after transfer and demonstrate reactivity to antigens of the bacterial flora.[86]

STAT-4 transgenic mice. As discussed, STAT-4 mediates IL-12 signaling and drives Th1 cell development. The phosphorylation of STAT-4 activates the expression of genes such as IFNγ. Overexpression of STAT-4 in T cells renders them more sensitive to IL-12 signaling and thus results in enhanced Th1 responses. Mice expressing STAT-4 as a transgene remain normal unless challenged by an immunization with DNP-KLH in CFA (this antigen is thought to mimic antigens of the bacterial flora in some way), which increased transgene expression in both spleen and colon and induced an unremitting severe colitis with dense infiltrates of $CD4^+$ T cells expressing nuclear STAT-4 and producing high levels of IFNγ and TNFα. $CD4^+$ T cells isolated from STAT-4 transgenic mice with colitis responded strongly to lysates of intestinal bacteria and were able to transfer colitis to SCID recipients, indicating a $CD4^+$ Th1 cell response to enteric bacterial antigens mediates the disease.[87]

An interesting feature in this model was the requirement of antigen-specific activation for colitis induction. STAT-4 transgenic mice do not develop colitis when given complete Freunds adjuvant IP without the DNP-KLH. Thus, these genetically susceptible mice do not develop excessive $CD4^+$ Th1 cell reactivity to the bacterial flora and subsequent colitis unless they sustained a triggering event that activated the pathogenic process. The exact relationship of DNP-KLH to the flora is unclear and other types of antigens were not tested. However, CFA contains mycobacterial antigens as an adjuvant and these mycobacterial antigens were not sufficient to trigger disease on their own, suggesting that some specificity was required for disease induction. Little is known about how pathogenic Th1 responses are triggered in genetically susceptible hosts and this seems to be an excellent model to explore such questions further.

Th2 Effectors in IBD

The $CD4^+$ subset responsible for disease pathogenesis in ulcerative colitis is not defined. Increased production of some Th2 cytokine such as IL-5 has been found in LP mononuclear cell isolates from ulcerative colitis patients, but IL-4, the signature cytokine associated with the Th2 responses, is not increased.[19] This discrepancy may be due to different roles of IL-4 and

IL-13 in Th2-mediated inflammation. Recent data suggests that IL-4 may be more important in disease induction and IL-13 acting later in disease progression and chronicity.[88] Based on these concepts and on a histopathologic resemblance of some Th2-mediated murine models of mucosal inflammation to human ulcerative colitis, some assert that UC is a 'Th2-like disease'.[14,15] Each of the three major models of Th2-mediated disease provide some insights into what this process entails.

The best characterized Th2 mediated model of IBD is the T cell receptor (TCR) α-chain-deficient mouse.[25] BALB.TCR$\alpha^{-/-}$.[89,90] α deficient mice demonstrate polyclonal expansion and activation of B cells with the production of multiple autoantibodies including anti-colon and anti-tropomyosin, antibodies which are present in patients with ulcerative colitis.[91] Mice also develop antibodies to multiple enteric bacterial antigens. BALB.TCRα deficient mice populate peripheral immune tissues with a unique population of $CD4^+$ T cells that express the TCRβ chain without TCRα chain. This novel TCRβ^+ T cell subset responds to polyclonal activators, is enriched in Peyer's patches and in colon lymphoid follicles early in life, and is abundant in the colon lamina propria and draining mesenteric lymph nodes of mice with colitis. This TCRβ^+ subset produces high levels of IL-4 and is the effector cell mediating disease in that treatment of mice with antibodies to the TCRβ chain or to IL-4 abrogates colitis and polyclonal B cell activation.[15] Analysis of the TCR repertoire indicates that the pathogenic TCRβ cell utilizes a restricted $Vb8.2^+$ chain with a conserved motif in the CDR3 region, and that this TCR might cross-react with both epithelial and bacterial antigens.[92] BALB.TCR α-deficient colitis appears to be a Th2-mediated inflammatory disease in which the pathogenic response to enteric bacterial antigens results in secondary immune response to intestinal autoantigens.[93] An interesting feature of this model is that a $CD1d^+$ B cell subset appears to have a regulatory role and ameliorates disease via the production of IL-10, which may be unique to this model.[94]

Some insight into the relative roles of IL-4 and IL-13 in Th2-mediated colitis comes from the Oxazolone-induced colitis model. Intrarectal administration of high doses of the haptenating agent oxazolone in ethanol leads to a colitis caused by a polarized Th2 T cell response. Colon cytokine production in oxazalone colitis is dominated by high levels of IL-4, IL-5, and IL-13, with normal or reduced IFNγ production. This Th2 response is the cause of the inflammation, in that administration of anti-IL-4 monoclonal antibodies ameliorates the disease, whereas anti-IL-12 administration exacerbates the disease.[14] Recent work indicates that IL-13 production by NK-T cells is a significant pathologic factor in oxazolone colitis in that neutralization of IL-13 by IL-13Ra2-Fc fusion protein prevents colitis. Oxazolone colitis does not occur in mice depleted of NK-T cells, in mice unable to present antigen to NK-T cells, or in mice lacking an NK-T cell-associated TCR, indicating that NK-T cells mediate oxazolone colitis through production of IL-13. There is preliminary data that IL-13 might be increased in LP of patients with ulcerative colitis.[88]

What are the factors that stimulate a Th-2 deviated response to bacterial associated antigens in the intestine? This answer to this question is not presently known, but is approachable experimentally in a novel model in which OVA-specific T cells from DO11.RAG-$2^{-/-}$ TCR transgenic mice are transferred into RAG-$2^{-/-}$ recipients whose intestinal tracts are colonized with OVA-expressing or control *Escherichia coli*. Transfer of antigen-naive DO11.RAG-$2^{-/-}$ T cells into recipients colonized with OVA-E. coli resulted in enhanced intestinal recruitment and cell cycling of OVA-specific T cells in the gut; however, there was no development of disease. In contrast, transfer of polarized Th2 or Th1 $CD4^+$ cells resulted in severe wasting and colitis in recipients colonized with OVA-expressing but not control *E. coli*. The histopathologic features of disease induced by Th2 and Th1 transfers were distinct, but disease severity was comparable. Induction of disease by both Th2 and Th1 transfers was dependent on bacterially associated OVA. This models shows that the same bacterially-associated antigen can induce either Th2 or Th1-mediated disease, but only after the $CD4^+$ T cells had been deviated to Th2 or Th1 responses in vitro.[16] This model will allow exploration of how such $CD4^+$ polarization might occur in the intestine in vivo.

The pathogenic potential of unopposed Th2 response in these murine models of Th2-mediated colitis strikes a cautionary note with respect to therapeutic strategies aimed at reversing Th1-mediated inflammatory activity by augmenting Th2 activity. These results cast doubt on the common notion that intestinal Th2 cells are beneficial and are able to regulate the pathogenic effects of the Th1 subset in the gut. There is no evidence that Th2 cells regulate Th1 cells in the intestine in any model. Instead, both $CD4^+$ Th1 and Th2 effector cells are under the control of various types of regulatory cells, including $CD4^+$ Tregs.

Intestinal T Regulatory Cells

Multiple types of Treg cells appear to exist (Table 2), including Tr1 cells, Th3 cells, $CD4^+CD25^+$ T cells, $CD4^+CD45RB^{lo}$ T cells, $CD8^+CD28-$ T cells, γδ T cells, and NKT cells. Different subsets of Tr cells show regulatory activities that are mediated either by immunosuppressive cytokines or by contact-dependent mechanisms.[36,95-98] Although the presence of Tregs in the intestine itself has not been proven for any of these Treg subsets, several of them have been shown to regulate intestinal inflammation in animal models of IBD.

$CD4^+CD45RB^{lo}$ T Cells

One of the most important findings in the CD45RB transfer model is that colitis is abrogated by cotransfer of $CD4^+$ $CD45RB^{lo}$ T cells along with $CD4^+$ $CD45RB^{hi}$ T cells, indicating that $CD4^+$ $CD45RB^{lo}$ T cells serve as regulatory cells and are present in the normal animals.[99] The protective effect of $CD45RB^{lo}$ T cells is not due to IL-4 production because $CD45RB^{lo}$ T cells from IL-4 deficient mice are protective and, secondly repeated administration of anti-IL-4 does not abrogate the regulatory function of normal $CD45RB^{lo}$ T cells. However, administration of either anti-TGFβ or anti-IL-10R1 can abrogate protection by $CD45RB^{lo}$ T cells. Consistent

Table 2. Subsets of regulatory T cells

Subset of Tr Cells	Possible Mechanisms Involved	Inhibition of IBD	References
Tr1 cells	IL-10 and TGFβ; Cognate interaction	Yes	Groux, 1997, Nature; Cong, 2002, JI
Th3 cells	TGFβ	ND	Chen, 1994, Science; Fukara, 1996, JCI
$CD4^+CD25^+$ T cells	TGFβ; CTLA-4; GITR	Yes	Sakaguchi, 1995, JI; Read, 2000, JEM; Thornton, 1998, JEM
$CD4^+CD45RB^{low}$ T cells	IL-10 and TGFβ	Yes	Morrissey, 1993, JEM; Powrie, 1993, II; Asseman, 1999, JEM
Immature DC induced Tr cells	Cognate interaction	Yes	Jonuleit, 2000, JEM; Dhodapkar, 2001,JEM
Vit 3/Dex Tr cells	IL-10	ND	Barrat, 2002, JEM
$\alpha_E\beta_7^+$ T cells	Unknown	Yes	Lehmann, 2002, PNAS
$CD8^+CD28^-$ T cells	Cognate interaction	Yes	Liu, 1998, II; Mayer, 1987, JEM
γδ T cells	KGF	Yes	Chen, 2002, PNAS
NK T cells	α-GalCer mediated activation	Yes	Cui, 1997, Science; Saubermann, 2002, Gastroenterolgy

with this, $CD4^+CD45RB^{lo}$ T cells from IL-10 deficient donors fail to protect, and instead cause colitis in scid recipients.[100,101] The $CD4^+CD45RB^{lo}$ T cell population is heterogeneous, consisting of both memory effector T cells as well as regulatory T cells. The Treg cells that are protective for colitis are enriched within the $CD25^+$ subset of $CD4^+CD45RB^{lo}$ T cells.[72,102]

T-Regulatory-1 Cells (Tr1)

Tr1 cells are induced in vitro by activating $CD4^+$ T cells with antigen in the presence of IL-10, in both human and mouse. Tr1 cells have a cytokine production profile and functional phenotype distinct from that of Th1 or Th2 cells,[36,98,103] producing high amounts of IL-10, and low levels of IFNγ, little or no IL-4, and no IL-2 upon TCR-mediated activation. Tr1 cells inhibit both naïve and memory T cells in an antigen-specific manner via a mechanism that is partially dependent on the production of IL-10 and TGFβ1 and partly on an ill-defined mechanism requiring cell contact. Tr1 cells produce IL-10 very rapidly, with significant levels of IL-10 detectable in supernatants as early as 4 hrs after activation, and peaking at 12-24 hrs, whereas control Th2 clones with the same antigen specificity produce IL-10 late after activation, peaking at 48-72 hrs. Production of TGFβ1 by Tr1 cells seems to be dependent on signaling through CTLA-4, in that secretion of TGF-β is blocked by a nonstimulatory anti-CTLA-4 mAb.[104] Tr1 cells proliferate poorly upon polyclonal TCR-mediated or antigen-specific activation compared to Th1 or Th2 cells with the same antigen specificity, and their activation requires costimulation signals provided by CD80 and CD86, but not CD40. Tr1 cells not only inhibit naïve as well as memory $CD4^+$ T cell responses, including both Th1 or Th2 responses, but also suppress $CD8^+$ T cell activation and the production of immunoglobulin by B cells.[104] When cotransferred with these pathogenic T cells, Tr1 cells inhibit Th1-mediated colitis development in scid recipients of naïve $CD4^+CD45RB^{hi}$ T cells[36] or of memory Th1 cells reactive to enteric bacterial antigens.[98] Tr1 cells also regulate Th2 cell mediated pathology, and suppress serum IgE responses and T cell responses in a model of immediate hypersensitivity via production of IL-10.[105]

$CD4^+$ T cells with Tr1 activity appear to be present in the intestinal LP in both humans and mice. Human LP $CD4^+$ T cells inhibit the autologous peripheral blood T cell proliferative response to *E. coli* antigens and this inhibition is abrogated by addition of anti-IL-10 or anti-TGFβ mAbs.[106] Cells that express IL-10 or IFNγ are frequent in mouse LPL as detected by in situ hybridization.[5] Murine LP $CD4^+$ T cells produce IL-10, but not IL-2, IL-4, and IFNg upon stimulation with enteric bacterial antigens. When anti-IL-10R1 mAb is added to such LP $CD4^+$ T cell cultures, IFNγ is detectable, but not IL-2 or IL-4. LP $CD4^+$ T cells inhibit a pathogenic Th1-cell in response to CBA stimulation, but not to anti-CD3 stimulation, consistent with a CBA-specific inhibition. Such inhibition is partially abrogated by addition of anti-IL-10 or anti-IL-10R1 mAb.[98] Thus, CBA-specific T cells with activity similar to Tr1 cells are present in normal lamina propria both in humans and mice, and can inhibit pathogenic T cell responses. These data indicate that $CD4^+$ T cells with Tr1 activity and reactive to enteric bacterial antigens are present in the lamina propria and most likely are involved in the intestinal immune homeostasis.

T-Helper-3 Cells (Th3)

Th3 cells are characterized by the production of high amounts of TGFβ1 when stimulated with their specific antigen. Th3 cells were originally identified in the peripheral blood of humans with multiple sclerosis and were subsequently identified as mediating oral tolerance to autoantigens fed to mice. Th3 cells appear to mediate the protective oral tolerance induced in mice fed TNP-colon proteins in the TNBS-ethanol colitis model.[107] Deficient mucosal TGFβ1 production has been found in IL-2-deficient mice, suggesting that this form of experimental colitis might be due to the lack of development of the Th3 subset.

It is unclear whether the Th3 subset is the same or different from the Tr1 subset. IL-10 stimulates TGFβ production and thus is needed for Th3 function,[107] which might explain why Th3 cells do not protect IL-10-deficient mice from colitis. It is possible that Tr1 and Th3 cells are the same subset, which can utilize either IL-10 or TGFβ as the predominant depending on the context or microenvironment.

$CD4^+CD25^+$ T Cells

$CD4^+CD25^+$ T cells are the most studied subset of Tregs. CD25 is the alpha chain of the IL2R and can be expressed on any activated T cell, however constitutive expression of CD25 is a characteristic of the $CD4^+CD25^+$ Treg subset. $CD4^+CD25^+$ T cells make up approximately 5 to 10% of peripheral $CD4^+$ T cells and have an anergic phenotype demonstrating little or no proliferation to a variety of stimuli such as anti-CD3. Instead, $CD4^+CD25^+$ Tr cells inhibit $CD4^+CD25^-$ T cell proliferation and IL-2 production to such stimulation.[108,109] This inhibition requires cognate interaction in vitro and does not require cytokines such as IL-10 or TGF[110] $CD4^+CD25^+$ Tregs express high levels of CTLA-4 and CTLA-4 appears critical for their regulatory function.[72,73] In addition, signaling through GITR plays an essential role in the function of $CD4^+CD25^+$ Tregs.[74,111] Although $CD4^+CD25^+$ Tregs do not proliferate to stimulation via TCR, such stimulation is required for $CD4^+CD25^+$ Tr cells to exert inhibition, and once so activated, the inhibition is antigen nonspecific, in that they can inhibit the responses of T cells with the same or different antigen specificity.[112,113] $CD4^+CD25^+$ Tregs can suppress differentiation and function of both Th1 and Th2 cells, as well as $CD8^+$ T cells and B cells, possibly through down-regulation of costimulatory molecule expression by APCs, such as CD40, CD80, and CD86 and cytokine production.[114-117] Removal of $CD4^+CD25^+$ T cells in vivo leads to spontaneous development of various autoimmune diseases as well as triggers excessive or misdirected immune responses to microbial antigens, thus causing immunopathology.[118] Reconstitution of such $CD4^+CD25^+$ cells prevents such autoimmunity. $CD4^+CD25^+$ T cells prevent colitis when cotransferred with $CD4^+CD45RB^{hi}$ T cells into $RAG^{-/-}$ mice and CTLA-4 plays an essential role in such inhibition. $CD4^+CD25^+$ Tregs inhibited both T cell-mediated and T cell-independent intestinal inflammation in a model of Helicobacter hepaticus infection of $RAG^{-/-}$ mice. The latter effect indicates that the innate immune system can be regulated by $CD4^+CD25^+$ Tregs. The inhibition of innate immune pathology was dependent on the production of IL-10 and TGFβ.[72,119]

$\alpha_E\beta_7^+$ T Cells

The integrin $\alpha_E\beta_7^+$ was initially described as a marker for IEL T cells. Recently, a new subset of Tregs cells expressing $a_Eb_7^+$ integrin has been identified.[120] These $\alpha_E\beta_7^+$ T cells express high levels of CTLA-4, and have a unique cytokine expression pattern with higher levels of IL-2, INFγ, IL-5, and IL-13, but lower levels of IL-10, compared to memory T cells. $\alpha_E\beta_7^+$ T cells inhibit naïve T cell responses and addition of anti-10 and anti-TGFβ mAbs does not abrogate this inhibition, indicating that these cytokines are not involved in its activity. If the $\alpha_E\beta_7^+$ T cells and the target cells are separated in culture by transwell chambers, $\alpha_E\beta_7^+$ T cells fail to suppress naïve T cell response, thus the regulatory function of $\alpha_E\beta_7^+$ T cells requires cell contact but not soluble factors. Transfer of $\alpha_E\beta_7^+$ T cells into scid mice does not cause colitis, rather, co-transfer of $\alpha_E\beta_7^+$ T cells with $CD4^+CD45RB^{hi}$ T cells prevented disease development in the scid recipients. The presence and role of this Treg subset in the normal intestine remains to be established.

$CD8^+CD28^-$ T Cells

$CD8^+$ T cells can suppress effector T cell function in several experimental systems as well as in humans. In humans, a subset of MHC-class I-restricted $CD8^+CD28^-$ T cells have regulatory activity. $CD8^+CD28^-$ T cells suppressed Th1 cell proliferation by a cell contact-dependent mechanism that involved interaction with an APC that coexpressed the MHC class I and class II antigens recognized by $CD8^+CD28^-$ and Th cells, respectively. $CD8^+$CD28- T cell culture supernatants are not inhibitory, and addition of anti-IL-4, anti-IL-10, anti-TGFβ, and anti-CTLA-4 failed to abrogate the suppressive activity of $CD8^+CD28^-$ T cells.[121,122] Interestingly, human intestinal epithelial cells (IECs) which express gp180, a CEA family member, can induce $CD8^+CD28^-$ T cells that express CD101 and CD103 and have inhibitory function.[123,124] IECs from IBD patients are defective in expression of gp180 and fail to expand this $CD8^+$CD28- T cell subset.

The regulatory function of $CD8^+$ T cells has also been demonstrated in several animal models of IBD. In a TNBS adoptive transfer model, the migration of sensitized $CD4^+$ T cells into the colon is inhibited by $CD8^+$ T cells. IL-2 deficient mice that are also deficient for $CD8^+$ T cells (β2M-/-) develop colitis earlier than IL-2 deficient mice with normal $CD8^+$ T cells.[9] $CD8^+$ T cells from C3H/HeJBir mice inhibit proliferation and IFNγ production of pathogenic Th1 cell reactive to enteric bacterial antigens; addition of anti-IL-10 mAb partially reverses such inhibition. Transfer of C3H/HeJBir $CD8^+$ T cells into scid mice does not cause colitis, rather these $CD8^+$ T cells can prevent colitis development when cotransferred with pathogenic Th1 cells reactive to enteric bacterial antigens.[125] The mechanism by which MHC class I-restricted $CD8^+$ T cells inhibit class II restricted $CD4^+$ T cell responses to enteric bacterial antigens remains unknown.

NK T Cells

NK T cells exist in low numbers in the peripheral blood (<1-2% of blood T cells) and most other tissues. They are CD1d-restricted and coexpress NK cell marker NK1.1 and an invariant TCR chain (TCRAV24AJQ in human and TCRAV14AJ281 in mice) paired with polymorphic TCRβ chains from a restricted number of TCR Vβ families. Upon stimulation, NK T cells produce high amounts of IFNγ, IL-4, as well as IL-10. A regulatory function of NK T cells has been demonstrated in various autoimmune diseases, as well as in IBD. In the DSS colitis model, administration of a-galactosylceramide (alpha-GalCer), a lipid ligand known to bind CD1d and activate NK T cells in a CD1d-restricted fashion, significantly ameliorates the severity of colitis. Depletion of NK T cells by anti-NK1.1 mAb abrogated the protection afforded by alpha-GalCer. Transfer of NK T cells into $RAG^{-/-}$ mice which are treated with DSS, protect against colitis development. These data indicate that the alpha-GalCer-activated NK T cells inhibit colitis development in DSS-treated mice.[126]

It is clear from the above that there are multiple subsets of regulatory T cells with various mechanisms present that control the intestinal immune response in normal hosts. The relationships among these regulatory T cells remains unknown. Some may belong to different lineages of T cells or may simply represent different stages of differentiation of the same lineage of regulatory T cells.

What Drives Effector and Regulatory T Cells in IBD?

It is clear now that $CD4^+$ T cells can be effector cells or regulatory cells in IBD development, however it is not clear what factors drive the $CD4^+$ T cells differentiate into effector cells or regulatory cells. Accumulating evidence indicates that enteric flora play a key role in triggering both effector cell and regulatory cell development.

Enteric Bacteria

The driving force generating effector T cells, either Th1 or Th2, and thus the induction of inflammation is the commensal microbiota. In almost all the models tested so far, such as IL-2 deficient mice, IL-10 deficient mice, CD45RB transfer model, and SAMP-1/Yit mice, disease develops in a specific pathogen-free or conventional environment but not in a germ-free environment, and in some instances colitis is ameliorated when the mice are treated with antibiotics.[44,85,127-130] Colitic mice showed an enhanced T cell response to enteric bacterial antigens in multiple models. More direct evidence supporting the concept that microbial antigens drive effector T cell function comes from an extensive series of studies on C3H/HeJBir mice that spontaneously develop colitis.[12,18] $CD4^+$ T cells of these mice develop a strong Th1 response to a small set of selected enteric bacterial antigens, and these T cells can transfer disease to SCID mice once activated by enteric bacterial antigens in vitro. Interestingly, enteric bacterial antigens do not only stimulate effector cells but also can stimulate regulatory T cells. When stimulated with enteric bacterial antigens in vitro in the presence of IL-10, a Tr1 cell line generated from C3H/HeJBir mice inhibited pathogenic Th1 cell response in vitro and in vivo through production of IL-10 and TGFβ, as well as cognate interactions in vitro, and prevented colitis in vivo induced by pathogenic Th1 cells when cotransferred with Th1 cells.[98] It is unknown whether the enteric bacterial antigens that stimulate T effector cells versus Treg cells are the same or different, but it is clear that enteric bacterial antigens can drive the development and function of both as cognate ligands of the TCR.

There appears to be a complex three-way interaction in the intestine between the microbiota, the epithelium, and mucosal immune cells, the latter involving cells of both the innate and acquired immune system. Each component is in dialogue with the other two.[131] Some microbial products likely target innate immune cells and serve an adjuvant function, by interacting with Toll-like receptors. Others microbial products are able to stimulate T cells and B cells. The leading candidates for passing these microbial signals to T cells, are intestinal epithelial cells and dendritic cells. In regard to the first, primary abnormalities of epithelial cells can result in an abnormal T cell reactivity to the microbiota sufficient to induce colitis. This is best illustrated by the *mdr1α-/-* mouse, in which a deficient expression of this transport protein in epithelial cells eventuates in increased Th1 reactivity to the microbiota and colitis.[132] Direct evidence that dendritic cells are transducing microbial signals comes from a novel system showing that the microbiota drive IL-12p40 and IL-23 expression by CD11c+CD11b-CD8- dendritic cells in the mouse ileum.[133] Another level of complexity in understanding these complex interactions derives from a multigenic regulation of the mucosal immune response to the microbiota, which is evident in the wide phenotypic differences among inbred mouse strains bearing the same induced mutation such as IL-10 deficiency.[134]

Microbial Pathogens Can Modulate Immune System

There is evidence that microbial pathogens can modulate host immune response. Recently, Tr1 clones specific for filamentous hemagglutinin (FHA) and pertactin have been generated from the lungs of mice during acute infection with B. pertussis. Such pathogen-specific Tr1 cells were induced at a mucosal surface during infection and local Th1 responses were suppressed. These Tr1 clones expressed T1/ST2 and chemokine receptor CCR5, secreted high levels of IL-10, but not IL-4 or IFNγ, and suppressed Th1 responses against B. pertussis or an unrelated pathogen. Furthermore, FHA inhibited IL-12 and stimulated IL-10 production by dendritic cells (DCs), and these DCs directed naive T cells into the regulatory subtype. The induction of Tr1 cells after interaction of a pathogen-derived molecule with cells of the innate immune system represents a novel strategy exploited by an infectious pathogen to subvert protective immune responses in vivo.[135]

FHA, a virulence factor of *Bordetella pertussis*, interacts directly with DCs to induce IL-10 and inhibit LPS-induced IL-12 and inflammatory chemokine production. FHA binds to leukocyte response integrin (Vβ3, CD61) and integrin-associated protein (CD47) complex and stimulates bacterial adherence to the β2-integrin CR3. Engagement of CR3 or CD47 on macrophages or DCs with specific antibodies or natural ligands has been shown to suppress IL-12 production, and this may represent a physiological phenomenon for normal control of immune responses in vivo. Although FHA-activated DCs may have features of immature DCs, they appear to represent a distinct subtype of DCs, with enhanced expression of CD86 and CD40 over that seen on immature DCs and in addition secrete high levels of IL-10 after appropriate stimulation.[135] Whether other bacteria can stimulate this pathway in the intestine, either pathogens or commensals, is an unknown but fascinating question.

Probiotic Microbes

Probiotics, including bacteria, such as *Lactobacilli sp* and *Bifidobacteria sp*, some *Escherichia coli, Enterococci*, and certain yeast, mainly *Saccharomyces* have been shown to provide beneficial effects on human health and on IBD.[136] Administration of certain strains of probiotics can prevent colitis in animal models, such as IL-10 deficient mice and HLAβ27 transgenic rats by inhibiting Th1 cytokines, such as IFNγ and IL-12, whereas maintaining production of the immunoregulatory cytokine TGF-β1.[137,138] Probiotics can act on various immune cell types, including B cell, DC and macrophages as well as T cells.

Different probiotic strains vary in their ability to modulate T cell responses. Some promote Th1 type responses, and others enhance both Th1 and Th2 type responses,[139,140] and some may induce Tr1 cells.[141] In regard to the last possibility, it has been shown that Lactobacillus paracasei induces high levels of IL-10. When L. paracasei bacteria are added to mixed lymphocyte cultures in which BALB/c $CD4^+$ T cells are stimulated weekly in the presence of irradiated allogeneic splenocytes, the L. paracasei strongly inhibited the proliferative response of $CD4^+$ T cells in a dose-dependent fashion. This is accompanied by a marked decrease of both Th1 and Th2 cytokines but maintenance of IL-10 and induction of TGF-β in a dose-dependent manner.[141] It seems possible that this probiotic is stimulating regulatory T cells similar to the Tr1 cells that are involved in maintenance of intestinal immune homeostasis.

Conclusion

Data from multiple experimental models to date provide strong support for the immunologic hypothesis that a dysregulated mucosal $CD4^+$ T cell response to antigens of the enteric bacteria in a genetically susceptible host results in chronic intestinal inflammation. The host interaction with the flora is complex, but there are a select number of cells and molecules that are critical to this effort. When these key pathways are impaired, the host response to the bacterial flora results in IBD. The key effector cell responsible for disease in most instances is the $CD4^+$ T cell. Although there is data in some systems that Th1 and Th2 subsets can reciprocally regulate one another, each of these CD4 T cell effector CD4 T cell subsets has been found to mediate colitis in various mouse models. There is no data at present that demonstrates that Th2 cells regulate Th1 cells in the intestine or vice versa, and thus experimental colitis is not explained as an imbalance between Th1 and Th2 subsets. At present the data are compatible with the concept that excessive responses of either the Th1 or Th2 effector subsets are detrimental and can result in inflammatory bowel disease, and that this is prevented in the normal host by the presence and function of regulatory cells, particularly $CD4^+$ Tregs. It is unclear how many distinct subsets of regulatory cells are present in the intestine, or the role that each may play in regulating the mucosal immune response to commensal bacterial antigens.

References

1. James SP. Mucosal T-cell function. Gastroenterol Clin N Am 1992; 20:597-612.
2. Lefrancois L. Basic aspects of intraepithelial lymphocyte immunobiology. In: Ogra PL, Mestecky J, Lamm ME et al, eds. Handbook of Mucosal Immunology. San Diego: Academic Press, 1994:287-298.
3. James SPaK H. Gastrointestinal lamona propria T cells. In: Ogra PL, Mestecky J, Lamm ME et al, eds. Mucosal Immunology. San Diego: Academic Press, 1999:381-396.
4. Targan SR, Deem RL, Liu M et al. Definition of a lamina propria T cell responsive state. Enhanced cytokine responsiveness of T cells stimulated through the CD2 pathway. J Immunol 1995; 154(2):664-75.
5. Saparov A, Elson CO, DevoreCarter D et al. Single-cell analyses of CD4+ T cells from alpha beta T cell receptor-transgenic mice: A distinct mucosal cytokine phenotype in the absence of transgene-specific antigen. Eur J Immunol 1997; 27(7):1774-81.
6. Bucy RP, Panoskaltsis-Mortari A, Huang GQ et al. Heterogeneity of single cell cytokine gene expression in clonal T cell populations. J Exp Med 1994; 180(4):1251-62.
7. Lefrancois L, Puddington L. Basic aspects of intraepithelial lymphocyte immunology. In: Ogra PL, Mestecky J, Lamm ME et al, eds. Mucosal Immunology. San Diego: Academic Press, 1999:413-428.
8. Elson CO, Sartor RB, Tennyson GS et al. Experimental models of inflammatory bowel disease. Gastroenterology 1995; 109(4):1344-67.
9. Simpson SJ, Mizoguchi E, Allen D et al. Evidence that CD4+, but not CD8+ T cells are responsible for murine interleukin-2-deficient colitis. Eur J Immunol 1995; 25(9):2618-25.
10. Okamoto S, Watanabe M, Yamazaki M et al. A synthetic mimetic of CD4 is able to suppress disease in a rodent model of immune colitis. Eur J Immunol 1999; 29(1):355-66.
11. Powrie F, Leach MW, Mauze S et al. Inhibition of Th1 responses prevents inflammatory bowel disease in scid mice reconstituted with CD45RBhi CD4+ T cells. Immunity 1994; 1(7):553-62.

12. Cong Y, Brandwein SL, McCabe RP et al. CD4+ T cells reactive to enteric bacterial antigens in spontaneously colitic C3H/HeJBir mice: Increased T helper cell type 1 response and ability to transfer disease. J Exp Med 1998; 187(6):855-64.
13. Simpson SJ, Shah S, Comiskey M et al. T cell-mediated pathology in two models of experimental colitis depends predominantly on the interleukin 12/Signal transducer and activator of transcription (Stat)-4 pathway, but is not conditional on interferon gamma expression by T cells. J Exp Med 1998; 187(8):1225-34.
14. Boirivant M, Fuss IJ, Chu A et al. Oxazolone colitis: A murine model of T helper cell type 2 colitis treatable with antibodies to interleukin 4. J Exp Med 1998; 188(10):1929-39.
15. Mizoguchi A, Mizoguchi E, Bhan AK. The critical role of interleukin 4 but not interferon gamma in the pathogenesis of colitis in T-cell receptor alpha mutant mice. Gastroenterology 1999; 116(2):320-6.
16. Iqbal N, Oliver JR, Wagner FH et al. T helper 1 and T helper 2 cells are pathogenic in an antigen-specific model of colitis. J Exp Med 2002; 195(1):71-84.
17. Mullin G, Maycon ZR, Sampat A et al. Intestinal mucosal lymphokine production in Crohn's disease has a Th1 profile. Gastroenterology 1993; 104:A750.
18. Sundberg JP, Elson CO, Bedigian H et al. Spontaneous, heritable colitis in a new substrain of C3H/HeJ mice. Gastroenterology 1994; 107(6):1726-35.
19. Fuss IJ, Neurath M, Boirivant M et al. Disparate CD4+ lamina propria (LP) lymphokine secretion profiles in inflammatory bowel disease. Crohn's disease LP cells manifest increased secretion of IFN-gamma, whereas ulcerative colitis LP cells manifest increased secretion of IL-5. J Immunol 1996; 157(3):1261-70.
20. Monteleone G, Biancone L, Marasco R et al. Interleukin 12 is expressed and actively released by Crohn's disease intestinal lamina propria mononuclear cells. Gastroenterology 1997; 112(4):1169-78.
21. Strober W, Fuss IJ, Ehrhardt RO et al. Mucosal immunoregulation and inflammatory bowel disease: New insights from murine models of inflammation. Scand J Immunol 1998; 48(5):453-8.
22. Kontoyiannis D, Pasparakis M, Pizarro TT et al. Impaired on/off regulation of TNF biosynthesis in mice lacking TNF AU-rich elements: Implications for joint and gut-associated immunopathologies. Immunity 1999; 10(3):387-98.
23. Kosiewicz MM, Nast CC, Krishnan A et al. Th1-type responses mediate spontaneous ileitis in a novel murine model of Crohn's disease. J Clin Invest 2001; 107(6):695-702.
24. Mizoguchi A, Mizoguchi E, Chiba C et al. Role of appendix in the development of inflammatory bowel disease in TCR-alpha mutant mice. J Exp Med 1996; 184(2):707-15.
25. Bhan AK, Mizoguchi E, Smith RN et al. Spontaneous chronic colitis in TCR alpha-mutant mice; an experimental model of human ulcerative colitis. Int Rev Immunol 2000; 19(1):123-38.
26. Bluestone JA. New perspectives of CD28-B7-mediated T cell costimulation. Immunity 1995; 2(6):555-9.
27. Boussiotis VA, Freeman GJ, Gribben JG et al. The role of B7-1/B7-2: CD28/CLTA-4 pathways in the prevention of anergy, induction of productive immunity and down-regulation of the immune response. Immunol Rev 1996; 153:5-26.
28. Cella M, Scheidegger D, Palmer-Lehmann K et al. Ligation of CD40 on dendritic cells triggers production of high levels of interleukin-12 and enhances T cell stimulatory capacity: T-T help via APC activation. J Exp Med 1996; 184(2):747-52.
29. Chen AI, McAdam AJ, Buhlmann JE et al. Ox40-ligand has a critical costimulatory role in dendritic cell: T cell interactions. Immunity 1999; 11(6):689-98.
30. McAdam AJ, Chang TT, Lumelsky AE et al. Mouse inducible costimulatory molecule (ICOS) expression is enhanced by CD28 costimulation and regulates differentiation of CD4+ T cells. J Immunol 2000; 165(9):5035-40.
31. Fiorentino DF, Bond MW, Mosmann TR. Two types of mouse T helper cell. IV. Th2 clones secrete a factor that inhibits cytokine production by Th1 clones. J Exp Med 1989; 170(6):2081-95.
32. Glimcher LH, Murphy KM. Lineage commitment in the immune system: The T helper lymphocyte grows up. Genes Dev 2000; 14(14):1693-711.
33. Murphy KM, Reiner SL. The lineage decisions of helper T cells. Nat Rev Immunol 2002; 2(12):933-44.
34. Robinson DS, O'Garra A. Further checkpoints in Th1 development. Immunity 2002; 16(6):755-8.
35. Farrar JD, Asnagli H, Murphy KM. T helper subset development: Roles of instruction, selection, and transcription. J Clin Invest 2002; 109(4):431-5.
36. Groux H, O'Garra A, Bigler M et al. A CD4+ T-cell subset inhibits antigen-specific T-cell responses and prevents colitis. Nature 1997; 389(6652):737-42.
37. O'Garra A. Cytokines induce the development of functionally heterogeneous T helper cell subsets. Immunity 1998; 8(3):275-83.

38. Barrat FJ, Cua DJ, Boonstra A et al. In vitro generation of interleukin 10-producing regulatory CD4(+) T cells is induced by immunosuppressive drugs and inhibited by T helper type 1 (Th1)- and Th2-inducing cytokines. J Exp Med 2002; 195(5):603-16.
39. Magram J, Connaughton SE, Warrier RR et al. IL-12-deficient mice are defective in IFN gamma production and type 1 cytokine responses. Immunity 1996; 4(5):471-81.
40. Wu C, Ferrante J, Gately MK et al. Characterization of IL-12 receptor beta1 chain (IL-12Rbeta1)-deficient mice: IL-12Rbeta1 is an essential component of the functional mouse IL-12 receptor. J Immunol 1997; 159(4):1658-65.
41. Kuhn R, Rajewsky K, Muller W. Generation and analysis of interleukin-4 deficient mice. Science 1991; 254(5032):707-10.
42. Kopf M, Le Gros G, Coyle AJ et al. Immune responses of IL-4, IL-5, IL-6 deficient mice. Immunol Rev 1995; 148:45-69.
43. Mohrs M, Ledermann B, Kohler G et al. Differences between IL-4- and IL-4 receptor alpha-deficient mice in chronic leishmaniasis reveal a protective role for IL-13 receptor signaling. J Immunol 1999; 162(12):7302-8.
44. Kuhn R, Lohler J, Rennick D et al. Interleukin-10-deficient mice develop chronic enterocolitis. Cell 1993; 75(2):263-74.
45. Shull MM, Ormsby I, Kier AB et al. Targeted disruption of the mouse transforming growth factor-beta 1 gene results in multifocal inflammatory disease. Nature 1992; 359(6397):693-9.
46. Kulkarni AB, Huh CG, Becker D et al. Transforming growth factor beta 1 null mutation in mice causes excessive inflammatory response and early death. Proc Natl Acad Sci USA 1993; 90(2):770-4.
47. Wang ZE, Reiner SL, Zheng S et al. CD4+ effector cells default to the Th2 pathway in interferon gamma-deficient mice infected with Leishmania major. J Exp Med 1994; 179(4):1367-71.
48. Barbulescu K, Becker C, Schlaak JF et al. IL-12 and IL-18 differentially regulate the transcriptional activity of the human IFN-gamma promoter in primary CD4+ T lymphocytes. J Immunol 1998; 160(8):3642-7.
49. Akira S. The role of IL-18 in innate immunity. Curr Opin Immunol 2000; 12(1):59-63.
50. Pien GC, Satoskar AR, Takeda K et al. Cutting edge: Selective IL-18 requirements for induction of compartmental IFN-gamma responses during viral infection. J Immunol 2000; 165(9):4787-91.
51. Grunig G, Warnock M, Wakil AE et al. Requirement for IL-13 independently of IL-4 in experimental asthma. Science 1998; 282(5397):2261-3.
52. Foster PS, Martinez-Moczygemba M, Huston DP et al. Interleukins-4, -5, and -13: Emerging therapeutic targets in allergic disease. Pharmacol Ther 2002; 94(3):253-64.
53. Wills-Karp M, Chiaramonte M. Interleukin-13 in asthma. Curr Opin Pulm Med 2003; 9(1):21-7.
54. O'Garra A. Commitment factors for T helper cells. Curr Biol 2000; 10(13):R492-4.
55. Rengarajan J, Szabo SJ, Glimcher LH. Transcriptional regulation of Th1/Th2 polarization. Immunol Today 2000; 21(10):479-83.
56. Thierfelder WE, van Deursen JM, Yamamoto K et al. Requirement for Stat4 in interleukin-12-mediated responses of natural killer and T cells. Nature 1996; 382(6587):171-4.
57. Heath VL, Kurata H, Lee HJ et al. Checkpoints in the regulation of T helper 1 responses. Curr Top Microbiol Immunol 2002; 266:23-39.
58. Takeda K, Tanaka T, Shi W et al. Essential role of Stat6 in IL-4 signalling. Nature 1996; 380(6575):627-30.
59. Ouyang W, Lohning M, Gao Z et al. Stat6-independent GATA-3 autoactivation directs IL-4-independent Th2 development and commitment. Immunity 2000; 12(1):27-37.
60. Kurata H, Lee HJ, O'Garra A et al. Ectopic expression of activated Stat6 induces the expression of Th2-specific cytokines and transcription factors in developing Th1 cells. Immunity 1999; 11(6):677-88.
61. Zhu J, Guo L, Watson CJ et al. Stat6 is necessary and sufficient for IL-4's role in Th2 differentiation and cell expansion. J Immunol 2001; 166(12):7276-81.
62. Szabo SJ, Kim ST, Costa GL et al. A novel transcription factor, T-bet, directs Th1 lineage commitment. Cell 2000; 100(6):655-69.
63. Das J, Chen CH, Yang L et al. A critical role for NF-kappa B in GATA3 expression and TH2 differentiation in allergic airway inflammation. Nat Immunol 2001; 2(1):45-50.
64. Szabo SJ, Sullivan BM, Stemmann C et al. Distinct effects of T-bet in TH1 lineage commitment and IFN-gamma production in CD4 and CD8 T cells. Science 2002; 295(5553):338-42.
65. Zheng W, Flavell RA. The transcription factor GATA-3 is necessary and sufficient for Th2 cytokine gene expression in CD4 T cells. Cell 1997; 89(4):587-96.
66. Lee HJ, Takemoto N, Kurata H et al. GATA-3 induces T helper cell type 2 (Th2) cytokine expression and chromatin remodeling in committed Th1 cells. J Exp Med 2000; 192(1):105-15.

67. Miaw SC, Choi A, Yu E et al. ROG, repressor of GATA, regulates the expression of cytokine genes. Immunity 2000; 12(3):323-33.
68. Usui T, Nishikomori R, Kitani A et al. GATA-3 suppresses Th1 development by downregulation of Stat4 and not through effects on IL-12Rbeta2 chain or T-bet. Immunity 2003; 18(3):415-28.
69. Ranger AM, Hodge MR, Gravallese EM et al. Delayed lymphoid repopulation with defects in IL-4-driven responses produced by inactivation of NF-ATc. Immunity 1998; 8(1):125-34.
70. Ho IC, Lo D, Glimcher LH. c-maf promotes T helper cell type 2 (Th2) and attenuates Th1 differentiation by both interleukin 4-dependent and -independent mechanisms. J Exp Med 1998; 188(10):1859-66.
71. Rengarajan J, Tang B, Glimcher LH. NFATc2 and NFATc3 regulate T(H)2 differentiation and modulate TCR-responsiveness of naive T(H)cells. Nat Immunol 2002; 3(1):48-54.
72. Read S, Malmstrom V, Powrie F. Cytotoxic T lymphocyte-associated antigen 4 plays an essential role in the function of CD25(+)CD4(+) regulatory cells that control intestinal inflammation. J Exp Med 2000; 192(2):295-302.
73. Takahashi T, Tagami T, Yamazaki S et al. Immunologic self-tolerance maintained by CD25(+)CD4(+) regulatory T cells constitutively expressing cytotoxic T lymphocyte-associated antigen 4. J Exp Med 2000; 192(2):303-10.
74. Shimizu J, Yamazaki S, Takahashi T et al. Stimulation of CD25(+)CD4(+) regulatory T cells through GITR breaks immunological self-tolerance. Nat Immunol 2002; 3(2):135-42.
75. Groux H. Type 1 T-regulatory cells: Their role in the control of immune responses. Transplantation 2003; 75(9 Suppl):8S-12S.
76. O'Garra A, Vieira P. Twenty-first century Foxp3. Nat Immunol 2003; 4(4):304-6.
77. Fontenot JD, Gavin MA, Rudensky AY. Foxp3 programs the development and function of CD4+CD25+ regulatory T cells. Nat Immunol 2003; 4(4):330-6.
78. Hori S, Nomura T, Sakaguchi S. Control of regulatory T cell development by the transcription factor Foxp3. Science 2003; 299(5609):1057-61.
79. Khattri R, Cox T, Yasayko SA et al. An essential role for Scurfin in CD4+CD25+ T regulatory cells. Nat Immunol 2003; 4(4):337-42.
80. Parronchi P, Romagnani P, Annunziato F et al. Type 1 T-helper cell predominance and interleukin-12 expression in the gut of patients with Crohn's disease. Am J Pathol 1997; 150(3):823-32.
81. Powrie F, Leach MW, Mauze S et al. Phenotypically distinct subsets of CD4+ T cells induce or protect from chronic intestinal inflammation in C. B-17 scid mice. Int Immunol 1993; 5(11):1461-71.
82. Brimnes J, Reimann J, Mogens MH et al. Enteric bacterial antigens activate CD4+ T cells from scid mice with inflammatory bowel disease. Eur J Immunol 2001; 31:23-31.
83. Leach MW, Bean AG, Mauze S et al. Inflammatory bowel disease in C.B-17 scid mice reconstituted with the CD45RBhigh subset of CD4+ T cells. Am J Pathol 1996; 148(5):1503-15.
84. Mackay F, Browning JL, Lawton P et al. Both the lymphotoxin and tumor necrosis factor pathways are involved in experimental murine models of colitis. Gastroenterology 1998; 115(6):1464-75.
85. Aranda R, Sydora BC, McAllister PL et al. Analysis of intestinal lymphocytes in mouse colitis mediated by transfer of CD4+, CD45RBhigh T cells to SCID recipients. J Immunol 1997; 158(7):3464-73.
86. Matsuda JL, Gapin L, Sydora BC et al. Systemic activation and antigen-driven oligoclonal expansion of T cells in a mouse model of colitis. J Immunol 2000; 164(5):2797-806.
87. Wirtz S, Finotto S, Kanzler S et al. Cutting edge: Chronic intestinal inflammation in STAT-4 transgenic mice: Characterization of disease and adoptive transfer by TNF- plus IFN-gamma-producing CD4(+) T cells that respond to bacterial antigens. J Immunol 1999; 162(4):1884-1888.
88. Heller F, Fuss IJ, Nieuwenhuis EE et al. Oxazolone colitis, a Th2 colitis model resembling ulcerative colitis, is mediated by IL-13-producing NK-T cells. Immunity 2002; 17(5):629-38.
89. Mombaerts P, Mizoguchi E, Grusby MJ et al. Spontaneous development of inflammatory bowel disease in T cell receptor mutant mice. Cell 1993; 75(2):1-20.
90. Mizoguchi E, Mizoguchi A, Bhan AK. Role of cytokines in the early stages of chronic colitis in TCR alpha-mutant mice. Lab Invest 1997; 76(3):385-97.
91. Mizoguchi A, Mizoguchi E, Chiba C et al. Cytokine imbalance and autoantibody production in T cell receptor-alpha mutant mice with inflammatory bowel disease. J Exp Med 1996; 183(3):847-56.
92. Mizoguchi A, Mizoguchi E, Saubermann LJ et al. Limited CD4 T-cell diversity associated with colitis in T-cell receptor alpha mutant mice requires a T helper 2 environment. Gastroenterology 2000; 119(4):983-95.
93. Mizoguchi A, Mizoguchi E, Tonegawa S et al. Alteration of a polyclonal to an oligoclonal immune response to cecal aerobic bacterial antigens in TCRa mutant mice with inflammatory bowel disease. Int Immunol 1996; 8(9):1387-1394.

94. Mizoguchi A, Mizoguchi E, Takedatsu H et al. Chronic intestinal inflammatory condition generates IL-10-producing regulatory B cell subset characterized by CD1d upregulation. Immunity 2002; 16(2):219-30.
95. Sakaguchi S. Regulatory T cells: Key controllers of immunologic self-tolerance. Cell 2000; 101(5):455-8.
96. Roncarolo MG, Levings MK, Traversari C. Differentiation of T regulatory cells by immature dendritic cells. J Exp Med 2001; 193(2):F5-9.
97. Shevach EM. Certified professionals: CD4(+)CD25(+) suppressor T cells. J Exp Med 2001; 193(11):F41-6.
98. Cong Y, Weaver CT, Lazenby A et al. Bacterial-reactive T regulatory cells inhibit pathogenic immune responses to the enteric flora. J Immunol 2002; 169(11):6112-9.
99. Powrie F, Correa-Oliveira R, Mauze S et al. Regulatory interactions between CD45RBhigh and CD45RBlow CD4+ T cells are important for the balance between protective and pathogenic cell-mediated immunity. J Exp Med 1994; 179(2):589-600.
100. Powrie F, Carlino J, Leach MW et al. A critical role for transforming growth factor-beta but not interleukin 4 in the suppression of T helper type 1-mediated colitis by CD45RB(low) CD4+ T cells. J Exp Med 1996; 183(6):2669-74.
101. Asseman C, Mauze S, Leach MW et al. An essential role for interleukin 10 in the function of regulatory T cells that inhibit intestinal inflammation. J Exp Med 1999; 190(7):995-1004.
102. Annacker O, Pimenta-Araujo R, Burlen-Defranoux O et al. CD25+ CD4+ T cells regulate the expansion of peripheral CD4 T cells through the production of IL-10. J Immunol 2001; 166(5):3008-18.
103. Bacchetta R, Sartirana C, Levings MK et al. Growth and expansion of human T regulatory type 1 cells are independent from TCR activation but require exogenous cytokines. Eur J Immunol 2002; 32(8):2237-45.
104. Kitani A, Chua K, Nakamura K et al. Activated self-MHC-reactive T cells have the cytokine phenotype of Th3/T regulatory cell 1 T cells. J Immunol 2000; 165(2):691-702.
105. Cottrez F, Hurst SD, Coffman RL et al. T regulatory cells 1 inhibit a Th2-specific response in vivo. J Immunol 2000; 165(9):4848-53.
106. Khoo UY, Proctor IE, Macpherson AJ. CD4+ T cell down-regulation in human intestinal mucosa: Evidence for intestinal tolerance to luminal bacterial antigens. J Immunol 1997; 158(8):3626-34.
107. Fuss IJ, Boirivant M, Lacy B et al. The interrelated roles of TGF-beta and IL-10 in the regulation of experimental colitis. J Immunol 2002; 168(2):900-8.
108. Asano M, Toda M, Sakaguchi N et al. Autoimmune disease as a consequence of developmental abnormality of a T cell subpopulation. J Exp Med 1996; 184(2):387-96.
109. Kuniyasu Y, Takahashi T, Itoh M et al. Naturally anergic and suppressive CD25(+)CD4(+) T cells as a functionally and phenotypically distinct immunoregulatory T cell subpopulation. Int Immunol 2000; 12(8):1145-55.
110. Suri-Payer E, Amar AZ, Thornton AM et al. CD4+CD25+ T cells inhibit both the induction and effector function of autoreactive T cells and represent a unique lineage of immunoregulatory cells. J Immunol 1998; 160(3):1212-8.
111. McHugh RS, Whitters MJ, Piccirillo CA et al. CD4(+)CD25(+) immunoregulatory T cells: Gene expression analysis reveals a functional role for the glucocorticoid-induced TNF receptor. Immunity 2002; 16(2):311-23.
112. Sakaguchi S. Policing the regulators. Nat Immunol 2001; 2(4):283-4.
113. Shevach EM. CD4+ CD25+ suppressor T cells: More questions than answers. Nat Rev Immunol 2002; 2(6):389-400.
114. Piccirillo CA, Shevach EM. Cutting edge: Control of CD8+ T cell activation by CD4+CD25+ immunoregulatory cells. J Immunol 2001; 167(3):1137-40.
115. Chai JG, Tsang JY, Lechler R et al. CD4+CD25+ T cells as immunoregulatory T cells in vitro. Eur J Immunol 2002; 32(8):2365-75.
116. Seo SJ, Fields ML, Buckler JL et al. The impact of T helper and T regulatory cells on the regulation of anti-double-stranded DNA B cells. Immunity 2002; 16(4):535-46.
117. Xu D, Liu H, Komai-Koma M et al. CD4+CD25+ regulatory T cells suppress differentiation and functions of Th1 and Th2 cells, Leishmania major infection, and colitis in mice. J Immunol 2003; 170(1):394-9.
118. Sakaguchi S, Sakaguchi N, Shimizu J et al. Immunologic tolerance maintained by CD25+ CD4+ regulatory T cells: Their common role in controlling autoimmunity, tumor immunity, and transplantation tolerance. Immunol Rev 2001; 182:18-32.
119. Maloy KJ, Salaun L, Cahill R et al. CD4+CD25+ T(R) cells suppress innate immune pathology through cytokine-dependent mechanisms. J Exp Med 2003; 197(1):111-9.

120. Lehmann J, Huehn J, de la Rosa M et al. Expression of the integrin alpha Ebeta 7 identifies unique subsets of CD25+ as well as CD25- regulatory T cells. Proc Natl Acad Sci USA 2002; 99(20):13031-6.
121. Liu Z, Tugulea S, Cortesini R et al. Specific suppression of T helper alloreactivity by allo-MHC class I-restricted CD8+CD28- T cells. Int Immunol 1998; 10(6):775-83.
122. Chang CC, Ciubotariu R, Manavalan JS et al. Tolerization of dendritic cells by T(S) cells: The crucial role of inhibitory receptors ILT3 and ILT4. Nat Immunol 2002; 3(3):237-43.
123. Campbell NA, Kim HS, Blumberg RS et al. The nonclassical class I molecule CD1d associates with the novel CD8 ligand gp180 on intestinal epithelial cells. J Biol Chem 1999; 274(37):26259-65.
124. Allez M, Brimnes J, Dotan I et al. Expansion of CD8+ T cells with regulatory function after interaction with intestinal epithelial cells. Gastroenterology 2002; 123(5):1516-26.
125. Cong Y, Weaver CT, Nguyen H et al. CD8+ T cells, but not B cells inhibit enteric bacterial antigen-specific CD4+ T cell-induced colitis. Gastroenterology 1999; 116:A690.
126. Saubermann LJ, Beck P, De Jong YP et al. Activation of natural killer T cells by alpha-galactosylceramide in the presence of CD1d provides protection against colitis in mice. Gastroenterology 2000; 119(1):119-28.
127. Morrissey PJ, Charrier K, Braddy S et al. CD4+ T cells that express high levels of CD45RB induce wasting disease when transferred into congenic severe combined immunodeficient mice. Disease development is prevented by cotransfer of purified CD4+ T cells. J Exp Med 1993; 178:237-244.
128. Sadlack B, Merz H, Schorle H et al. Ulcerative colitis-like disease in mice with a disrupted interleukin-2 gene. Cell 1993; 75(2):253-61.
129. Contractor NV, Bassiri H, Reya T et al. Lymphoid hyperplasia, autoimmunity, and compromised intestinal intraepithelial lymphocyte development in colitis-free gnotobiotic IL-2-deficient mice. J Immunol 1998; 160(1):385-94.
130. Bamias G, Marini M, Moskaluk CA et al. Down-regulation of intestinal lymphocyte activation and Th1 cytokine production by antibiotic therapy in a murine model of Crohn's disease. J Immunol 2002; 169(9):5308-14.
131. McCracken VJ, Lorenz RG. The gastrointestinal ecosystem: A precarious alliance among epithelium, immunity and microbiota. Cell Microbiol 2001; 3(1):1-11.
132. Panwala CM, Jones JC, Viney JL. A novel model of inflammatory bowel disease: Mice deficient for the multiple drug resistance gene, mdr1a, spontaneously develop colitis. J Immunol 1998; 161(10):5733-44.
133. Becker C, Wirtz S, Blessing M et al. Constitutive p40 promoter activation and IL-23 production in the terminal ileum mediated by dendritic cells. J Clin Invest 2003; 112(5):693-706.
134. Bristol IJ, Farmer MA, Cong Y et al. Heritable susceptibility for colitis in mice induced by IL-10 deficiency. Inflammatory Bowel Disease 2000; 6(4):290-302.
135. McGuirk P, McCann C, Mills KH. Pathogen-specific T regulatory 1 cells induced in the respiratory tract by a bacterial molecule that stimulates interleukin 10 production by dendritic cells: A novel strategy for evasion of protective T helper type 1 responses by Bordetella pertussis. J Exp Med 2002; 195(2):221-31.
136. Cong Y, Konrad A, Iqbal N et al. Probiotics and immune regulation of inflammatory bowel diseases. Curr Drug Targets Inflamm Allergy 2003; 2(2):145-154.
137. Rath HC, Herfarth HH, Ikeda JS et al. Normal luminal bacteria, especially bacteroides species, mediate chronic colitis, gastritis, and arthritis in HLA-B27/human beta2 microglobulin transgenic rats. J Clin Invest 1996; 98(4):945-53.
138. Schultz M, Veltkamp C, Dieleman LA et al. Lactobacillus plantarum 299V in the treatment and prevention of spontaneous colitis in interleukin-10-deficient mice. Inflamm Bowel Dis 2002; 8(2):71-80.
139. Aattour N, Bouras M, Tome D et al. Oral ingestion of lactic-acid bacteria by rats increases lymphocyte proliferation and interferon-gamma production. Br J Nutr 2002; 87(4):367-73.
140. Shida K, Makino K, Morishita A et al. Lactobacillus casei inhibits antigen-induced IgE secretion through regulation of cytokine production in murine splenocyte cultures. Int Arch Allergy Immunol 1998; 115(4):278-87.
141. von der Weid T, Bulliard C, Schiffrin EJ. Induction by a lactic acid bacterium of a population of CD4(+) T cells with low proliferative capacity that produce transforming growth factor beta and interleukin-10. Clin Diagn Lab Immunol 2001; 8(4):695-701.
142. Chen Y, Kuchroo VK, Inobe J et al. Regulatory T cell clones induced by oral tolerance: suppression of autoimmune encephalomyelitis. Science 1994; 265(5176):1237-40.

Chapter 10

The B-Cell System in Inflammatory Bowel Disease

Per Brandtzaeg, Hege S. Carlsen and Trond S. Halstensen

Abstract

Secretory immunity is the best-defined part of the mucosal immune system. This adaptive humoral defense mechanism depends on a fine-tuned cooperation between secretory epithelia and local plasma cells. Such mucosal immunocytes produce preferentially dimers and larger polymers of immunoglobulin A (collectively called pIgA), which contain J chain and therefore can bind to the epithelial secretory component (SC). This transmembrane glycoprotein functions as pIg receptor (pIgR) that also translocates pentameric IgM to the epithelial surface. B cells with a high level of J-chain expression and pIg-pIgR interactions at mucosal effector sites are thus necessary for the generation of secretory antibodies (SIgA and SIgM).

Secretory antibodies perform immune exclusion in a first-line defense, thereby counteracting microbial colonization and mucosal penetration of soluble antigens. However, local production of pIgA is significantly down-regulated in inflammatory bowel disease (IBD), as revealed by strikingly decreased J-chain expression. Although the total increase of the immunocyte population in IBD lesions probably compensates for the relatively reduced pIgA production, decreased pIgR/SC expression in regenerating and dysplastic epithelium signifies that the SIgA system is topically deficient. There is, moreover, a significant shift from IgA2 to IgA1 production, the latter subclass being less resistant to proteolytic degradation. These changes—together with activation of mucosal macrophages and a dramatic increase of IgG-producing cells—may reflect local establishment of a second defense line which, however, is unsuccessful in its attempt to eliminate antigens derived from the indigenous microbial flora. Such a 'frustrated' local humoral immune system results in altered immunological homeostasis and jeopardized mucosal integrity.

Complement activation observed in relation to epithelium-bound IgG1 in ulcerative colitis indicates, moreover, that the surface epithelium is subjected to immunological attack by an autoimmune reaction. These luminal deposits regularly contain terminal cytotoxic complement, and often also C3b as a sign of persistent activation. Comparison of identical twins, discordant with regard to ulcerative colitis, suggests that the markedly skewed local IgG1 response seen in this IBD entity may be genetically determined.

The initial event(s) eliciting B-cell driven immunopathology in IBD remains unknown. Abrogation of oral tolerance to certain antigens from commensal bacteria has been suggested as a putative early mechanism, and lymphoid neogenesis and hyperplasia in the lesions most likely signify massive microbial overstimulation of the local B-cell system. Such ectopic lymphoid microcompartments may contribute substantially to the proinflammatory systemic-type of B-cell responses occurring in established IBD lesions.

Immune Mechanisms in Inflammatory Bowel Disease, edited by Richard S. Blumberg and Markus F. Neurath.

Introduction

Studies on the pathogenesis of inflammatory bowel disease (IBD) should be related to current knowledge about the immunophysiology of normal human intestinal mucosa and organized gut-associated lymphoid tissue (GALT). The prevailing adaptive immune effector mechanism throughout the gut is an immunoglobulin A (IgA)-producing B-cell system that basically provides noninflammatory first-line defense functions within the mucosa and, in particular, gives rise to secretory IgA (SIgA) antibodies that perform 'immune exclusion'.[1-3] This term is coined for antibody functions at the mcosal surface—aiming to inhibit both microbial colonization and noxious antigen penetration through the epithelial barrier.

B Cell-Mediated Mucosal Homeostasis in the Normal Gut

The generation of SIgA antibodies in the gut depends on mucosal IgA-producing immunocytes (B-cell blasts and plasma cells) that accumulate in the mucosal lamina propria by selective homing mechanisms after being primed in GALT—including Peyer's patches, isolated lymphoid follicles (ILFs) and the appendix.[2,4,5] At least 80% of the body's Ig-producing cells are located in the intestinal mucosa, which therefore constitutes the largest effector organ providing adaptive humoral immunity.[1,2]

Secretory Immunity and Immune Exclusion

Most mucosal immunocytes produce dimers and larger polymers of IgA (collectively called pIgA), which contain a disulfide-linked 15-kD polypeptide called the 'joining' or J chain. Notably, the J chain is a prerequisite for active translocation of pIgA through secretory elements such as the intestinal crypts of Lieberkühn.[6,7] This external transport is mediated by the polymeric Ig receptor (pIgR), also named membrane secretory component (SC).[2] J chain-containing pentameric IgM is externally translocated by the same mechanism.[2,6]

In the human gut, pIgR is basolaterally expressed mainly by the columnar crypt cells, but some expression is also seen in the surface epithelium of the large bowel.[2,10] Cleavage of the extracellular portion of pIgR enables release of SIgA and SIgM from the apical epithelial cell membrane to the lumen. In this manner the ectodomain of pIgR is 'sacrificed' to become bound SC, which stabilizes the quaternary structure of the secretory antibodies—particularly by covalent bonding in SIgA.[2,9] Immune exclusion performed by SIgA and secretory IgM (SIgM) antibodies thus depends on an intimate cooperation between the mucosal B-cell system and the pIgR-expressing epithelium (Fig. 1), although serum-derived and locally produced IgG antibodies reaching the lumen by paracellular diffusion may also contribute.[10] IgG is rapidly degraded in the gut lumen, but the hepatic superantigen (protein Fv) may form large complexes with degraded antibodies of different specificities, thereby reinforcing their immune exclusion function.[11]

IgA-Mediated Local Homeostasis

Most of the mucosal immunocytes (70%-80%) are located in a 200-µm zone around the crypt openings, both in the small and large bowel—that is, near epithelium with a high level of pIgR/SC expression.[8] The numerous pIgA-producing cells are supposedly not only essential for secretory immunity, but may also be important for homeostasis within the mucosa by several anti-inflammatory mechanisms. IgA lacks ordinary complement-activating properties and may therefore block nonspecific biological amplification mechanisms triggered by locally produced or serum-derived IgG antibodies.[12,13] This is important in view of the fact that immune complexes are probably formed even within the normal lamina propria due to some influx of soluble antigens, particularly following food intake.[14] In vitro and in vivo experiments[15,16] have suggested that soluble antigens—after pIgA-mediated noninflammatory trapping in immune complexes—may be cleared by the secretory epithelium via pIgR-mediated translocation to the lumen (Fig. 1). Similar experiments have suggested that pIgA-containing immune complexes also can form within virus-infected secretory epithelial cells, which are able to take up corresponding antibodies via pIgR;[15] in vivo data have confirmed that this mechanism may participate in the intestinal defense against rotavirus infection.[17]

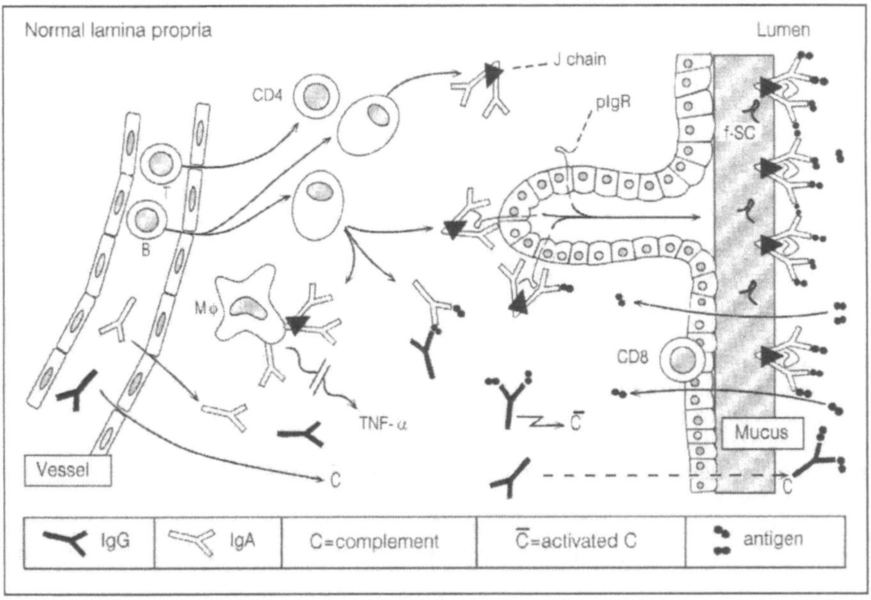

Figure 1. Normal humoral homeostasis in intestinal mucosa. Contributing variables are represented as a balance between various Ig classes (for simplicity, only IgG and IgA are indicated). Secretory IgA is generated from dimeric IgA (with associated J chain) produced by local plasma cells and transported by the polymeric Ig receptor (pIgR) to the lumen together with unoccupied cleaved receptor, or free SC (f-SC). Secretory IgA antibodies act in first-line defense by performing antigen exclusion in mucus on epithelial surface (to the right). Antigens circumventing this barrier meet corresponding serum-derived IgG antibodies in lamina propria. The formed immune complexes may activate complement, and the resulting inflammatory mediators probably cause temporarily increased paracellular leakage of IgG antibodies through the epithelium (broken arrow). Sustained inflammation is normally inhibited by blocking antibody activities exerted in the lamina propria by serum-derived and locally produced IgA (competition for antigen depicted), which independent of antibody specificity also may inhibit mediator release (TNF-α depicted) from activated phagocytic cells such as macrophages (Mφ). Moreover, antigens bound to dimeric IgA antibody may be returned in a noninflammatory way to the gut lumen by the pIgR-mediated transport mechanism as indicated. Adapted from reference 120.

Locally produced IgA may further influence mucosal homeostasis by interacting with the Fcα receptor (CD89) on leukocytes in the lamina propria. First, monomeric IgA—and particularly pIgA or IgA-containing immune complexes—are able to suppress attraction of neutrophils, eosinophils and monocytes—thereby potentially reducing the availability of the many proinflammatory activities of such cells.[13] Second, IgA can apparently downregulate the secretion of proinflammatory cytokines such as tumor necrosis factor (TNF)-α from activated monocytes.[18] However, it is uncertain whether this homeostatic mechanism operates in the normal gut (Fig. 1), because mucosal macrophages do not express detectable surface CD89—at least not in the small intestine.[19,20] Third, neutrophil and monocyte activation that results in generation of reactive oxygen metabolites ('respiratory burst'), is also reportedly inhibited by IgA.[21] Conversely, pIgA or aggregated monomeric IgA may temporarily trigger monocytes to enhanced activity—including TNF-α secretion[22]—and IgA (particularly SIgA) appears to be a potent activator of eosinophils.[23,24] These in vitro results suggest that the participation of IgA in mucosal homeostasis is quite fine-tuned—although perhaps being partly skewed towards a proinflammatory potential in IBD lesions, which contain increased numbers of CD89$^+$granulocytes[20] and also recently recruited monocyte-like macrophages with expression of CD14, the lipopolysaccharide (LPS) receptor,[25] as well as TLR2 and TLR4 (Toll-like receptors).[26]

Humoral Immunity and Immune Exclusion in IBD Lesions

Established IBD lesions are dominated by an excessive number of immunocytes—remarkably skewed towards IgG production—depending on the severity of inflammation.[1,8,13,27] This shift from the normal mucosal pIgA predominance may initially be beneficial as a powerful second line of defense because IgG antibodies can efficiently mediate immune elimination via phagocytosis and antibody-dependent cell-mediated cytotoxicity (ADCC) against microorganisms. However, the chronicity of IBD lesions signifies failure of antigen elimination with persistence of various nonspecific biological amplification mechanisms. This adverse development involves sustained inflammation and tissue damage, featured as severely altered homeostasis and defective mucosal barrier function (Fig. 2). Thus, whereas application of fluorescent in situ hybridization (FISH) on tissue sections from normal colon reportedly reveals no microorganisms, 83% of UC and 25% of colonic CD specimens show mucosal invasion of a large variety of commensal bacteria.[28] In UC a proinflammatory antimicrobial reaction is additionally promoted by a significant shift of the local IgG overproduction towards the highly complement-activating IgG1 subclass[27,29]—apparently reflecting a genetic impact as revealed by comparing identical twins—healthy or afflicted with UC.[30]

In parallel with the increased IgG production, the J-chain expression of mucosal immunocytes is decreased both in CD and UC.[31] In both types of active inflammatory colonic lesions there is also a shift from the IgA2 to the less stable IgA1 subclass;[27,32] more than 50% of the increased mucosal IgA1 immunocyte subset is J chain-deficient—therefore

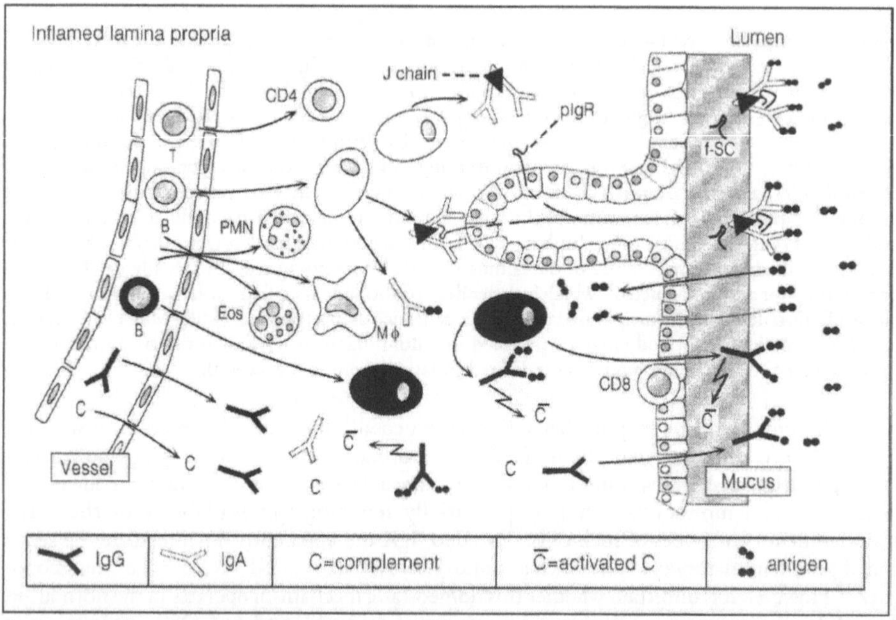

Figure 2. Altered humoral homeostasis in IBD lesion. For comparison with normal mucosal homeostasis, see Figure 1. Immunological imbalance develops when there is distorted local B-cell differentiation because of unselective leukocyte extravasation and increased epithelial permeability with excessive antigen/mitogen exposure of the immune system. To eliminate foreign material from the lamina propria, a second line of mucosal defense is generated, including exudation and local production of IgG as well as extravasation of polymorphonuclear granulocytes (PMN), eosinophils (Eos), and monocyte-derived macrophages (Mφ). Because IgG antibodies exhibit proinflammatory properties, a vicious circle develops with further increase of antigen penetrability, massive attraction of inflammatory cells, and release of proteolytic enzymes as well as other inflammatory mediators (not shown). Adapted from reference 120.

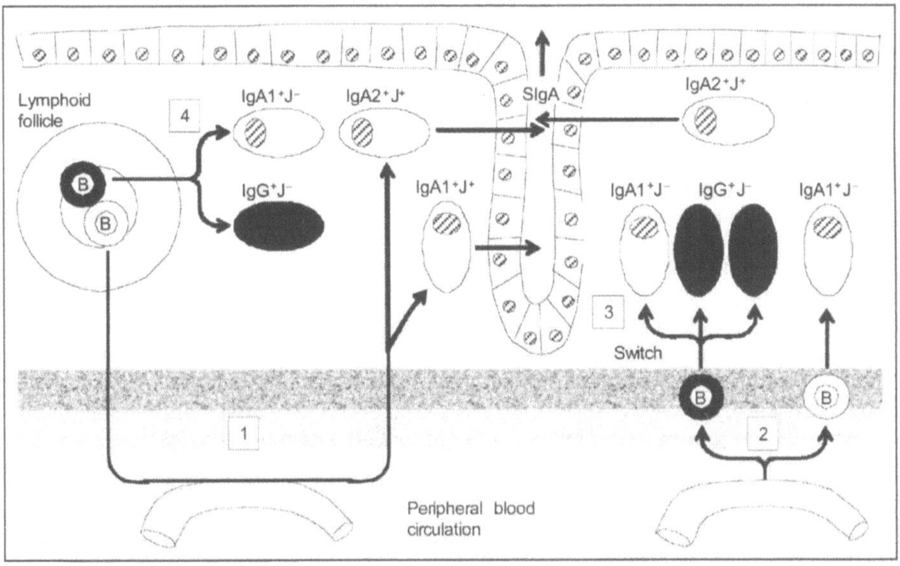

Figure 3. Putative mechanisms that might explain the observed skewing of local plasma cells from a mucosal immunophenotype (J^+) in normal lamina propria (Fig. 1) to a systemic immunophenotype (J^-) in IBD lesions. (1) Normal endothelial recognition mechanisms allow extravasation of B cells belonging to the mucosal immune system ($IgA2^+J^+$ and $IgA1^+J^+$). (2) In IBD, endothelial recognition mechanisms are altered from a mucosal to an inflammatory type, which allows extravasation of B cells with systemic immunophenotype ($IgA1^+J^-$ and IgG^+J^-). (3) Switching of heavy-chain genes may generate $IgA1^+J^-$ cells locally from IgG^+J^- precursors of systemic immunophenotype. (4) Sustained stimulation of lymphoid follicles may enhance local generation of mature B-cell clones with a systemic immunophenotype shown by downregulated J-chain expression ($IgA1^+J^-$ and IgG^+J^-), in contrast to the normally dominating generation of early memory clones in GALT with a high level of J-chain expression and mucosal homing properties (1). Modified from reference 1.

producing monomers that cannot be translocated externally by pIgR.[27,33] The same is true for a substantial fraction (25%-35%) of the expanded IgA2-producing subset.[33] These adverse local B-cell alterations supposedly reflect less restricted leukocyte extravasation due to a changed profile of adhesion molecules on the mucosal microvascular endothelium, allowing B cells expressing characteristics of systemic immunity to enter the lesion (Fig. 3).

The overall local pIgA production is, nevertheless, quantitatively maintained in IBD because of a great increase in the total number of immunocytes in the lesions.[33] However, reactive and dysplastic epithelial elements show reduced pIgR/SC expression,[34] which results in patchy defects of SIgA and SIgM secretion and thereby reduced secretory immunity.[13] The consequences of such topically decreased immune exclusion capacity for the chonicity of IBD remain elusive, but a deficient mucosal barrier function may promote bacterial invasion.[28] It is of further note that rectal exposure to an array of food antigens demonstrated increased mucosal uptake and sensitization in CD patients compared with normal controls.[35] This finding harmonizes with the increased mucosal leakiness exhibited by pIgR/SC knockout mice, which lack SIgA and SIgM.[36] Finally, a strikingly increased frequency of CD is observed in subjects with selective IgA deficiencies (L. Hammarström, personal communication).

Local Production of Antibodies to the Indigenous Gut Flora in IBD

Some knowledge exists about the antimicrobial specificities of local humoral immunity in IBD. Antibody production against fecal anaerobic bacteria has been reported, and the IgG response was found to be especially prominent in UC lesions.[37] No local response was detected against fecal aerobic bacteria, although corresponding antibodies were present in serum. Electroblotting has also demonstrated a much more prominent IgG activity against fecal microbial antigens in mucosal IBD extracts than in corresponding serum samples.[38] Isolated intestinal mononuclear cells have been shown to produce antibodies to common fecal *Escherichia coli* strains, and this IgG response was particularly elevated in CD.[39] Moreover, the serum titers to Gram-positive anaerobic coccoid rods such as *Eubacterium, Peptostreptococcus* and *Coprococcus* spp. are generally found to be higher in CD than in UC and other diarrheal diseases.[40,41]

Break of Oral Tolerance against Indigenous Bacteria

The above findings have more recently been extended at the mucosal level for apparently locally produced IgG against cytoplasmic antigens from a range of Gram-positive as well as Gram-negative commensal fecal bacteria; the antibody activities were reportedly higher in CD than in UC, and higher in UC than in other types of intestinal inflammation.[42] Thus, nonspecific mucosal damage and bacterial invasion alone[28] did not seem to explain the intensified local IgG response to the commensal gut flora. Studies in rodents have shown that indigenous intestinal bacteria normally are poorly stimulatory for the B-cell system.[43,44] One explanation might be that the enteric flora induces waves of self-limiting SIgA responses in GALT while permanently colonizing the gut.[45] Such intermittent immune exclusion could contribute to the partial tolerance that apparently exists towards autologous normal resident bacteria in the gut.[46] This hyporesponsiveness or oral tolerance phenomenon is clearly abrogated in human IBD,[47,48] which agrees with several experimental models for intestinal inflammation in rodents (see below).

At the local level, such break of putative tolerance against the indigenous microbiota is apparently signified by increased in vivo antibody coating of luminal bacteria. In healthy controls, approximately 40% of fecal anaerobic bacteria are coated with IgA, 12% with IgG and 12% with IgM.[49] In IBD patients, these figures are raised to 65%, 45% and 50%, respectively, with no difference between UC and CD.[50] In addition to some leakage of antibodies from serum (Fig. 2), this result most likely reflects the markedly elevated local production of these antibody classes known to take place in active IBD[1,8,13,27]—with the relative average increase being more prominent for IgG (x30) and IgM (x2.5) than for IgA (x1.7-2). In fact, adjacent to CD ulcers the number of immunocytes is increased 100- to 200-fold for the IgG class and 8- to 12-fold for the IgM class, compared with 1.2- to 6.7-fold for the IgA class.[27] Based on analysis of serum antibodies, however, there seems to be considerable heterogeneity in microbial specificities among IBD patients; rather than a global loss of immunological tolerance against intestinal bacteria, individual subsets of patients with varying immune responses to selected microbial antigens can reportedly be identified, and the same appears to be true in relation to autoantigens.[51]

Local Production of Anti-Epithelial IgG1 Autoantibodies in Ulcerative Colitis

Hybridomas of antibody-producing cells obtained from mesenteric lymph nodes draining IBD lesions reportedly often exhibited specificities for *E. coli* and mycobacteria (heat-shock proteins?) but generally not for food and various autoantigens.[52] However, one clone showed reactivity against the cytoplasm of human colonic enterocytes although the authors did not specify whether it came from an UC or CD patient.[52] Further evidence for local production of autoantibodies to the colonic epithelium in UC was obtained with dispersed mononuclear cells from the lesions.[53] Some of these antibodies could be a sole epiphenomenon, as suggested in the IL-2 knockout mouse,[54] while others might depend on a certain genetic background.[55]

Characterization of Local Autoantibody Response

Spontaneous production of autoantibodies to colonic epithelium by mononuclear cells from active UC lesions showed that the frequency of active B cells was much higher than among peripheral blood lymphocytes.[53] Such local antibodies of the IgG1 subclass were found to be directed both against a colonic 40-kD protein fraction (p40) and against purified tropomyosin (TM) isoforms (TM5 and TM1); the antibodies were produced at significantly higher levels in UC than in CD or nonIBD lesions.[56,57] A remarkably high percentage (42%) of B cells from UC lesions has subsequently been reported to produce IgG against TM5, in contrast to a much lower figure (2.5%) in colonic CD and negative results in nonIBD controls.[58]

TM is a family of cytoskeletal microfilamental proteins with tissue-specific isoforms—TM5 being particularly expressed by the intestinal epithelium.[57] As reviewed elsewhere,[59] Das et al[60] had earlier described considerable amino-acid homology and immunological crossreactivity between TM and the colonic p40 protein fraction, which initially was proposed to contain the major autoantigen involved in UC. However, Das and coworkers more recently identified a novel colonic epithelial glycoprotein with a molecular weight of at least 200 kD, called colon epithelial protein (CEP), which appears to be responsible for much of the immunoreactivity in the p40 fraction.[59,61] Selectively expressed in the large bowel, CEP is a plasma membrane-associated protein that may act as a chaperon in bringing complexed TM5 to the surface of the epithelial cells.[59] Altogether, therefore, autoantibodies are apparently produced locally in UC both against CEP and TM5, and a biased B-cell usage of Ig heavy chain genes in the lesion may reflect preferential stimulation with an epithelial antigen.[62,63]

Putative Pathogenic Role of Anti-Epithelial IgG1 Antibodies

Suggestive evidence for a pathogenic role of antibodies against epithelial autoantigen(s) was obtained in our laboratory when we found the early complement activation product C3b to be apically localized on colonic surface epithelium in at least 90% of UC patients, usually together with the cytolytic terminal complement complex (TCC)—both deposits being quantitatively related to the topical degree of inflammation.[64] Many of the patients also showed apically bound IgG1, often together with C3b and TCC. This suggested that complement activation had been induced from the luminal side by colon-specific IgG1 autoantibodies (Fig. 4)—a possibility that was further supported by the presence of components such as C1q and C4c, which are involved in the classical antibody-dependent complement activation pathway.[65] Moreover, the immune deposition was not merely a result of inflammation because neither IgG1 nor activated complement from the classical pathway was observed on the epithelium in intensely inflamed CD lesions.[65] Instead, subepithelial C3b and TCC deposition observed in about 50% of the patients with ileal or colonic CD presumably reflected secondary alternative pathway activation.[65] Also, aggregates of bacteria coated with C3b were observed in inflamed lamina propria, mainly within CD lesions.[65]

When the apical epithelial distribution of IgG1 and activated complement on large bowel epithelium was further studied in relation to epithelial expression of CEP (presumably in complex with TM5), this antigen was observed in goblet cells of normal terminal ileum and proximal colon but not in rectal goblet cells.[66] By contrast, large bowel columnar enterocytes showed increasing reactivity in distal direction, expanding to intense cytoplasmic expression in the rectum.[66] This autoantigen distribution might explain several features of UC if the epithelium indeed is a target of pathogenic autoimmunity (Fig. 4). Thus, both 'pouchitis' and 'backwash ileitis' could depend on ileal goblet cell expression of the antigen, while its distribution in the large bowel—from being mainly found in goblet cells of the cecum to appearing exclusively in the cytoplasm and at the apical face of rectal enterocytes—could explain why UC begins, and generally is most severe, in the left colon. Furthermore, the extraintestinal distribution of the same antigen(s) might have bearing on certain complications of UC beyond the gut as reviewed elsewhere.[55,59]

Luminal availability of IgG1 autoantibodies and complement factors may be envisioned even in an early phase of IBD (Fig. 4) because the initial response of an irritated epithelium is to open up its tight junctions with a temporary profuse bulk leakage of plasma proteins.[10] Although the gut epithelium apparently can protect itself against a cytotoxic attack by producing complement regulatory proteins[67,68] such as decay-accelerating factor (DAF, CD55),

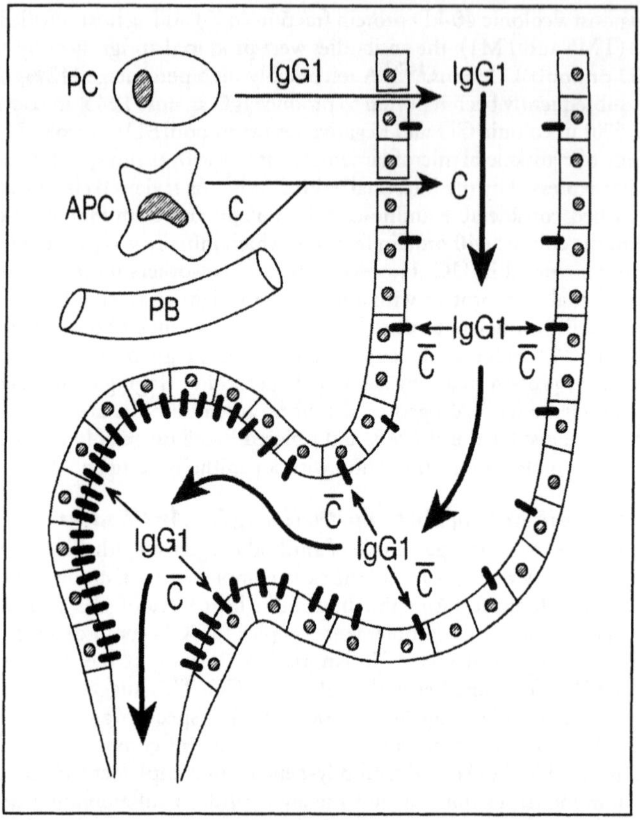

Figure 4. Schematic depiction of putative autoimmune pathogenic mechanism in ulcerative colitis. The antibody-mediated attack is directed against the luminal face of the large bowel epithelium by complement-activating IgG1 specific for apical enterocyte antigens (▬) whose expression increases in distal direction. It is indicated that IgG1 antibodies produced by lamina propria plasma cells (PC), as well as complement factors (C) derived from antigen-presenting cell (APC) or peripheral blood (PB), reach the gut lumen by paracellular leakage through the surface epithelium. $\bar{C}$ = activated complement. Adapted from reference 55.

protectin (CD59), and particularly membrane cofactor protein (MCP, CD46), antiepithelial autoimmune reactions might intensify ongoing tissue damage and explain the wide superficial ulcerations characteristic of UC. Of note, however, recent observations in IL-2 knockout mice emphasize the primary enterocolitis-driving pathogenic role of commensal bacteria; this mouse strain does not develop UC-like lesions in the distal gut when being devoid of its intestinal microflora, despite the occurrence of generalized autoimmune reactions and extraintestinal lymphoid hyperplasia.[69] Autoimmunity is most likely a bystander phenomenon also in human UC, but may nevertheless significantly intensify the pathology of this IBD entity by undermining the epithelial barrier function as alluded to above (Fig. 4).

Local Production and Diagnostic Aid of Other (Auto)Antibodies in IBD

Measurements of circulating antineutrophil cytoplasmic antibodies (ANCAs), especially of the perinuclear variety (pANCA), are considered to represent a useful diagnostic adjunct in

UC patients. These tests should preferably be combined with determinations of antibodies against baker's yeast and brewer's yeast (anti-*Saccharomyces cerevisiae* antibodies, ASCAs) as well as against anaerobic coccoid rods—both reactivities being more frequently found in CD than in UC.[70,71] A subgroup of UC patients with predisposition to sclerosing cholangitis appears, for unknown reason(s), to be most prone to pANCA production.[72] It is likewise unknown why ASCAs are preferentially produced in CD patients. These antibodies are directed against oligomannosidic epitopes present in the yeast cell wall, and may signify crossreacting antimicrobial antibodies or a response against excessive uptake of food products through a leaky intestinal barrier. In so-called indeterminate colitis, when a distinction between UC and colonic CD remains unsolved, application of tests for pANCAs and ASCAs has resulted in differentiation of CD from UC or CD-like UC cases, although no differentiation was possible between UC and UC-like CD cases.[73]

Regrettably, however, many patients (~48%) with indeterminate colitis do not have detectable pANCAs or ASCAs in serum.[73] The reason may be insufficient spillover from the IBD lesions, which probably represent the major production site. Thus, spontaneous synthesis of ANCAs by mononuclear cells from UC lesions has been reported, particularly IgG antibodies of the pANCA variety.[74] Interestingly, lymphocytes from peripheral blood of UC patients did not secrete such antibodies spontaneously or after activation, and the frequency of positive mucosal samples was only 11% in CD compared with 70% in UC.[74] This result suggested that the UC lesion favors local accumulation of B-cell clones with pANCA specificity, but a possible pathogenic role of ANCAs remains controversial.

Mechanisms for Induction of Local Proinflammatory B-Cell Responses

Based on the observations reviewed above, it may be concluded that B cells featuring a systemic type of immune response gain undue access to—and/or are generated locally in—the intestinal mucosa of IBD patients (Fig. 3), and that the accumulation of selected antimicrobial and autoimmune antibody-producing clones is determined by genetic factors as well as by local availability of corresponding stimulatory antigens. However, very limited knowledge exists about the state of the mucosal immune system prior to clinical IBD presentation. In theory, either overactivation or inadequate stimulation of GALT might predispose to later disease development in the gut.

Putative Role of the Appendix

No clear answers with regard to immunological predisposition to IBD have emerged from international case control studies, except for a decreased incidence of prior appendectomy in UC patients.[75-77] Interestingly, a recent report suggested that appendectomy at an early age (before 20 years) also protects against CD, at least by delaying its onset.[78] It remains uncertain how such clinical observations should be interpreted,[79] but it appears that the beneficial effect of appendectomy is enhanced by the presence of an inflammatory condition in the appendix or mesenteric lymph nodes,[80] implying overstimulation at these immune-inductive sites. Both the appendix and mesenteric lymph nodes are functionally part of GALT[81] and contain activated B-cell follicles that generate a substantial contingent of IgG-producing cells, often with downregulated J-chain expression as a sign of clonal maturity.[2,4] Thus, in the appendix there are approximately equal proportions of IgG and IgA immunocytes in follicle-associated lamina propria including the dome areas, and the same is true for ILFs in the colon and ileum.[82-84] Perhaps B cells derived from a similar follicular, relatively mature B-cell subset—after homing to the large bowel mucosa—could develop a proinflammatory IgG response against indigenous bacteria (and autoantigens?) that predisposes to IBD (Fig. 3).

This possibility is corroborated by the strikingly reduced gut inflammation observed in T-cell receptor-α mutant mice (TCR-$\alpha^{-/-}$) after early (<5 weeks of age) 'appendectomy' (removal of the cecal patch).[85] Interestingly, the proliferative response of B cells from the appendix of TCR-$\alpha^{-/-}$ mice is reportedly quite strong after stimulation with *E. coli* antigens, and increased levels of autoantibodies to tropomyosin are also produced.[85] Appendectomy (but not splenectomy) has likewise been shown to reduce disease severity in a dextran sulfate-induced murine model of colitis.[86] Together, these findings strongly support the suggested importance of the appendix as a site for activation of B cells involved in production of pathogenic proinflammatory antimicrobial IgG antibodies. Similar experiments with removal of the mesenteric lymph nodes would be of considerable interest, but are in practice not feasible.

Putative Role of Isolated Lymphoid Follicles

As mentioned above, ILFs in the colon and ileum (like in the appendix) generate a remarkably large number of IgG-producing cells appearing adjacent to these follicles.[84] Their contribution to proinflammatory responses even in distant lamina propria is therefore an interesting possibility (Fig. 3). The density of ILFs is normally quite small, however—only one follicle reportedly being present per 269 villi in the jejunum and one per 28 villi in the ileum.[87] In the normal large bowel, the density of ILFs is likewise relatively small—enumerated in tissue sections to increase from 0.02/mm muscularis mucosae in the ascending colon to 0.06/mm muscularis mucosae in the rectosigmoid.[88] Taking the diameter of each follicle to be ~2 mm, these data harmonize with figures obtained by inspection of the normal large bowel from the luminal side.[89] Importantly, the density is increased approximately 4.5-fold in UC and 7-fold in CD of the colon.[88]

In other histological studies, the number of ILFs per mm^2 of section area of normal colonic mucosa has been determined to be <0.1, a figure that increases to 1.3-4.9 in moderately or severely inflamed UC lesions.[90] In addition, the lymphoid structures occurring in IBD tend to be large and irregular—often being situated between the muscularis mucosae and the crypts, therefore referred to as basal lymphoid aggregates (BLAs). Such BLAs are considered as a significant discriminator of IBD *versus* acute self-limited colitis in rectal biopsies.[91] BLAs are a purely histopathological feature, and to our knowledge they can not be distinguished functionally from other types of lymphoid aggregates in IBD or from ILFs in normal colon mucosa. Thus, BLAs generally appear to be parts of a single lymphoid structure with both submucosal and mucosal extensions.[87,88] Therefore, we prefer to use the term 'irregular lymphoid aggregates' to include all organized lymphoid elements of IBD lesions. The size of the aggregates increases with the disease intensity, and in severely inflamed UC lesions they may constitute up to 45% of the section area.[90] Some 40%-80% of these aggregates contain detectable follicular dendritic cells (FDCs), suggesting that the lymphoid hyperplasia represents a true germinal center reaction.[25,90] However, as discussed above, the number of irregular lymphoid aggregates is much larger in IBD lesions than the number of ILFs in normal colon mucosa—implying that lymphoid neogenesis contributes significantly to the aberrant local B-cell system in the diseased gut.

Role of Chemokines in Formation of Lymphoid Structures

Chemokines have recently emerged as crucial orchestrators for lymphocyte migration and activation.[92] Such 'chemoattractant cytokines' exert their function by binding to cell surface receptors, and can be divided into two main categories: homeostatic and inflammatory. The former type is constitutively expressed in a certain tissue or organ, which suggests a specific function that involves normal leukocyte migration. Conversely, the inflammatory chemokines are strongly upregulated by proinflammatory or immunological stimuli and most likely participate in reactive processes by targeting effector leukocytes.[93]

Constitutively expressed in organized lymphoid tissue, several chemokines have been identified as major cues for homeostatic lymphocyte trafficking and positioning.[94,95] The CXC chemokine BCA-1 (B cell-attracting chemokine-1)/ CXCL13 (CXC chemokine ligand 13) is an efficacious attractant of naive B cells in vitro and has been shown to be produced in follicles of human lymph nodes.[96] This chemokine was concurrently described in mice[97] and called BLC (B-lymphocyte chemoattractant). Several lines of evidence suggest that BLC is directly involved in the formation of organized lymphoid tissue. Thus, ectopic expression of BLC can induce a series of events leading to the formation of extranodal B-cell follicles.[98] Also of note, mice deficient in the corresponding receptor, CXCR5 (BLR1), show impaired development of Peyer's patches, inguinal lymph nodes, and splenic follicles. Furthermore, B cells from mice lacking CXCR5 fail to enter lymphoid follicles of Peyer's patches and the spleen after transfer to wild-type littermates, despite their normal extravasation in the T-cell zones.[99] BLC-deficient mice likewise show severe lack of Peyer's patches.[100] Together, these findings document a fundamental requirement for the BLC-CXCR5 ligand-receptor pair in the development of murine GALT.

Expression of BCA-1 and CXCR5 in Normal Human Gut and IBD Lesions

In the original work of Legler et al,[96] BCA-1 mRNA transcripts were found in the human appendix and stomach but neither in the colon nor small intestine. Such disparity suggested that this homeostatic chemokine—although being crucial for the development of murine GALT—might be restricted to certain parts of human GALT, perhaps having a different and less important role than in mice. To address this possibility, we recently examined normal human GALT specimens for BCA-1 and CXCR5 expression.[101] We found that BCA-1 and CXCR5 are expressed in normal GALT structures both in the ileum (Peyer's patches) and colon (ILFs), as well as in irregular lymphoid aggregates in UC and CD lesions. These findings provided the first evidence to suggest an important role of this chemokine-receptor pair in the formation of both normal and aberrant lymphoid structures in the human gut.

The general expression of CXCR5 seen in follicular mantle zones,[101] agreed with the notion that this chemokine is a selective and highly efficacious chemoattractant for circulating naive human B lymphocytes.[96] The level of CXCR5 on these cells was relatively low, however, apparently paralleling the moderate potency of BCA-1 on in vitro migration of peripheral blood B cells.[96] Conversely, we found T cells strongly positive for CXCR5 in the B-cell follicles of GALT (Fig. 5). Such $CD4^+$ $CXCR5^+$ T cells have been functionally described as 'follicular B-helper T cells' (T_{FH} cells) because they show all the characteristics required for efficient B-cell help within the lymphoid follicles.[102-104]

A T_{FH} cell subset identified as $CD57^+$ $CXCR5^+$, has been termed germinal center T-helper (GC-Th) cells because they appear to be essential for B-cell differentiation and antibody production; this subset is localized only within the germinal centers.[105] Because we observed particularly strong expression of CXCR5 on scattered T cells restricted to the germinal centers of both normal GALT and irregular lymphoid aggregates in IBD, they most likely belonged to this subset.[101] Thus, FACS analysis of isolated lymphoid cells from murine Peyer's patches[99] and human tonsils[102,103,105] has revealed a much higher proportion of $CXCR5^+$ T cells, implicating the presence of a substantial fraction of such cells in the T-cell zones of these tissues. Our observation of $CXCR5^+$ T cells restricted to germinal centers in both GALT[101] and tonsils (unpublished observation) most likely reflects the lower sensitivity of the applied in situ immunostaining method. The prominent CXCR5 expression on GC-Th cells is probably a result of local upregulation, which could explain that this subset reportedly shows a better chemotactic response to BCA-1 than other $CXCR5^+$ T cells isolated from tonsils.[105]

Distribution and Source of BCA-1 in Normal and Inflamed Human Gut

A fibrillar meshwork positive for BCA-1 protein was detected at the periphery of lymphoid follicles in normal GALT structures including Peyer's patches.[101] Notably, we observed extensive colocalization with the extracellular matrix protein fibronectin, implying that BCA-1 was deposited on reticular fibers in the follicles where the fibrillar network is looser than in other compartments of lymphoid tissue[106]. However, in the inner mantle zone some colocalization of BCA-1 protein and the FDC markers CD21, CD23, CD35, DRC and MAdCAM-1 was

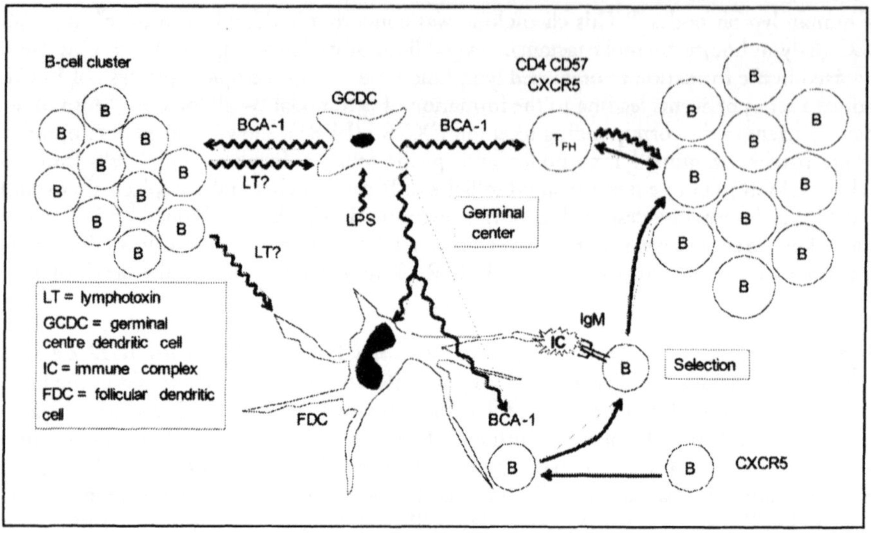

Figure 5. Simplified model for lymphoid neogenesis in IBD lesions, showing the progression from a small B-cell cluster to an organized follicle with germinal center and FDCs. As discussed in the text, the major source of BCA-1 (CXCL13) in germinal centres appears to be GCDCs, which most likely represent activated macrophages. B cells (and probably T cells or other cell types) produce LT, which may exert a reciprocal effect on germinal center and FDC formation. Both follicular B-helper T (T_{FH}) cells and B cells with a high level of CXCR5 expression are attracted to the follicle by BCA-1, but germinal center B cells downregulate CXCR5 after proliferation and selection for high antigen affinity on IC-bearing FDCs via their surface receptor (IgM). Such affinity maturation is followed by Ig class switching (not shown).

found. Also, in situ hybridization showed that BCA-1 mRNA was expressed in a pattern that strongly suggested a dendritic cellular morphology.[101] Likewise, BLC has been reported to be produced by nonlymphoid cells present in murine lymphoid follicles,[97] and its origin was taken to be FDCs, although this was not explicitly shown to be the case.[107] The partial overlapping seen for our BCA-1 immunostaining and several traditional FDC markers, might result from chemokine secretion by another cell type and deposition on peripherally located FDCs. Therefore, although an FDC-related subset of dendritic cells remains a potential source of BCA-1, no conclusive evidence exists for this possibility.

We believe that our observations in IBD lesions elucidated the relationship between BCA-1 production and FDCs in the formation of lymphoid follicles; chemokine-expressing relatively large cells did apparently develop concurrently with lymphoid neogenesis, and before the detection of classical FDCs in aberrant follicles.[108] Also notably, single cells clearly containing cytoplasmic BCA-1 protein in normal GALT germinal centers did not show the characteristics of FDCs but were rather scattered within the FDC meshwork. This immunohistochemical finding was supported by the distribution of specific mRNA signals in the same compartment.[101] Although some colocalization of BCA-1 protein and the FDC markers was seen in the inner mantle zone of IBD follicles, the BCA-1-bearing dendritic elements were mainly overlapping with extracellular fibrils as in normal GALT.[101] Moreover, large BCA-1 protein and mRNA-expressing cells occurred not only in the irregular lymphoid aggregates, but also within or near smaller collections of B cells where no FDCs could be detected.[108]

Macrophages May Drive Aberrant B-Cell Responses

It is of further note that we were able to phenotype as macrophages the scattered BCA-1-expressing cells seen in germinal centers of normal GALT as well as within and outside of irregular lymphoid aggregates in IBD lesions and in synovial tissue afflicted with rheumatoid arthritis.[108] We believe these cells are identical to the previously described germinal center dendritic cells reported to stimulate T cells in this compartment.[109] Outside of the lymphoid aggregates some of these macrophages were shown to be CD14$^+$, suggesting recent extravasation. Moreover, we found that monocytes from healthy blood donors—particularly after in vitro maturation towards macrophages—could be stimulated by LPS (endotoxin) from *E. coli* to secrete BCA-1 protein.[108] Our findings therefore imply that newly recruited macrophages play an early and important role in lymphoid neogenesis. Induction of BCA-1 production in these cells, as well as lymphotoxin (LT) release from B cells and/or other cell types,[110-113] could together provide the microenvironment required for formation of follicles and FDCs (Fig. 5).

BCA-1 expression has likewise been studied in some other human diseases such as rheumatoid arthritis[114,115] and Sjögren's syndrome.[116,117] Contrary to our findings in normal GALT and IBD lesions,[101] and more recently in rheumatoid arthritis,[108] immunostaining for BCA-1 in the latter lesion has previously been reported to be associated only with FDCs of germinal centers, while the mantle zones were negative.[114,115] In situ transcripts for BCA-1 and the cytokine LTβ suggested that both were necessary (but not sufficient) for the occurrence of FDCs in rheumatoid arthritis.[115] This would be in keeping with our results[108] consistently demonstrating that FDCs were surrounded by a reticular meshwork bearing BCA-1, which most likely in the induction phase was mainly derived from activated macrophages (Fig. 5).

As discussed above, findings in experimental animal models strongly support the notion that BLC-CXCR5 interactions are critical in the formation of organized lymphoid tissue. Our detection of both BCA-1 and CXCR5 in lymphoid aggregates formed in IBD lesions strengthens this notion,[101] and we propose a mechanism for induction of BCA at sites of chronic inflammation (Fig. 5) . It has been suggested that local production of BCA-1 and other lymphoid tissue-inducing chemokines could convert an inflammatory lesion from an acute to a chronic state, and that blocking of the chemokine activity therefore might be of clinical value.[118] IBD may thus become a potential candidate for such chemokine blockade in the future. Moreover, therapeutic targeting of monocyte extravasation could also turn out to be advantageous to this end by inhibiting local BCA-1 induction.

IBD Apparently Reflects Break of Oral Tolerance against Indigenous Bacteria

Although the initial IBD event has not been defined, it seems justified to consider the established lesions as varying clinical presentations of nonspecific inflammatory amplification due to overactivation of innate and adaptive immune mechanisms—operating on a polygenetic background in response to the commensal microbial flora.[119,120] There is increasing evidence that multiple genes are involved in the disease predisposition.[119] Genetically determined autoimmunity may, moreover, contribute to the ulcerative mucosal destruction in UC as previously discussed.[55] Confounding exogenous cofactors in the pathogenesis could be variables such as smoking, which appears to increase the risk of acquiring CD but to protect against UC,[119] and even against associated or nonassociated primary sclerosing cholangitis.[121]

Putative Pathways to Chronic Inflammation in IBD

No fundamental differences have been identified in the immunopathology of the two disease entities except for more intensive Th1-cell responses in CD and more overt humoral autoimmune responses in UC patients.[120] Although various IBD models developed in knockout and transgenic rodents strongly suggest that the major pathogenic effector mechanisms triggered by the indigenous microflora are cell-mediated,[119,122,123] humoral (B cell-driven) immunity may nevertheless play a crucial role.[27,55,120] In fact, antibody-mediated complement activation appears to be essential even for the recruitment of T cells that induce cell-mediated hypersensitivity,[124] and IgE-dependent mast cell degranulation could be involved in the pathology onset as well.[125] Moreover, in mice with deficiency of SIgA due to lack of Ig class switching, a striking hypertrophy of ILFs develops over time, apparently in response to an overgrowth of anaerobic commensal bacteria.[126] This development has some resemblance to the irregular lymphoid aggregates seen in long-standing human IBD lesions as discussed above.

Altogether, two chief theories exist to explain the immunopathological development and perpetuation of IBD: (a) some unknown (probably bacterial) antigen(s) gains undue access to, and is unsuccessfully eliminated from, the mucosa—therefore triggering sustained ('frustrated') local immune responses with resulting immunopathology; and (b) an inherent immunoregulatory defect gives rise to overreaction—or abrogation of oral tolerance—against constituents of the indigenous gut microflora. These two possibilities are difficult, if not impossible, to distinguish and could in fact be part of the same scenario because of the great concentration of bacteria in the distal gut lumen. Indeed, the take-home lesson from most disease models in gene-manipulated animals is that a predilection exists for immunopathology to occur in the distal gut when either the regulation of adaptive immunity or the intestinal barrier function is compromised.[119,122,123]

Macrophages Linking Innate to Adaptive Immunity in IBD

The association of certain mutations in the *NOD2 (CARD15)* gene with clinical subsets of CD patients,[127-132] provides strong support for the possibility that aberrant 'sensing' of the intestinal microbiota is an early pathogenic event—perhaps involving deficient signalling for downregulatory mechanisms in the gut.[133] Imbalanced triggering of innate immunity may in fact be a common theme on a polygenetic background in the pathogenesis of a spectrum of clinical IBD entities. As discussed above, activated macrophages or dendritic cells most likely represent an important link between innate immunity and an overstimulated adaptive immune system in the gut, which in various ways could contribute to mucosal pathology. Generation of aberrant lymphoid aggregates in persistent IBD lesions via interactions between BCA-1 and CXCR5-expressing B and T_{FH} cells is one of these possibilities (Fig. 5). Chronically activated B-cell follicles give rise to a contingent of mature immunocytes that not only produce proinflammatory IgG antibodies (Fig. 3), but also might contribute to immunopathology by TNF-α secretion.[134]

On the other hand, it can not be excluded that ectopic lymphoid neogenesis as seen in persistent IBD lesions, signifies a reactive mechanism aiming at homeostatic immune control. It is of note that B cell-deficient (mMT) mice show poor low-dose oral tolerance induction,[135] and regulatory B cells with distinct cytokine profiles apparently exist.[136] Interestingly, in a T cell-mediated murine colitis model (TCR-$\alpha^{-/-}$ mice), induction of a CD1d-expressing regulatory B-cell phenotype occurs both in colonic lamina propria and reactive (hyperplastic) mesenteric lymph nodes; these B cells reportedly suppress the progression of chronic gut inflammation by their IL-10 secretion.[137] Further studies are needed to investigate whether regulatory B cells likewise emerge in human IBD lesions.

Acknowledgements

Studies in the Brandtzaeg laboratory are supported by the Research Council of Norway, the Norwegian Cancer Society, Anders Jahre's Fund, and Rakel and Otto Kr. Bruun's Legacy. We thank the technical staff at LIIPAT for expert assistance. Hege Eliassen and Erik K. Hagen are gratefully acknowledged for excellent help with the manuscript.

References

1. Brandtzaeg P, Halstensen TS, Kett K et al. Immunobiology and immunopathology of human gut mucosa: Humoral immunity and intraepithelial lymphocytes. Gastroenterology 1989; 97:1562-1584.
2. Brandtzaeg, Farstad IN, Johansen F-E et al. The B-cell system of human mucosae and exocrine glands. Immunol Rev 1999; 171:45-87.
3. Brandtzaeg P, Baekkevold ES, Farstad IN et al. Regional specialization in the mucosal immune system: What happens in the microcompartments? Immunol Today 1999; 20:141-151.
4. Brandtzaeg P, Farstad IN, Haraldsen G. Regional specialization in the mucosal immune system: Primed cells do not always home along the same track. Immunol Today 1999; 20:267-277.

5. Kunkel EJ, Butcher EC. Chemokines and the tissue-specific migration of lymphocytes. Immunity 2002; 16:1-4.
6. Brandtzaeg P, Prydz H. Direct evidence for an integrated function of J chain and secretory component in epithelial transport of immunoglobulins. Nature 1984; 311:71-73.
7. Johansen F-E, Braathen R, Brandtzaeg P. The J chain is essential for polymeric Ig receptor-mediated epithelial transport of IgA. J Immunol 2001; 167:5185-5192.
8. Brandtzaeg P, Baklien K. Immunohistochemical studies of the formation and epithelial transport of immunoglobulins in normal and diseased human intestinal mucosa. Scand J Gastroenterol 1976; 11(Suppl 36):1-45.
9. Norderhaug IN, Johansen F-E, Schjerven H et al. Regulation of the formation and external transport of secretory immunoglobulins. Crit Rev Immunol 1999; 19:481-508.
10. Persson CG, Erjefalt JS, Greiff L et al. Contribution of plasma-derived molecules to mucosal immune defence, disease and repair in the airways. Scand J Immunol 1998; 47:302-313.
11. Bouvet JP, Pires R, Iscaki S et al. Nonimmune macromolecular complexes of Ig in human gut lumen. Probable enhancement of antibody functions. J Immunol 1993; 151:2562-2571.
12. Russell MW, Reinholdt J, Kilian M. Anti-inflammatory activity of human IgA antibodies and their Fab α fragments: Inhibition of IgG-mediated complement activation. Eur J Immunol 1989; 19:2243-2249.
13. Brandtzaeg P, Baklien K, Bjerke K et al. Nature and properties of the human gastrointestinal immune system. In: Miller K, Nicklin S, eds. Immunology of the Gastrointestinal Tract. Boca Raton, Florida: CRC Press, 1987:1:1-85.
14. Brandtzaeg P. Mechanisms of gastrointestinal reactions to food. Environ. Toxicol Pharmacol 1997; 4:9-24.
15. Mazanec MB, Nedrud JG, Kaetzel CS et al. A three-tiered view of the role of IgA in mucosal defense. Immunol Today 1993; 14:430-435.
16. Robinson JK, Blanchard TG, Levine AD et al. A mucosal IgA-mediated excretory immune system in vivo. J Immunol 2001; 166:3688-3692.
17. Burns JW, Siadat-Pajouh M, Krishnaney AA et al. Protective effect of rotavirus VP6-specific IgA monoclonal antibodies that lack neutralizing activity. Science 1996; 272:104-107.
18. Wolf HM, Fischer MB, Puhringer H et al. Human serum IgA downregulates the release of inflammatory cytokines (tumor necrosis factor-α, interleukin-6) in human monocytes. Blood 1994; 83:1278-1288.
19. Smith PD, Smythies LE, Mosteller-Barnum M et al. Intestinal macrophages lack CD14 and CD89 and consequently are down-regulated for LPS- and IgA-mediated activities. J Immunol 2001; 167:2651-2656.
20. Hamre R, Farstad IN, Brandtzaeg P et al. Expression and modulation of the human immunoglobulin A Fc receptor (CD89) and the FcR γ chain on myeloid cells in blood and tissue. Scand J Immunol 2003; 57:506-16.
21. Wolf HM, Vogel E, Fischer MB et al. Inhibition of receptor-dependent and receptor-independent generation of the respiratory burst in human neutrophils and monocytes by human serum IgA. Pediatr Res 1994; 36:235-243.
22. Deviere J, Vaerman JP, Content J et al. IgA triggers tumor necrosis factor α secretion by monocytes: A study in normal subjects and patients with alcoholic cirrhosis. Hepatology 1991; 13:670-675.
23. Lamkhioued B, Gounni AS, Gruart V et al. Human eosinophils express a receptor for secretory component. Role in secretory IgA-dependent activation. Eur J Immunol 1995; 25:117-125.
24. Motegi Y, Kita H. Interaction with secretory component stimulates effector functions of human eosinophils but not of neutrophils. J Immunol 1998; 161:4340-4346.
25. Rugtveit J, Bakka A, Brandtzaeg P. Differential distribution of B7.1 (CD80) and B7.2 (CD86) costimulatory molecules on mucosal macrophage subsets in human inflammatory bowel disease (IBD). Clin Exp Immunol 1997; 110:104-113.
26. Hausmann M, Kiessling S, Mestermann S et al. Toll-like receptors 2 and 4 are up-regulated during intestinal inflammation. Gastroenterology 2002; 122:1987-2000.
27. Brandtzaeg P, Halstensen TS, Kett K. Immunopathology of inflammatory bowel disease. In: MacDermott RP, Stenson WF, eds. Inflammatory Bowel Disease. Current Topics in Gastroenterology (Series: Zakim D, ed.). London: Elsevier, 1992:95-136.
28. Kleessen B, Kroesen AJ, Buhr HJ et al. Mucosal and invading bacteria in patients with inflammatory bowel disease compared with controls. Scand J Gastroenterol 2002; 37:1034-1041.
29. Kett K, Rognum TO, Brandtzaeg P. Mucosal subclass distribution of immunoglobulin G-producing cells is different in ulcerative colitis and Crohn's disease of the colon. Gastroenterology 1987; 93:919-924.
30. Helgeland L, Tysk C, Järnerot G et al. The IgG subclass distribution in serum and rectal mucosa of monozygotic twins with or without inflammatory bowel disease. Gut 1992; 33:1358-1364.

31. Brandtzaeg P, Korsrud FR. Significance of different J-chain profiles in human tissues: Generation of IgA and IgM with binding site for secretory component is related to the J-chain expressing capacity of the total local immunocyte population, including IgG- and IgD-producing cells, and depends on the clinical state of the tissue. Clin Exp Immunol 1984; 58:709-718.
32. Kett K, Brandtzaeg P. Local IgA subclass alterations in ulcerative colitis and Crohn's disease of the colon. Gut 1987; 28:1013-1021.
33. Kett K, Brandtzaeg P, Fausa O. J-chain expression is more prominent in immunoglobulin A2 than in immunoglobulin A1 colonic immunocytes and is decreased in both subclasses associated with inflammatory bowel disease. Gastroenterology 1988; 94:1419-1425.
34. Rognum TO, Elgjo K, Fausa O et al. Immunohistochemical evaluation of carcinoembryonic antigen, secretory component, and epithelial IgA in ulcerative colitis with dysplasia. Gut 1982; 23:123-133.
35. Van Den Bogaerde J, Cahill J, Emmanuel AV et al. Gut mucosal response to food antigens in Crohn's disease. Aliment Pharmacol Ther 2002; 16:1903-1915.
36. Johansen F-E, Pekna M, Norderhaug IN et al. Absence of epithelial immunoglobulin A transport, with increased mucosal leakiness, in polymeric immunoglobulin receptor/secretory component-deficient mice. J Exp Med 1999; 190:915-921.
37. Monteiro E, Fossey J, Shiner M et al. Antibacterial antibodies in rectal and colonic mucosa in ulcerative colitis. Lancet 1971; 1:249-250.
38. Folkersen J, Sofeldt S, Svehag SE. Application of electroblotting technique to studies of the intestinal antibody response to extractable fecal antigens. Scand J Gastroenterol 1985; 20:247-253.
39. Heddle RJ, La Brooy JT, Shearman DJ. Escherichia coli antibody-secreting cells in the human intestine. Clin Exp Immunol 1982; 48:469-476.
40. Auer IO, Röder A, Wensinck F et al. Selected bacterial antibodies in Crohn's disease and ulcerative colitis. Scand J Gastroenterol 1984; 18:217-223.
41. Howells B, Matthews N, Mayberry JF et al. Agglutinins to anaerobic bacteria in Crohn's disease and in Indian patients with diarrhoea. J Med Microbiol 1984; 17:207-209.
42. Macpherson A, Khoo UY, Forgacs I et al. Mucosal antibodies in inflammatory bowel disease are directed against intestinal bacteria. Gut 1996; 38:365-375.
43. Foo MC, Lee A. Immunological response of mice to members of the autochthonous intestinal microflora. Infect Immun 1972; 6:525-532.
44. Berg RD, Savage DC. Immune responses of specific pathogen-free and gnotobiotic mice to antigens of indigenous and nonindigenous microorganisms. Infect Immun 1975; 11:320-329.
45. Shroff KE, Meslin K, Cebra JJ. Commensal enteric bacteria engender a self-limiting humoral mucosal immune response while permanently colonizing the gut. Infect Immun 1995; 63:3904-3913.
46. MacDonald TT. Breakdown of tolerance to the intestinal bacterial flora in inflammatory bowel disease (IBD). Clin Exp Immunol 1995; 102:445-447.
47. Duchmann R, Kaiser I, Hermann E et al. Tolerance exists towards resident intestinal flora but is broken in active inflammatory bowel disease (IBD). Clin Exp Immunol 1995; 102:448-455.
48. Duchmann R, Neurath MF, Meyer zum Buschenfelde KH. Responses to self and nonself intestinal microflora in health and inflammatory bowel disease. Res Immunol 1997; 148:589-594.
49. van der Waaij LA, Limburg PC, Mesander G et al. In vivo IgA coating of anaerobic bacteria in human faeces. Gut 1996; 38:348-354.
50. van der Waaij LA, Kroese FGM, Jansen PLM et al. IBD patients have a very high percentage of colonic anaerobic bacteria that are in vivo coated with IgA, IgG or IgM, irrespective of clinical activity. Gut 1997; 41(Suppl 3):A117.
51. Landers CJ, Cohavy O, Misra R et al. Selected loss of tolerance evidenced by Crohn's disease-associated immune responses to auto- and microbial antigens. Gastroenterology 2002; 123:689-699.
52. Chao LP, Steele J, Rodrigues C et al. Specificity of antibodies secreted by hybridomas generated from activated B cells in the mesenteric lymph nodes of patients with inflammatory bowel disease. Gut 1988; 29:35-40.
53. Hibi T, Ohara M, Toda K et al. In vitro anticolon antibody production by mucosal or peripheral blood lymphocytes from patients with ulcerative colitis. Gut 1990; 31:1371-1376.
54. Sadlack B, Merz H, Schorle H et al. Ulcerative colitis-like disease in mice with a disrupted interleukin-2 gene. Cell 1993; 75:253-261.
55. Brandtzaeg P. Autoimmunity and ulcerative colitis: Can two enigmas make sense together? (Editorial). Gastroenterology 1995; 109:307-312.
56. Biancone L, Mandal A, Yang H et al. Production of immunoglobulin G and G1 antibodies to cytoskeletal protein by lamina propria cells in ulcerative colitis. Gastroenterology 1995; 109:3-12.
57. Geng X, Biancone L, Dai HH et al. Tropomyosin isoforms in intestinal mucosa: Production of autoantibodies to tropomyosin isoforms in ulcerative colitis. Gastroenterology 1998; 114:912-922.

58. Onuma EK, Amenta PS, Ramaswamy K et al. Autoimmunity in ulcerative colitis (UC): A predominant colonic mucosal B cell response against human tropomyosin isoform 5. Clin Exp Immunol 2000; 121:466-471.
59. Das KM. Relationship of extraintestinal involvements in inflammatory bowel disease: New insights into autoimmune pathogenesis. Dig Dis Sci 1999; 44:1-13.
60. Das KM, Dasgupta A, Mandal A et al. Autoimmunity to cytoskeletal protein tropomyosin. A clue to the pathogenetic mechanism for ulcerative colitis. J Immunol 1993; 150:2487-2493.
61. Hassan T, Kanisawa Y, Meyers S et al. Expression of a unique protein on colon cancer cells that reacts with a novel monoclonal antibody and ulcerative colitis serum. Clin Exp Immunol 1995; 100:457-462.
62. Dunn-Walters DK, Boursier L, Hackett M et al. Biased JH usage in plasma cell immunoglobulin gene sequences from colonic mucosa in ulcerative colitis but not in Crohn's disease. Gut 1999; 44:382-386.
63. Inoue N, Watanabe M, Sato T et al. Restricted V(H) gene usage in lamina propria B cells producing anticolon antibody from patients with ulcerative colitis. Gastroenterology 2001; 121:15-23.
64. Halstensen TS, Mollnes TE, Garred P et al. Epithelial deposition of immunoglobulin G1 and activated complement (C3b and terminal complement complex) in ulcerative colitis. Gastroenterology 1990; 98:1264-1271.
65. Halstensen TS, Mollnes TE, Garred P et al. Surface epithelium related activation of complement differs in Crohn's disease and ulcerative colitis. Gut 1992; 33:902-908.
66. Halstensen TS, Das K, Brandtzaeg P. Epithelial deposits of immunoglobulin G1 and activated complement colocalise with the M_r 40 kD putative autoantigen in ulcerative colitis. Gut 1993; 34:650-657.
67. Berstad AE, Brandtzaeg P. Expression of cell membrane complement regulatory glycoproteins along the normal and diseased human gastrointestinal tract. Gut 1998; 42:522-529.
68. Ueki T, Mizuno M, Uesu T et al. Distribution of activated complement, C3b, and its degraded fragments, iC3b/C3dg, in the colonic mucosa of ulcerative colitis (UC). Clin Exp Immunol 1996; 104:286-292.
69. Contractor NV, Bassiri H, Reya T et al. Lymphoid hyperplasia, autoimmunity, and compromised intestinal intraepithelial lymphocyte development in colitis-free gnotobiotic IL-2-deficient mice. J Immunol 1998; 160:385-394.
70. Linskens RK, Mallant-Hent RC, Groothuismink ZM et al. Evaluation of serological markers to differentiate between ulcerative colitis and Crohn's disease: PANCA, ASCA and agglutinating antibodies to anaerobic coccoid rods. Eur J Gastroenterol Hepatol 2002; 14:1013-1018.
71. Bartunkova J, Kolarova I, Sediva A et al. Antineutrophil cytoplasmic antibodies, anti-Saccharomyces cerevisiae antibodies, and specific IgE to food allergens in children with inflammatory bowel diseases. Clin Immunol 2002; 102:162-168.
72. Seibold F, Slametschka D, Gregor M et al. Neutrophil autoantibodies: A genetic marker in primary sclerosing cholangitis and ulcerative colitis. Gastroenterology 1994; 107:532-536.
73. Joossens S, Reinisch W, Vermeire S et al. The value of serologic markers in indeterminate colitis: A prospective follow-up study. Gastroenterology 2002; 122:1242-1247.
74. Targan SR, Landers CJ, Cobb L et al. Perinuclear anti-neutrophil cytoplasmic antibodies are spontaneously produced by mucosal B cells of ulcerative colitis patients. J Immunol 1995; 155:3262-3267.
75. Gilat T, Hacohen D, Lilos P et al. Childhood factors in ulcerative colitis and Crohn's disease. An international cooperative study. Scand J Gastroenterol 1987; 22:1009-1024.
76. Rutgeerts P, D'Haens G, Hiele M et al. Appendectomy protects against ulcerative colitis. Gastroenterology 1994; 106:1251-1253.
77. Smithson JE, Radford-Smith G, Jewell GP. Appendectomy and tonsillectomy in patients with inflammatory bowel disease. J Clin Gastroenterol 1995; 21:283-286.
78. Mitchell SA, Thyssen M, Orchard TR et al. Cigarette smoking, appendectomy, and tonsillectomy as risk factors for the development of primary sclerosing cholangitis: A case control study. Gut 2002; 51:567-573.
79. Logan R. Appendectomy and ulcerative colitis: What connection? Gastroenterology 1994; 106:1382-1384.
80. Andersson RE, Olaison G, Tysk C et al. Appendectomy and protection against ulcerative colitis. N Engl J Med 2001; 344:808-814.
81. Campbell DJ, Butcher EC. Rapid acquisition of tissue-specific homing phenotypes by $CD4^+$ T cells activated in cutaneous or mucosal lymphoid tissues. J Exp Med 2002; 195:135-141.
82. Bjerke K, Brandtzaeg P, Rognum TO. Distribution of immunoglobulin producing cells is different in normal human appendix and colon mucosa. Gut 1986; 27:667-674.
83. Bjerke K, Brandtzaeg P. Terminally differentiated human intestinal B cells. J chain expression of IgA and IgG subclass-producing immunocytes in the distal ileum compared with mesenteric and peripheral lymph nodes. Clin Exp Immunol 1990; 82:411-415.

84. Bjerke K, Brandtzaeg P. Immunoglobulin- and J chain-producing cells associated with lymphoid follicles in the human appendix, colon and ileum, including Peyer's patches. Clin Exp Immunol 1986; 64:432-441.
85. Mizoguchi A, Mizoguchi E, Chiba C et al. Role of appendix in the development of inflammatory bowel disease in TCR-α mutant mice. J Exp Med 1996; 184:707-715.
86. Krieglstein CF, Cerwinka WH, Laroux FS et al. Role of appendix and spleen in experimental colitis. J Surg Res 2001; 101:166-175.
87. Moghaddami M, Cummins A, Mayrhofer G. Lymphocyte-filled villi: Comparison with other lymphoid aggregations in the mucosa of the human small intestine. Gastroenterology 1998; 115:1414-1425.
88. O'Leary AD, Sweeney EC. Lymphoglandular complexes of the colon: Structure and distribution. Histopathology 1986; 10:267-283.
89. Langman JM, Rowland R. The number and distribution of lymphoid follicles in the human large intestine. J Anat 1986; 149:189-194.
90. Yeung MM-W, Melgar S, Baranov V et al. Characterisation of mucosal lymphoid aggregates in ulcerative colitis: Immune cell phenotype and TcR-γδ expression. Gut 2000; 47:215-227.
91. Surawicz CM, Belic L. Rectal biopsy helps to distinguish acute self-limited colitis from idiopathic inflammatory bowel disease. Gastroenterology 1984; 86:104-113.
92. Mackay CR. Chemokines: Immunology's high impact factors. Nature Immunol 2001; 2:95-101.
93. Zlotnik A, Yoshie O. Chemokines: A new classification system and their role in immunity. Immunity 2000; 12:121-127.
94. Cyster JG. Chemokines and cell migration in secondary lymphoid organs. Science 1999; 286:2098-2102.
95. Moser B, Loetscher P. Lymphocyte traffic control by chemokines. Nature Immunol 2001; 2:123-128.
96. Legler DF, Loetscher M, Roos RS et al. B cell-attracting chemokine 1, a human CXC chemokine expressed in lymphoid tissues, selectively attracts B lymphocytes via BLR1/CXCR5. J Exp Med 1998; 187:655-660.
97. Gunn MD, Ngo VN, Ansel KM et al. A B-cell-homing chemokine made in lymphoid follicles activates Burkitt's lymphoma receptor-1. Nature 1998; 391:799-803.
98. Luther SA, Lopez T, Bai W et al. BLC expression in pancreatic islets causes B cell recruitment and lymphotoxin-dependent lymphoid neogenesis. Immunity 2000; 12:471-481.
99. Forster R, Mattis AE, Kremmer E et al. A putative chemokine receptor, BLR1, directs B cell migration to defined lymphoid organs and specific anatomic compartments of the spleen. Cell 1996; 87:1037-1047.
100. Ansel KM, Ngo VN, Hyman PL et al. A chemokine-driven positive feedback loop organizes lymphoid follicles. Nature 2000; 406:309-314.
101. Carlsen HS, Baekkevold ES, Johansen FE et al. B cell attracting chemokine 1 (CXCL13) and its receptor CXCR5 are expressed in normal and aberrant gut associated lymphoid tissue. Gut 2002; 51:364-371.
102. Schaerli P, Willimann K, Lang AB et al. CXC chemokine receptor 5 expression defines follicular homing T cells with B cell helper function. J Exp Med 2000; 192:1553-1562.
103. Breitfeld D, Ohl L, Kremmer E et al. Follicular B helper T cells express CXC chemokine receptor 5, localize to B cell follicles, and support immunoglobulin production. J Exp Med 2000; 192:1545-152.
104. Moser B, Schaerli P, Loetscher P. $CXCR5^+$ T cells: Follicular homing takes center stage in T-helper-cell responses. Trends Immunol 2002; 23:250-254.
105. Kim CH, Rott LS, Clark-Lewis I et al. Subspecialization of $CXCR5^+$ T Cells. B helper activity is focused in a germinal center-localized subset of $cxcr5^+$ T cells. J Exp Med 2001; 193:1373-1382.
106. Gretz JE, Kaldjian EP, Anderson AO et al. Sophisticated strategies for information encounter in the lymph node: The reticular network as a conduit of soluble information and a highway for cell traffic. J Immunol 1996; 157:495-499.
107. Cyster JG, Ansel KM, Reif K et al. Follicular stromal cells and lymphocyte homing to follicles. Immunol Rev 2000; 176:181-193.
108. Carlsen HS, Baekkevold ES, Morton HC et al. Monocyte-like and mature macrophages produce CXCL13 (B cell-attracting chemokine 1) in inflammatory lesions with lymphoid neogenesis. Blood 2004; 104:3021-7.
109. Grouard G, Durand I, Filgueira L et al. Dendritic cells capable of stimulating T cells in germinal centres. Nature 1996; 384:364-367.
110. Fu Y-X, Huang G, Wang Y et al. B lymphocytes induce the formation of follicular dendritic cell clusters in a lymphotoxin α-dependent fashion. J Exp Med 1998; 187:1009-1018.

111. Gonzalez M, Mackay F, Browning JL et al. The sequential role of lymphotoxin and B cells in the development of splenic follicles. J Exp Med 1998; 187:997-1007.
112. Endres R, Alimzhanov MB, Plitz T et al. Mature follicular dendritic cell networks depend on expression of lymphotoxin β receptor by radioresistant stromal cells and of lymphotoxin β and tumor necrosis factor by B cells. J Exp Med 1999; 189:159-168.
113. Tumanov A, Kuprash D, Lagarkova M et al. Distinct role of surface lymphotoxin expressed by B cells in the organization of secondary lymphoid tissues. Immunity 2002; 17:239-250.
114. Shi K, Hayashida K, Kaneko M et al. Lymphoid chemokine B cell-attracting chemokine-1 (CXCL13) is expressed in germinal center of ectopic lymphoid follicles within the synovium of chronic arthritis patients. J Immunol 2001; 166:650-655.
115. Takemura S, Braun A, Crowson C et al. Lymphoid neogenesis in rheumatoid synovitis. J Immunol 2001; 167:1072-1080.
116. Xanthou G, Polihronis M, Tzioufas AG et al. "Lymphoid" chemokine messenger RNA expression by epithelial cells in the chronic inflammatory lesion of the salivary glands of Sjogren's syndrome patients: Possible participation in lymphoid structure formation. Arthritis Rheum 2001; 44:408-418.
117. Amft N, Curnow SJ, Scheel-Toellner D et al. Ectopic expression of the B cell-attracting chemokine BCA-1 (CXCL13) on endothelial cells and within lymphoid follicles contributes to the establishment of germinal center-like structures in Sjogren's syndrome. Arthritis Rheum 2001; 44:2633-2641.
118. Ansel KM, Cyster JG. Chemokines in lymphopoiesis and lymphoid organ development. Curr Opin Immunol 2001; 13:172-179.
119. Fiocchi C. Inflammatory bowel disease: Etiology and pathogenesis. Gastroenterology 1998; 115:182-205.
120. Brandtzaeg P, Haraldsen G, Helgeland L et al. New insights into the immunopathology of human inflammatory bowel disease. Drugs Today 1999; 35(Suppl A):33-70.
121. Mitchell SA, Thyssen M, Orchard TR et al. Cigarette smoking, appendectomy, and tonsillectomy as risk factors for the development of primary sclerosing cholangitis: A case control study. Gut 2002; 51:567-573.
122. Blumberg RS, Saubermann LJ, Strober W. Animal models of mucosal inflammation and their relation to human inflammatory bowel disease. Curr Opin Immunol 1999; 11:648-656.
123. MacDonald TT, Monteleone G, Pender SL. Recent developments in the immunology of inflammatory bowel disease. Scand J Immunol 2000; 51:2-9.
124. Tsuji RF, Szczepanik M, Kawikova I et al. B cell-dependent T cell responses: IgM antibodies are required to elicit contact sensitivity. J Exp Med 2002; 196:1277-1290.
125. Dombrowicz D, Nutten S, Desreumaux P et al. Role of the high affinity immunoglobulin E receptor in bacterial translocation and intestinal inflammation. J Exp Med 2001; 193:25-34.
126. Fagarasan S, Muramatsu M, Suzuki K et al. Critical roles of activation-induced cytidine deaminase in the homeostasis of gut flora. Science 2002; 298:1424-1427.
127. Hugot JP, Chamaillard M, Zouali H et al. Association of NOD2 leucine-rich repeat variants with susceptibility to Crohn's disease. Nature 2001; 411:599-603.
128. Ogura Y, Bonen DK, Inohara N et al. A frameshift mutation in NOD2 associated with susceptibility to Crohn's disease. Nature 2001; 411:603-606.
129. Hampe J, Grebe J, Nikolaus S et al. Association of NOD2 (CARD 15) genotype with clinical course of Crohn's disease: A cohort study. Lancet 2002; 359:1661-1665.
130. Ahmad T, Armuzzi A, Bunce M et al. The molecular classification of the clinical manifestations of Crohn's disease. Gastroenterology 2002; 122:854-866.
131. Cuthbert AP, Fisher SA, Mirza MM et al. The contribution of NOD2 gene mutations to the risk and site of disease in inflammatory bowel disease. Gastroenterology 2002; 122:867-874.
132. Abreu MT, Taylor KD, Lin YC et al. Mutations in NOD2 are associated with fibrostenosing disease in patients with Crohn's disease. Gastroenterology 2002; 123:679-688.
133. Judge T, Lichtenstein GR. The NOD2 gene and Crohn's disease: Another triumph for molecular genetics. Gastroenterology 2002; 122:826-828.
134. Di Girolamo N, Visvanathan K, Lloyd A et al. Expression of TNF-α by human plasma cells in chronic inflammation. J Leukoc Biol 1997; 61:667-678.
135. Gonnella PA, Waldner HP, Weiner HL. B cell-deficient (mu MT) mice have alterations in the cytokine microenvironment of the gut-associated lymphoid tissue (GALT) and a defect in the low dose mechanism of oral tolerance. J Immunol 2001; 166:4456-4464.
136. Harris DP, Haynes L, Sayles PC et al. Reciprocal regulation of polarized cytokine production by effector B and T cells. Nature Immunol 2000; 1:475-482.
137. Mizoguchi A, Mizoguchi E, Takedatsu H et al. Chronic intestinal inflammatory condition generates IL-10-producing regulatory B cell subset characterized by CD1d upregulation. Immunity 2002; 16:219-230.

Chapter 11

Alterations of Mesenchymal and Endothelial Cells in Inflammatory Bowel Diseases

Claudio Fiocchi, Kenji Ina, Silvio Danese, André Z.A. Leite and Jon D. Vogel

Abstract

The pathogenesis of complex chronic diseases like inflammatory bowel disease (IBD) can no longer be viewed as a one-way street in which classical immune cells have exclusive control over the initiation, duration and outcome of the disease. There is enough experimental evidence to demonstrate that nonimmune cells, among which are mucosal mesenchymal and endothelial cells, also play a decisive role by interacting with immune cells and establishing a two-way reciprocal exchange of signals and responses that dictate the ultimate outcome of inflammation. Smooth muscle cells and fibroblasts/myofibroblasts display a variety of immune functions and modulate the activity and survival of T-cells. Mucosal microvascular cells, through the expression of adhesion molecules and secretion of chemokines, regulate the quantity and quality of leukocytes transmigrating into the interstitial space. A number of receptor-ligand pairs are expressed by immune and nonimmune cells that control their functional interplay, but the CD40/CD40 ligand system may be the most effective because CD40 is expressed by activated muscle and endothelial cells, while the CD40 ligand is expressed by activated T-cells and platelets. The activation of this system in IBD can lead to the establishment of a continuous cycle of nonimmune cell-dependent, antigen-independent interactions that perpetuates gut inflammation.

Introduction

The abnormalities underlying IBD pathogenesis are not restricted to those mediated by classical immune cells such as T and B lymphocytes, macrophages and dendritic cells, but also nonimmune cells. In addition to epithelial cells, all other nonimmune cellular elements present in the bowel may also be involved, depending on the type, stage and duration of IBD. The structural abnormalities found in a tissue or organ affected by chronic inflammation are not simply the mere reflection of a passive response by nonimmune cells to local insults, but represent changes resulting from their active participation in the disease process. In fact, many of the functions traditionally attributed to classical immune cells, such as cytokine production or expression of MHC class II antigens, can also be exerted by non lymphoid cells, blurring the definition of "immune" and "nonimmune". The crucial contribution of nonimmune cells to chronic inflammation has long been recognized in other conditions like rheumatoid arthritis,[1] and this notion is finally being appreciated in IBD, where a complex interplay of immune and nonimmune cell interactions is involved in gut inflammation.[2] Since the contribution of epithelial and neuroimmune cells to IBD is discussed in other chapters, this review will

Immune Mechanisms in Inflammatory Bowel Disease, edited by Richard S. Blumberg and Markus F. Neurath.

specifically focus on the role of two distinct types of mesenchymal cells, smooth muscle cells and fibroblasts/myofibroblasts, and mucosal microvascular endothelial cells.

Mesenchymal Cells

Mesenchymal cells represent a large class of fibroblastic-looking cells whose classification is complicated by the varied and adaptive characteristic of each cell type. On one hand there are fully differentiated α-actin-positive smooth muscle cells that, in the bowel wall, are found in the ***muscularis mucosae*** and the longitudinal and circular ***muscularis propria*** layers. On the other hand there is a poorly characterized group of phenotypically related cells in the lamina propria and submucosa that include fibroblasts and myofibroblasts, each with variable amounts of α-actin, vimentin and desmin.[3,4] The heterogeneous and variable phenotype of fibroblasts and myofibroblasts reflect their remarkable functional plasticity, which becomes particularly evident in response to inflammation.[5] For purpose of clarity and to reflect the evolution of knowledge in this field, smooth muscle cells and fibroblasts/myofibroblasts will be discussed separately.

Smooth Muscle Cells

Smooth muscle cells have been traditionally viewed as submissive targets of chronic inflammation in IBD. In ulcerative colitis (UC) this behavior would result in a remarkable thickening of the muscularis mucosa, and in Crohn's disease (CD) thickening of the whole intestinal wall with subsequent stricture formation and eventually bowel obstruction. In addition, in both UC and CD, abnormalities of bowel motility would result from an altered contractile response of muscle cells to inflammation. This view was first modified by studies showing that smooth muscle cells actively contribute to IBD pathogenesis through an augmented collagen and metalloproteinase synthesis in response to locally produced cytokines and growth factors, particularly interleukin (IL) -1β and transforming growth factor (TGF) -β1.[6,7] The discovery that cultured muscle cells (myoblasts) display immunological functions, such as expression of MHC class II antigens and activation of $CD4^+$ T-cells,[8] prompted additional investigation of intestinal muscle cells in IBD. In a series of studies utilizing the rat model of *Trichinella spiralis* infection and cultured murine intestinal smooth muscle cells, Collins and collaborators showed that changes in muscle cell function following *T. spiralis* infection were T-cell-dependent, muscle cells could produce cytokines in response to mediators produced during gut inflammation, and interferon (IFN) -γ-pretreated muscle cells could activate syngeneic T-cells in a MHC class II-dependent antigen-specific fashion, providing convincing evidence for an immunomodulatory activity of muscle cells in gut inflammation.[9,10] These and subsequent studies helped to consolidate not only the immune function of intestinal muscle cells as active players in inflammation, but also to integrate this cell type into complex enteric neuromuscular circuits with important implications for chronicity of inflammation, IBD-associated dysmotility, and related functional disorders.

Fibroblasts and Myofibroblasts

Fibroblasts are classically defined as α-actin-negative mesenchymal cells, such as those typically found in human skin, whereas the definition of myofibroblasts, also called smooth muscle-like fibroblasts, stromal cells, lipocytes, pericytes or stellate cells, is still uncertain and confusing because they most likely represent a continuum between the two ends of the spectrum, fibroblasts and fully differentiated smooth muscle cells.[11,12] An in depth discussion of differences between fibroblasts and myofibroblasts is beyond the scope of this review, and both terms will be used under an umbrella encompassing all functional variants relevant to gut inflammation. The concept that fibroblasts/myofibroblasts have multiple immune functions and contribute to regulate inflammation is now firmly established.[13,14] Under physiological conditions one of the fundamental functions they carry out is protecting T-cells from apoptosis in specific tissue microenvironments through both contact-mediated mechanisms and secretion of soluble factors.[15] This is also true in the intestine, where human intestinal fibroblasts (HIF) prevent the death of mucosal T-cells,[16] prolonging the survival of memory T-cells for future antigen-specific interactions and protection against injury. On the other hand, when this protective activity goes awry, fibroblasts may exert an opposite and detrimental activity because failure of turning off their immune-inflammatory program leads to release of inflammatory molecules and inappropriate retention of leukocytes within inflamed tissue.[17] Thus, mucosal fibroblasts may regulate the transition from acute resolving to persistent inflammation, a key step that determines whether an immune response becomes irreversibly chronic like in IBD.

Fibroblasts from different compartments of the body display specialized functions, but they also produce a large number of molecules that are common to all of them and contribute to regulate immunity and inflammation, including cytokines, chemokines, eicosanoids, cell adhesion and costimulatory molecules. This is also the case for intestinal fibroblasts/myofibroblasts regardless of small or large bowel origin. These cells produce cytokines at the mRNA or protein level, like IL-1α and β, IL-6, IL-8, IL-10 and growth factors like granulocyte-macrophage colony stimulating factor (GM-CSF) and TGF-β1,[18,19] all of which can initiate or regulate inflammation. They also express cell surface molecules that enhance cell interactions between immune cells or immune-nonimmune cells, such as ICAM-1, VCAM-1 and CD40,[20,21] and promote or amplify immune and inflammatory responses. Contrasting with the production of several molecules with an injurious potential due to their pro-inflammatory activity, intestinal fibroblasts/myofibroblasts also produce molecules that have protective and anti-inflammatory activity. Subepithelial myofibroblasts produce extracellular matrix (ECM) proteins like collagen type IV, laminin-β1 and -γ1, and fibronectin that promote epithelial cell proliferation and restitution.[22] Even more interesting is the spontaneous and continuous production of cyclooxygenase-2-dependent prostaglandin E_2, which is independent of exogenous stimuli and may contribute to maintenance of immune tolerance by inducing a state of hyporesponsiveness in the mucosa.[23] Few studies have directly investigated functional differences between fibroblasts/myofibroblasts isolated from normal and IBD mucosa. A preliminary report on the isolation and characterization of human intestinal mucosal stellate cells shows that there is increased growth, proliferation and differentiation with early expression of α-actin when these cells are isolated from CD or UC mucosa,[24] suggesting a local conditioning of fibroblast/myofibroblast precursors by chronic inflammation.

In addition to producing various molecules that modulate the response of immune cells in their vicinity, intestinal fibroblasts/myofibroblasts in turn respond to multiple immune, nonimmune cell products, as well as enteric bacterial components by modifying their proliferation, cytokine and chemokine secretory patterns, cell adhesion and costimulatory molecule expression, and production of ECM molecules.[19,20,25,26] Thus, intestinal fibroblasts/myofibroblasts are strategically situated underneath a layer of antigen-sampling epithelial cells and surrounded by antigen-responding immune cells. In this unique location they interact with both epithelial and immune compartments, and generate protective or deleterious signals through the production of down- or up-regulatory molecules that determine whether mucosal homeostasis or inflammation will prevail.

Endothelial Cells

In contrast to mesenchymal cells, the notion that endothelial cells are intimately involved in inflammation is firmly established.[27] Endothelial cells are often referred to as "gatekeepers of inflammation", since they control both the type and number of leukocytes that extravasate from the intravascular into the interstitial compartment, a pivotal step that represents the *sine qua non* for tissue inflammation. The prime function of endothelial cells is obviously a physiological one, e.g., the distribution of leukocytes throughout the body to maintain immune homeostasis.[28] This is a complex process that displays specialized features depending on the specific tissue or organ where lymphoid cells are destined to reside. Such process is tightly regulated by a large number of homing and adhesion molecules expressed in a "key-and-lock" fashion (receptor and counter-receptor pairs) on the surface of microvascular and lymphoid cells, forming the basis of a multistep extravasation cascade which includes tethering/rolling, activation, adhesion, spreading, and transmigration.[29] One of the most important homing and adhesion molecular pairs controlling lymphocyte migration to the gut is formed by MAd-CAM-1, which belongs to the immunoglobulin family and is expressed on high endothelial venules in the mucosa, and the integrin α4β7, which is expressed on circulating lymphocytes destined to the gut.[29] Several other molecules participate in the adhesion and homing process, including a number of chemokines produced by the endothelial cells themselves.

In pathological conditions, including inflammation, the behavior of endothelial cells is drastically modified, inducing multiple changes in the expression of genes whose products control coagulation, ECM remodeling, angiogenesis and contact with platelets and leukocytes.[30] In the inflamed intestine these changes occur particularly at the level of the microcirculation, profoundly affecting the function of mucosal endothelial cells and their interaction with leukocytes.[31,32] Microvascular mucosal endothelial cells can be studied in fixed tissue sections of human intestine, or live cultures named human intestinal microvascular endothelial cells (HIMEC). The characteristics of these cells from normal and IBD-involved mucosa will be discussed in greater detail in the following paragraph.

Human Intestinal Microvascular Endothelial Cells

Microvascular cells in IBD mucosa respond to inflammation by altering the expression pattern of cell adhesion molecules. Immunohistochemical analysis of intestinal tissue sections shows that the expression of L-selectin, normally absent in non inflamed mucosa, is abundant in vascular cells of both CD and UC intestine.[33] The expression of ICAM-1, found on capillaries and venules in the normal mucosa, is upregulated in parallel with the histological severity of IBD.[34] Especially noteworthy is the augmented expression of MAd-CAM-1 in inflammatory foci of the lamina propria in CD and UC tissues, a phenomenon believed to contribute to the heightened recruitment of inflammatory cells into the mucosa.[35]

Studies with isolated HIMEC cultures have greatly expanded our understanding of the distinctive characteristics of these cells in both normal and inflamed intestine. Some basic characteristics are shared with other endothelial cells, as HIMEC also contain Weibel-Palade bodies, express von Willebrand factor, CD31 and VE-cadherin.[36] In the unstimulated state HIMEC express mRNA for a large number of immunoregulatory and pro-inflammatory cytokines and chemokines, including IL-3, IL-7, IL-8, IL-11, IL-14, IL-15, tumor necrosis factor (TNF) -α, TGF-β1, and GM-CSF, and after stimulation with IL-1β or TNF-α they produce IL-1α and upregulate secretion of IL-6 and IL-8.[37] In addition, HIMEC exposed to IL-1β, TNF-α or lipopolysaccharide de novo express E-selectin and VCAM-1 while enhancing ICAM-1 expression.[38] This indicates that HIMEC can contribute to the network of cytokines and adhesion molecules found in the normal as well as inflamed mucosa. Some of their functional characteristics may be unique because, unlike human umbilical vein endothelial cells (HUVEC), they contain IL-8 prestored in intracellular granules that can be quickly released in response to histamine or thrombin, providing a rapid pathway for activation and retention of neutrophils at sites of inflammation.[39] Another important function of HIMEC is the capacity to upregulate HLA-DR, -DP and -DQ upon exposure to IFN-γ treatment and present antigen in a MHC class II-dependent fashion.[40]

Particularly important to IBD pathogenesis are the results of studies performed with HIMEC derived from actively involved CD and UC mucosa. A key observation is that IBD HIMEC display a remarkably greater binding capacity for different classes of leukocyte, including monocytic cells, T cells and neutrophils, than HIMEC from normal mucosa.[41] This is particularly important because this distinctive feature persists in cells cultured for multiple passages, suggesting that prolonged exposure to an inflammatory milieu has transformed the functional phenotype of HIMEC into a hyper-adhesive one that supports an increased leukocyte recruitment and promotes persistence of inflammation. This hypothesis was confirmed by a subsequent study showing that the enhanced leukocyte binding capacity of IBD HIMEC is an acquired defect, since HIMEC from uninvolved intestinal segments of the same IBD patients bind the same number of leukocytes as control normal HIMEC do.[42] Some of the possible mechanisms underlying the hyperadhesiveness of IBD HIMEC for leukocytes have been explored in a series of reports by Binion and collaborators. Based on the fact that nitric oxide (NO) suppresses endothelial activation and leukocyte adhesion, these authors initially showed that inhibitors of inducible nitric oxide synthase (iNOS) increased binding of leukocytes to activated HIMEC and antioxidant compounds reversed the effect of NOS inhibitors on HIMEC-leukocyte interaction, and concluded that iNOS-derived NO acts as an endogenous antioxidant that downregulates leukocyte binding.[43] Then, using IBD HIMEC, they showed that these cells have significantly decreased levels of iNOS and NO production following activation, and that addition of NO restores a normal leukocyte binding pattern in IBD HIMEC, suggesting that loss of iNOS is a feature of chronically inflamed mucosal microvascular endothelial cells that contributes to sustain intestinal inflammation.[44] More recently, they reported that cyclosporine A enhances HIMEC leukocyte binding capacity through the inhibition of iNOS, a pro-inflammatory effect that may explain the lack of efficacy of this immunosuppressive drug in the long-term treatment of IBD.[45]

Integration of Intestinal Immune-Nonimmune Cell Interactions by the CD40/CD40 Ligand System

Although nonimmune cells can exert a variety of immunoregulatory and pro-inflammatory activities, this almost invariably happens in the context of a close physical and functional interaction with immune cells. The same is true in IBD, where T-cells, B-cells, macrophages and dendritic cells constantly exchange signals with epithelial, mesenchymal, vascular and neural cells.[2] There are multiple molecular requirements for optimal cell interactions during an immune/inflammatory response. For instance, when T-cells recognize a specific antigen, this occurs through the engagement of the T-cell receptor and MHC class II antigens on the surface of antigen-presenting cells (APC). The formation of this trimolecular complex is stabilized by additional molecular pairs such as ICAM-1 on the APC binding LFA-1 on the T-cell, and the resulting T-cell response is magnified by costimulatory signals triggered by the binding of B7.1 on the APC to CD28 on the T-cell. All these events must be amplified by additional signaling pathways for an immune response to become effective, or for them to persist like in autoimmune or chronic inflammatory diseases. Among others, the CD40/CD40 ligand (CD40L) system appears ideally suited for the amplification of an immune response.[46,47] CD40 is a protein homologous to members of the TNF receptor family and can be expressed by essentially all cells, including nonimmune cells such as fibroblasts, endothelial, epithelial and nerve cells. In contrast the CD40L, a homologous to members of the TNF family, has an extremely restricted expression pattern, practically being found only on the surface of activated T-cells and platelets, from where it is shed in a soluble form that is biologically active.[46-48] Thus, the CD40/CD40L system can very effectively amplify immune responses in specific organs and tissues by serving as a biological bridge linking activated T-cells and many other cells types, leading to the production of cytokines, chemokines, adhesion molecules, eicosanoids, metalloproteinases, and NO in addition to modulating cell survival and apoptosis (Fig. 1). There is strong evidence that the CD40/CD40L system is activated in IBD. Expression of CD40 and CD40L is increased in both CD and UC tissues,[49,50] and blockade of CD40-CD40L interaction downregulates inflammation in various animal models of experimental colitis.[51,52]

Because of our interest in the role of nonimmune cells in IBD pathogenesis, we carried out a series of experiments aimed at evaluating the expression of CD40 by intestinal mesenchymal

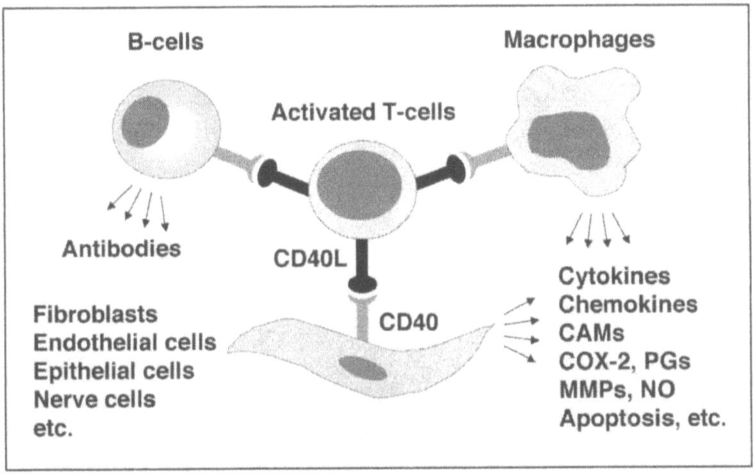

Figure 1. Pleiotropic effects resulting from the interaction of activated CD40L-positive T-cells with immune and nonimmune cells. Engagement of CD40 on B-cells leads to antibody production, whereas a much wider range of products are released by monocytes and macrophages activated through the CD40 pathway, including cytokines, chemokines, cell adhesion molecules (CAMs), cyclooxygenases (COX), prostaglandins (PGs), metalloproteinases (MMPs), nitric oxide (NO), etc. Activation of nonimmune cells through the CD40 pathway induces a spectrum of mediators similar to that secreted by monocytes and macrophages.

(HIF) and endothelial cells (HIMEC), and the outcome of the interaction with CD40L-positive T-cells and platelets. The results of these studies are herein summarized and the potential functional implications for IBD pathogenesis briefly discussed. Both HIF and HIMEC upregulate the surface expression of CD40 when exposed to IFN-γ and, after coculture with CD40L-positive T-cells or soluble CD40L, they produce substantial levels of the chemokine IL-8 and RANTES. Moreover, HIF and HIMEC baseline expression of ICAM-1 and VCAM-1 is also augmented after engagement of the CD40 receptor.[21] Thus, the CD40/CD40L system-mediated increase in chemokine and adhesion molecule production may result in the attraction of and interaction with additional T-cells leading, in turn, to further chemokine production. In addition to becoming activated and increasing chemokine production upon CD40 engagement, HIF downregulate their collagen synthetic capacity after contact with membrane-bound or soluble CD40L, pointing to a potentially beneficial antifibrotic effect of the CD40/CD40L system in IBD.[26] Since activated platelets express CD40L,[48] we investigated whether this type of nonimmune cell can also engage CD40 on HIMEC and lead to their activation. As assessed by flow cytometry, IBD platelets express significantly higher CD40L levels than those of normal platelets, and confocal microscopy detects CD40L-positive platelets in IBD-involved mucosa.[53] Furthermore, after direct physical contact, activated platelets upregulate ICAM-1 and VCAM-1 expression as well as IL-8 production by HIMEC in a CD40-dependent fashion.

Taken together, these preliminary results confirm that the CD40/CD40L dyad is a very efficient system leading to the activation of potent signaling pathways in intestinal mesenchymal and endothelial cells. These cells, in turn, secrete soluble mediators and upregulate adhesion molecules that cause more T cell to bind and thus sustain a continuous cycle of immune-nonimmune cell interactions (Fig. 2). This provides the conceptual framework for an antigen-independent recruitment of T cells to the inflamed intestinal mucosa, and highlights the role of nonimmune cells as crucial elements in the maintenance of IBD.

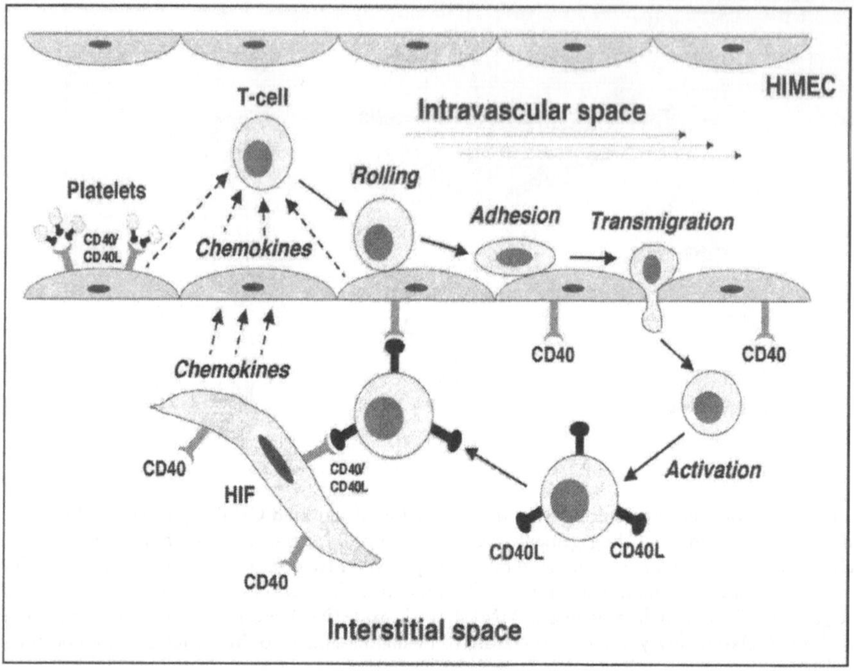

Figure 2. Hypothetical framework for continuous recruitment of T cells into the gut mucosa through a CD40/CD40L-dependent mechanism. In the early stages of mucosal inflammation local T cells become activated and express CD40L, allowing them to bind to and activate mesenchymal (HIF) and endothelial (HIMEC) cells through the CD40 pathway. HIF and HIMEC start secreting chemokines that attract more T cells that transmigrate into the interstitial space, become activated and express CD40L, initiating a new cycle of chemokine secretion by nonimmune cells which results in additional T-cell recruitment. Activated platelets in the circulation of IBD patients contribute to this process through the expression of CD40L on their surface. The end result is a nonimmune cell-mediated, antigen-independent mechanism perpetuating inflammation in IBD.

References

1. Smith JB, Haynes MK. Rheumatoid arthritis-A molecular understanding. Ann Intern Med 2002; 136:908-922.
2. Fiocchi C. Intestinal inflammation: A complex interplay of immune-non immune cell interactions. Am J Physiol 1997; 273:G769-G775.
3. Pucilowska JB, Williams KL, Lund PK. Fibrogenesis. IV. Fibrosis and inflammatory bowel disease: Cellular mediators and animal models. Am J Physiol 2000; 279:G653-G659.
4. Lund PK, Zuniga CC. Intestinal fibrosis in human and experimental inflammatory bowel disease. Curr Opin Gastroenterol 2001; 17:318-323.
5. Schmitt-Graff A, Desmouliere A, Gabbiani G. Heterogeneity of myofibroblast phenotypic features: An example of fibroblast cell plasticity. Virchows Archiv 1994; 425:3-24.
6. Matthes H, Herbst H, Schuppan D et al. Cellular localization of procollagen gene transcripts in inflammatory bowel disease. Gastroenterology 1992; 102:431-442.
7. Graham MF, Willey A, Adams J et al. Interleukin 1β down-regulates collagen and augments collagenase expression in human intestinal smooth muscle cells. Gastroenterology 1996; 110:344-350.
8. Hohlfeld R, Engel AG. The immunobiology of muscle. Immunol Today 1994; 15:269-274.
9. Vermillion DE, Ernst PB, Collins SM. T-lymphocyte modulation of intestinal muscle function in the Trichinella-infected rat. Gastroenterology 1991; 101:31-38.
10. Hogaboam CM, Snider DP, Collins SM. Activation of T lymphocytes by syngeneic murine intestinal smooth muscle cells. Gastroenterology 1996; 110:1456-1466.

11. Powell DW, Mifflin RC, Valentich JD et al. Myofibroblasts. I. Paracrine cells important in health and disease. Am J Physiol 1999; 277:C1-C19.
12. Powell DW, Mifflin RC, Valentich JD et al. Myofibroblasts. II. Intestinal subepithelial myofibroblasts. Am J Physiol 1999; 277:C183-C210.
13. Jordana M, Sarnstrand B, Sime PJ et al. Immune-inflammatory functions of fibroblasts. Eur Respir J 1994; 7:2212-2222.
14. Smith RS, Smith TJ, Blieden TM et al. Fibroblasts as sentinel cells. Synthesis of chemokines and regulation of inflammation. Am J Pathol 1997; 151:317-322.
15. Akbar AN, Salmon M. Cellular environments and apoptosis: Tissue microenvironments control activated T-cell death. Immunol Today 1997; 18:72-76.
16. Ina K, Binion DG, West GA et al. Secretion of soluble factors and phagocytosis by intestinal fibroblasts regulate T-cell apoptosis. Gastroenterology 1995; 108:A841.
17. Buckley CD, Pilling D, Lord JM et al. Fibroblasts regulate the switch from acute resolving to chronic persistent inflammation. Trends Immunol 2001; 22:199-204.
18. Pang G, Couch L, Batey R et al. GM-CSF, IL-1α, IL-1β, IL-6, IL-8, IL-10, ICAM-1 and VCAM-1 gene expression and cytokine production in human duodenal fibroblasts stimulated by lipolysaccharide, IL-1α and TNF-α. Clin Exp Immunol 1994; 96:437-443.
19. Strong SA, Pizarro TT, Klein JS et al. Proinflammatory cytokines differentially modulate their own expression in human intestinal mucosal mesenchymal cells. Gastroenterology 1998; 114:1244-1256.
20. Musso A, Condon TP, West GA et al. Regulation of ICAM-1-mediated fibroblast-T-cell reciprocal interaction: Implications for modulation of gut inflammation. Gastroenterology 1999; 117:546-556.
21. Vogel JD, West GA, Sturm A et al. Essential role of the CD40 pathway in T-cell-mediated induction of chemokines and cell adhesion molecules by human intestinal fibroblasts (HIF) and microvascular endothelial cells (HIMEC). Gastroenterology 2001; 120:A192.
22. Mahida YR, Beltinger J, Makh S et al. Adult human colonic subepithelial myofibroblasts express extracellular matrix proteins and cyclooxygenase-1 and -2. Am J Physiol 1997; 273:G1341-G1348.
23. Newberry RD, McDonough JS, Stenson WF et al. Spontaneous and continuous cyclooxygenase-2-dependent prostaglandin E2 production by stromal cells in the murine small intestine lamina propria: Directing the tone of the intestinal immune response. J Immunol 2001; 166:4465-4472.
24. Leite AZ, De la Motte C, Strong SA et al. Isolation and functional characterization of human intestinal mucosal stellate cells, Gastroenterology 2002; 122:A-111. (Abstract).
25. Van Tol EA, Holt L, Li FL et al. Bacterial cell wall polymers promote intestinal fibrosis by direct stimulation of myofibroblasts. Am J Physiol 1999; 277:G245-G255.
26. Vogel JD, Majors A, Sturm A et al. Collagen synthesis by human intestinal fibroblasts (HIF) is modulated by T-cells through the CD40/CD40 ligand (CD40L) pathway. Gastroenterology 2001; 120:A135.
27. Pober JS, Cotran RS. The role of endothelial cells in inflammation. Transplantation 1990; 50:537-544.
28. Butcher EC, Williams M, Youngman K et al. Lymphocyte trafficking and regional immunity. Adv Immunol 1999; 72:209-253.
29. Salmi M, Jalkanen S. Molecules controlling lymphocyte migration to the gut. Gut 1999; 45:148-153.
30. Gerritsen ME, Bloor CM. Endothelial cell gene expression in response to injury. FASEB J 1993; 7:523-532.
31. Panes J, Granger DN. Leukocyte-endothelial interactions: Molecular mechanisms and implications in gastrointestinal disease. Gastroenterology 1998; 114:1066-1090.
32. Laroux FS, Grisham MB. Immunological basis of inflammatory bowel disease: Role of the microcirculation. Microcirculation 2001; 8:283-301.
33. Koizumi M, King N, Lobb R et al. Expression of vascular adhesion molecules in inflammatory bowel disease. Gastroenterology 1992; 103:840-847.
34. Nakamura S, Ohtani H, Watanabe Y et al. In situ expression of the cell adhesion molecules in inflammatory bowel disease. Lab Invest 1993; 69:77-85.
35. Briskin M, Winsor-Hines D, Shyjan A et al. Human mucosal addressin cell adhesion molecule-1 is preferentially expressed in intestinal tract and associated lymphoid tissue. Am J Pathol 1997; 151:97-110.
36. Haraldsen G, Rugveigt J, Kvale D et al. Isolation and longterm culture of human intestinal microvascular endothelial cells. Gut 1995; 37:225-234.
37. Nilsen EM, Johansen F-E, Jahnsen FL et al. Cytokine profiles of cultured microvascular endothelial cells from human intestine. Gut 1998; 42:635-642.
38. Haraldsen G, Kvale D, Lien B et al. Cytokine-regulated expression of E-selectin, intercellular adhesion molecule-1 (ICAM-1), and vascular cell adhesion molecule-1 (VCAM-1) in human intestinal microvascular endothelial cells. J Immunol 1996; 156:2559-2565.

39. Utgaard JO, Jahnsen FL, Bakka A et al. Rapid secretion of prestored interleukin 8 from Weibel-Palade bodies of microvascular endothelial cells. J Exp Med 1998; 188:1751-1756.
40. Haraldsen G, Sollid LM, Bakke O et al. Major histocompatibility complex call II-dependent antigen presentation by human intestinal endothelial cells. Gastroenterology 1998; 114:649-656.
41. Binion DG, West GA, Ina K et al. Enhanced leukocyte binding by intestinal microvascular endothelial cells in inflammatory bowel disease. Gastroenterology 1997; 112:1895-1907.
42. Binion DG, West GA, Volk EE et al. Acquired increase in leucocyte binding by intestinal microvascular endothelium in inflammatory bowel disease. Lancet 1998; 352:1742-1746.
43. Binion DG, Fu S, Ramanujam KR et al. iNOS expression in human intestinal microvascular endothelial cells inhibits leukocyte adhesion. Am J Physiol 1998; 275:G592-G603.
44. Binion DG, Rafiee P, Ramunujam KS et al. Deficient iNOS in inflammatory bowel disease intestinal microvascular endothelial cells results in increased leukocyte adhesion. Free Radical Biol Med 2000; 29:881-888.
45. Rafiee P, Johnson CP, Li MS et al. Cyclosporine A enhances leukocyte binding by human intestinal microvascular endothelial cells through inhibition of p38 MAPK and iNOS. J Biol Chem 2002; 277:35605-35615.
46. van Kooten C, Banchereau J. CD40-CD40 ligand. J Leukoc Biol 2000; 67:2-17.
47. Schonbeck U, Libby P. The CD40/CD154 receptor/ligand dyad. Cell Molec Life Sci 2001; 58:4-43.
48. Henn V, Slupsky JR, Grafe M et al. CD40 ligand on activated platelets triggers an inflammatory reaction of endothelial cells. Nature 1998; 391:591-594.
49. Liu Z, Colpaert S, D'Haens GR et al. Hyperexpression of CD40 ligand (CD154) in inflammatory bowel disease and its contribution to pathogenic cytokine production. J Immunol 1999; 163:4049-4057.
50. Polese L, Angriman I, Cecchetto A et al. The role of CD40 in ulcerative colitis: Histochemical analysis and clinical correlation. Eur J Gastroenterol Hepatol 2002; 14:237-241.
51. Stuber E, Strober W, Neurath M. Blocking the CD40L-CD40 interaction in vivo specifically prevents the priming of helper T helper 1 cells through the inhibition of interleukin 12 secretion. J Exp Med 1996; 183:693-698.
52. de Jong Y, Comiskey M, Kalled SL et al. Chronic murine colitis is dependent on the CD154/CD40 pathway and can be attenuated by anti-CD154 administration. Gastroenterology 2000; 119:715-723.
53. Danese S, Katz J, Sturm A et al. Activated platelets in the circulation of inflammatory bowel disease (IBD) patients express CD40ligand and induce an inflammatory response in the intestinal microvasculature. Gastroenterology 2002; 122:A-108, (Abstract).

Chapter 12

Role of Mast Cells and Eosinophils in Neuroimmune Interactions Regulating Mucosal Inflammation in Inflammatory Bowel Disease

Stephan C. Bischoff and Thomas Gebhardt

Introduction

The nervous and the immune systems are the body's "supersystems" permanently sensing, processing and responding to changes of the external and internal micro- and macro-environment. Together, both systems provide function, integrity, and homeostasis of the organism, and it is supposed that they do so by bi-directionally interacting with each other at different levels. Indeed, there is a growing body of evidence for neural regulation of immune functions and, vice versa, for impact of immune activation on the function of the nervous system.[1] Mechanistically, this is mainly brought about by the shared use of several ligands and receptors in cell types of both neural and immunological origin. Furthermore, it is supposed that dysregulation of neuroimmune cross-talk occurs during or even might cause various chronic inflammatory conditions.[1,2]

Based on several morphologic and functional studies, tissue mast cells (MC) and, albeit to lesser extent, eosinophils (Eo) are known as key players in neuroimmune interactions at barrier locations such as gut, lung, or skin.[1,3-5] Whereas MC-nerve interactions have been reported in many different studies, evidence for Eo-nerve interaction in the gut is rather limited. Intestinal MCs, for instance, interact with nerves and epithelial cells to regulate various physiologic processes such as intestinal motility, intestinal permeability, or ion and fluid secretion. Furthermore, intestinal MCs and Eos exert proinflammatory and immuno-modulatory properties and are involved in intestinal inflammation, e.g., in the course of allergic enteritis or inflammatory bowel disease (IBD). These observations led to the hypothesis that activation of MCs and possibly also Eos may be closely related to a dysregulated neural coordination of gut function as observed in chronically inflammed tissue.

In this chapter, we will review current knowledge about the regulation and function of human intestinal MCs and Eos and their role in intestinal inflammation, particularly in ulcerative colitis and Crohn's disease. Moreover, we will try to summarize several studies providing evidence for a crosstalk between nerves and intestinal MCs and Eos, respectively, and we will discuss the possible role of such interaction in the pathogenesis of IBD.

Immune Mechanisms in Inflammatory Bowel Disease, edited by Richard S. Blumberg and Markus F. Neurath.

Intestinal Mast Cells and Eosinophils

Distribution of Mast Cells and Eosinophils in the Gut

The normal human gastrointestinal tract contains numerous Eos and MCs, the latter preferentially located in close association with endothelial cells, fibroblasts, and neurons. Based on the content of neutral serine proteases tryptase and chymase, human MCs are classified in subtypes thought to differ in distribution, cytokine expression pattern, and responsiveness towards certain agonists.[6-8] The highest number of MCs is found in the lamina propria, where 2-3% of the cells are MCs, predominantly of the tryptase positive, chymase negative subtype (MC_T), resembling the mucosal MC subtype in rodents. In the submucosa, MC density is lower compared to the lamina propria (about 1% of all cells), and the frequency of the tryptase, chymase double positive MC subtype (MC_{TC}), resembling the connective tissue MCs subtype in rodents, is higher than that of the MC_T subtype. In accordance with MC distribution, highest numbers of Eos are found in the lamina propria where they represent approximately 4% of the cells in healthy human tissue of adults, whereas Eo numbers in the submucosa are markedly lower.[7-10] The gut is thought to be the most important Eo host, since, normally, several hundred times as many Eos are present in tissue such as gut, spleen, lymph nodes or thymus compared to the blood.[11] Both, Eos and MCs are rare or absent in the epithelial compartment, the lamina muscularis, and the serosa of the human gut under normal conditions.[7-10] However, under certain pathological conditions such as chronic inflammation or fibrotic tissue transformation, these distribution patterns are found to be profoundly altered with respect to absolute or relative cell numbers and subtype distribution.[9,12]

Regulation of Intestinal Mast Cell and Eosinophil Density

MC and Eo densities in the intestine are regulated by changing the influx of progenitor or readily differentiated cells, and the growth factor-dependent survival and proliferation rate within the tissue, respectively.

MCs originate from immature, bone marrow-derived CD34+ hematopoietic stem cells circulating in the peripheral blood as committed progenitors before homing to distinct anatomical sites where they achieve full maturation regulated by the local cytokine milieu and by the interaction with connective tissue cells such as fibroblasts or endothelial cells.[8] Human MC survival, growth, and development in tissue is crucially dependent on stem cell factor (SCF) which is expressed by many tissue cells such as fibroblasts, endothelial cells, and stromal cells.[8,13-15] Accordingly, in mice, loss of SCF function due to a mutation of either SCF or its receptor ***c-kit*** leads to MC deficiency, whereas gain of function mutations of ***c-kit*** in humans are associated with mastocytosis.[16-18] Moreover, SCF stimulates adhesion of cultured human intestinal MCs to extracellular matrix proteins (ECM) and to endothelial cells expressing similar adhesion molecules as ECM.[14,19] The MC-endothalial cell adhesion involves the interaction of adhesion molecule family members VLA-4, VCAM-1, LFA-1, ICAM-1, E-selectin, and sialylated Lewis-X antigen.[14,19]

The crucial role of $\alpha 4\beta 7$ integrin expressed on MC precursors, and the mucosal addressin cell adhesion molecule-1 (MAdCAM-1) expressed on high endothelial venules in the intestinal lamina propria for the gut homing of MC precursors has been demonstrated recently. $\beta 7$-deficient mice were found to virtually lack the MC population in the small intestine, whereas other organs such as lung, spleen, or large intestine displayed normal MC numbers.[20,21] Our own data show that fully maturated intestinal MCs are capable of proliferating in vitro in the presence of SCF, providing evidence that expansion of MC numbers in tissue is not only regulated by progenitor cell influx but also by local cell proliferation. Cytokines such as IL-4 and IL-3 were found to promote in vitro growth of isolated mature human intestinal MCs in synergism with SCF whereas TGF-β_1 inhibited it (refs. 22, 23 and our own unpublished observations). Chemotactic responses of human MCs can be elicited by a variety of stimuli including SCF, IL-8, TGF-β_1, eotaxins, RANTES, or complement components, therefore, possibly contributing to MC tissue distribution under normal and pathologic conditions.[24-26]

In contrast to MCs, Eos leave the bone marrow as mature cells circulating in peripheral blood and subsequently migrating into tissues such as intestine, spleen, or thymus.[11] In addition, peripheral blood also contains a small population of CD34+ progenitors of the basophilc/eosinophilic linage that vary in numbers e.g., during pollen season in sensitized individuals.[27] Eo survival, differentiation, and mobilization from the bone marrow is stimulated by particular cytokines such as IL-3, IL-5, and GM-CSF, and is inhibited by TGF-β_1.[11,28] Moreover, IL-1, IL-4, IL-13, and TNF-α regulate Eo trafficking by promoting adhesive interaction with endothelial cells which involves members of integrin and selectin families such as VLA-4, VCAM-1, LFA-1, ICAM-1, E-selectin, and sialylated Lewis-X antigen.[29] In analogy to its important role in the gut homing of intestinal lymphocytes and MC precursors, the $\alpha4\beta7$ integrin also critically regulates intestinal recruitment of Eos. Accordingly, mice with targeted disruption of the gene encoding for $\alpha4\beta7$ showed delayed influx and reduced magnitude of intestinal eosinophilia in response to helminth infection.[20] Among other Eo chemoattractants such as chemokines (e.g., RANTES, MCP-3, MIP-1α, IL-8), complement factors (e.g., C3a, C5a), and lipid metabolites (e.g., platelet-activating factor, leukotrienes, prostaglandins), the chemokine eotaxin seems to critically regulate Eo localization to the intestinal lamina propria under normal and pathologic conditions.[30-32] This was highlighted by a selective reduction of gastrointestinal Eos in eotaxin-deficient mice, whereas hematopoietic compartments showed normal levels of Eos in these animals.[33]

An overview on the origin and development of intestinal MCs and Eos is given in Figure 1.

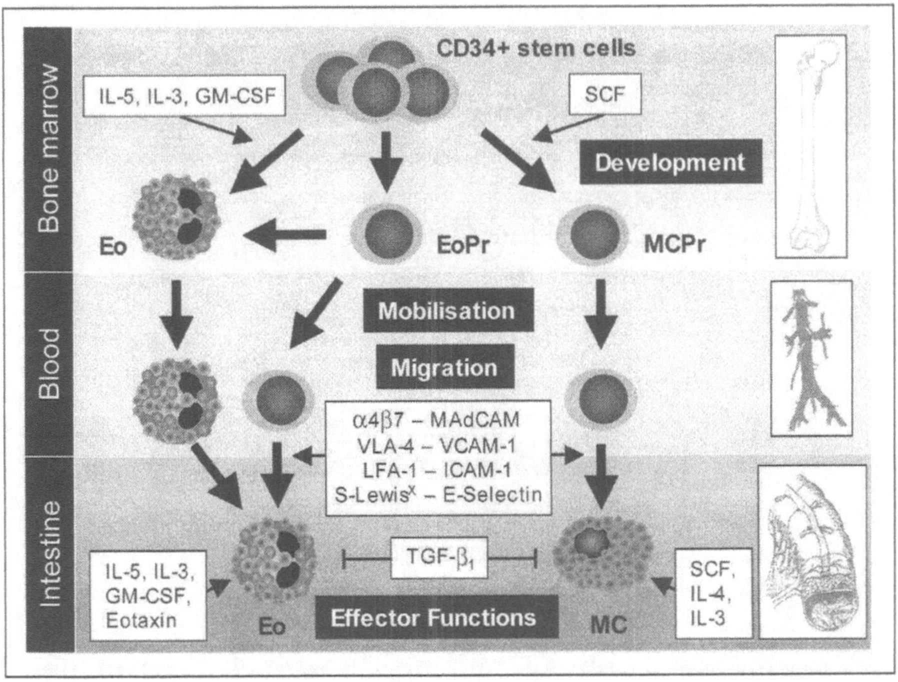

Figure 1. Origin and development of intestinal MCs and Eos. MC and Eo progenitors derive from CD34+ stem cells in the bone marrow. Eos primarily mature in the marrow and migrate via peripheral blood to the intestine. Additionally, Eo progenitors (EoPr) are also found in the peripheral blood and are recruited to the intestine where they mature. MC progenitors (MCPr) leave the marrow and migrate via peripheral blood to the intestine where they mature under the influence of local micro milieu. Important cytokines and adhesion molecules regulating development, growth, migration, and effector functions of both cell types are indicated.

Effector Functions of Human Intestinal Mast Cells

Both, MCs and Eos exert proinflammatory and immunoregulatory functions by releasing a broad variety of preformed and newly synthesized mediators upon stimulation (Figs. 2, 3).

In the gut, MC mediators such as histamine, leukotrienes, or prostaglandins mediate smooth muscle contraction, vasodilatation, leukocyte extravasation, epithelial ion secretion, and mucus secretion thought to be of importance for host defense, volume regulation and tissue homeostasis. The neutral proteases tryptase and chymase are known to mediate biological effects by degradation of peptides such as fibronectin, vasoactive intestinal peptide (VIP), or other neuropeptides, by regulating fibroblast or endothelial cell function, or by activating the complement, kalikrein-kinin, and angiotensin-aldosterone systems.[34] Beyond these well known biological effects, new important roles of such MC mediators are now emerging. For instance, histamine has recently been shown to regulate cytokine expression in murine T helper cells depending on their differential expression of histamine receptor subclasses.[35] Moreover, prostaglandin D_2 has been identified as a specific chemoattractant for cells implicated in allergic inflammation, namely type 2 T helper and cytotoxic cells, Eos, and basophils, binding to the CRTH2 receptor known to be selectively expressed in these cells.[36] In addition to these mediators, MCs produce multiple cytokines, growth factors, and neuropeptides such as TNF-α, IL-1, IL-3, IL-5, IL-6, IL-8, IL-13, IL-16, IL-18, TGF-β_1, basic fibroblast growth factor (βFGF), nerve growth factor (NGF), VIP, and others, emphasizing their important role in

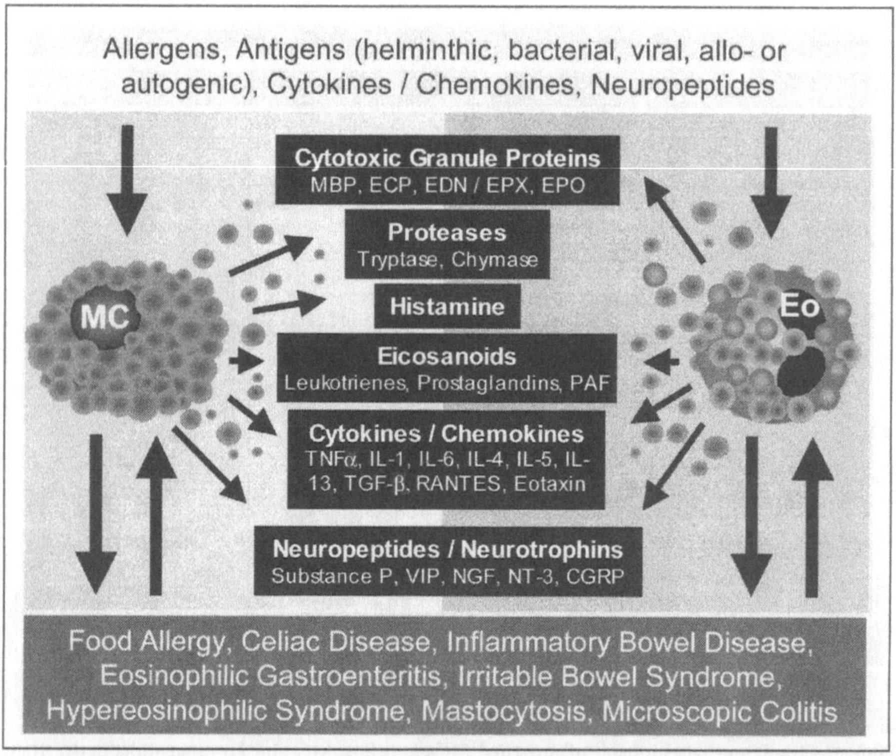

Figure 2. MC and Eo agonists, mediators, and associated intestinal diseases. Abbreviations: MBP, major basic protein; ECP, eosinophil cationic protein; EDN/EPX, eosinophil-derived neurotoxin; EPO, eosinophil peroxidase; PAF, platelet-activating factor; VIP, vasoactive intestinal peptide; NGF, nerve growth factor; NT-3, neurotrophin-3; CGRP, calcitonin-gene related peptide.

immunoregulation, tissue homeostasis, host defense against infection, or fibrotic tissue transformation.[8,37-42]

Crosslinking of cell surface-bound IgE by antigen or, experimentally, of IgE receptors by anti-IgE receptor antibodies, is the most potent and best characterized stimulus for MC activation. In addition, several IgE-independent triggering agents have been described acting on human skin or rodent MCs such as C5a, substance P (SP), VIP, morphine, f-Met, compound 48/80, or chemokines.[43,44] However, using human intestinal MCs freshly isolated from surgery specimens or cultured in the presence of SCF, we found no mediator release following challenge with these IgE-independent agonists, thus supporting the concept of MC heterogeneity at distinct anatomical locations (ref. 45 and our own unpublished observations). So far, only a few IgE-independent triggers causing mediator release in human intestinal MCs have been identified. Gram-negative bacteria cause histamine and TNF-α release.[38] SCF triggers for histamine release and leukotriene C_4 production under certain conditions.[45] and IL-4 induces IL-5 production in MCs cultured in medium supplemented with SCF.[22,39] Mediator release following IgE receptor-crosslinking in cultured human intestinal MCs is altered depending on the presence of cytokines such as IL-4, IL-3, and TGF-β_1 in addition to SCF. IL-4 and IL-3 support the release of histamine and production of leukotriene C_4, but not of prostaglandin D_2, in intestinal MCs following activation.[22,23] In contrast, TGF-β_1 inhibits release of histamine and production of leukotriene C_4, but, interestingly, primes intestinal

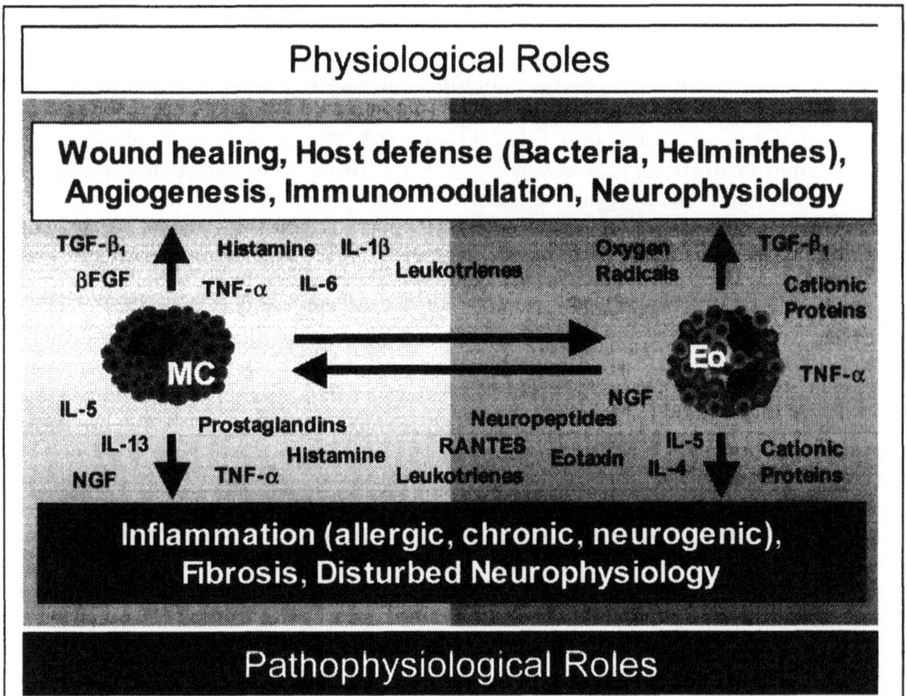

Figure 3. Dual role of intestinal MCs and Eos. On the one hand, both cell types exert physiological functions such as regulation of wound healing, defense against microorganisms, immune functions, and intestinal neurophysiology. On the other hand, both cell types are also implicated in a variety of pathological processes such as inflammation, tissue destruction, and disturbance of intestinal neurophysiology. Depending on the context they are released in, MC and Eo-derived mediators and cytokines either exert beneficial or harmful effects.

MCs for enhanced prostaglandin D_2 production (our own unpublished observations). Moreover, IL-4 profoundly alters the cytokine expression pattern in cultured human intestinal MCs, leading to an increased production of Th2-like cytokines such as IL-3, IL-5, and IL-13 following IgE-dependent activation.[39]

Effector Functions of Human Intestinal Eosinophils

The secondary or specific granules of Eos contain major basic protein (MPB), eosinophil cationic protein (ECP), eosinophil-derived neurotoxin (EDN, EPX), eosinophil peroxidase (EPO), some other other enzymes of uncertain significance, and proteins including a broad range of preformed cytokines and chemokines[30] (Fig. 2). MBP is cytotoxic for helminthic parasites and mammalian cells, activates the complement cascade, and leads to increased smooth muscle reactivity by causing dysfunction of vagal muscarinic M_2 receptors.[30,46] Moreover, MBP was shown to stimulate SP release from neonatal rat dorsal root ganglia neurons.[47] ECP, EPO, and EDN cause cytotoxic effects in helminthic parasites and mammalian cells by exerting ribonuclease activity or generation of unstable oxygen radicals, respectively. Among the cytokines and chemokines produced by eosinphils are IL-1, IL-2, IL-3, IL-4, IL-5, IL-6, IL-9, IL-10, IL-12, IL-16, TGF-β_1, IFN-γ, GM-CSF, TNF-α, eotaxin, IL-8, MIP-1α, RANTES, NGF, SCF, platelet-derived growth factor (PDGF) and others.[48]

Most of the studies investigating the ability of human Eos to produce cytokines were performed using cells isolated from peripheral blood and it is unclear to which extent these data can be easily transferred to tissue residing Eos. In this respect, it is interesting to note phenotypic modulation of Eos that include e.g., induced expression of activation markers CD69, ICAM-1, and CD25 as Eos move into tissue. However, few studies were performed using human intestinal tissue. These studies could demonstrate in situ expression of IL-3, IL-5, GM-CSF, IL-16, and TGF-β_1 in Eos residing in the intestinal mucosa.[49-51] In addition, human blood and intestinal Eos were found to store VIP, SP, calcitonin-gene related peptide (CGRP) and somatostatin, respectively.[52,53] When activated, Eos release these preformed mediators and, additionally, generate reactive oxygen species and lipid mediators such as leukotrienes, platelet-activating factor, or prostaglandins.[30] Under normal conditions, i.e., in noninflamed tissue, Eos are in a resting state to prevent tissue damage from Eo-derived mediators. In contrast to the outstanding role of IgE receptor crosslinking for activation of MCs, activation of Eos can be mediated by a broad range of stimuli, most of them also mediating survival or chemotaxis. Activation of Eos involves initial priming and subsequent triggering of the cells, and is mediated by cytokines such as IL-5, IL-3, GM-CSF, or TNF-α, by chemokines such as RANTES, MIP-1α, MCP-3, MCP-4, eotaxin, or IL-8, by complement components, by aggregated immunoglobulins, by lipid mediators, or by histamine.[54] Interestingly, many of these agonists are known as MC secretory products, thus MC activation might be involved in the rapid activation of Eos observed during allergic reactions.[55]

Apart from the production of these mediators, other immunological effector functions of human MCs and Eos such as phagocytosis and antigen presentation are poorly characterized. Recent studies in the murine system indicate that both cell types express MHC and accessory costimulatory molecules and are able to induce expansion of antigen-specific T cells.[56,57] However, the significance of this finding is unclear to date, and whether human MCs and Eos that also express MHC and accessory molecules are able to activate lymphocytes in an antigen-specific fashion in vivo has to be confirmed.[58,59]

Role of Mast Cells and Eosinophils in Intestinal Inflammation

The high numbers of MCs and Eos found in the gastrointestinal tract and their potency to release proinflammatory and immunoregulatory mediators lead to the speculation that they are involved in gastrointestinal pathologies. Indeed, altered numbers and activation states of MCs and Eos have been observed in several intestinal diseaes[30,34,60] (Fig. 2).

The pathogenesis of IBD is still an enigma. Obviously, not only a single etiologic agent is involved, but rather different cofactors that individually interact with each other leading to the chronic intestinal inflammation causing symptoms such as abdominal cramps, diarrhea, or rectal bleeding, respectively. Among the factors discussed are environmental factors such as diet, smoking, and intestinal flora, as well as genetic factors determining intestinal permeability, MHC-alleles, and mucosal immune functions.[61]

The speculation of MC and Eo involvement in IBD is mainly based on morphologic findings of altered cell numbers and elevated mediator concentrations found in these patients. Enhanced numbers of Eos have repeatedly been reported for both, ulcerative colitis and Crohn's disease, associated with ultrastructual evidence for activation, degranulation and deposition of granule-associated proteins.[9,62-67] Moreover, enhanced levels of ECP, EDN, and EPO in feces and gut perfusates, and increased release of ECP from activated peripheral blood Eos or from colorectal tissue samples during active phase of disease were consistently reported by several investigators.[65-77] Similar observations have been made with respect to MC accumulation and activation during the course of IBD. Consistently, accelerated MC numbers were observed during active ulcerative colitis, whereas for Crohn's disease the data are conflicting.[9,12,62,78-84] Using light microscopy to investigate tissue sections from Crohn's disease patients stained with May-Grünwald-Giemsa or with antibodies against MC tryptase, decreased MC numbers were reported in actively inflamed regions. However, this finding might be related to difficulties in detecting readily degranulated MCs since approaches using electron microscopy revealed markedly increased MC numbers with evidence of focal and complete degranulation.[62,83]

Of particular interest is the observation that MCs may not only act as pro-inflammatory effector cells but also as regulatory cells promoting wound healing following tissue destruction. In this respect, it is interesting to note that MC accumulation was observed in particular at the demarcation line of active ulcerative colitis and at sites of strictures in Crohn's disease.[12,82] Since strictures may be regarded as an abnormal healing process, this further supports the hypothesis that MCs are involved in tissue remodeling. Likely, distinct MC subpopulations are involved in inflammatory and healing processes in IBD since the MC_T subtype is thought to exert predominantly pro-inflammatory effects, whereas the MC_{TC} subtype mainly found in deeper layers such as submucosa and muscularis might be specialized to interact with fibroblasts by releasing anti-inflammatory and tissue remodeling factors such as TGF-β1 and basic fibroblast growth factor. Furthermore, elevated levels of tryptase, histamine, or methyl-histamine measured in feces, gut perfusates, tissue samples, or urine, or increased in vitro mediator release from tissue samples following challenge with anti-IgE antibodies or SP from IBD patients strongly indicate a role of MCs in IBD.[71,84-89] In addition, in vitro studies with MCs isolated from intestinal tissue demonstrated increased release of histamine, leukotrienes, prostaglandins, and TNF-α in cells derived from IBD patients compared to controls, suggesting that they were primed or preactivated by the local inflammatory tissue condition.[38,45,81,90]

Since Eos and MCs are regulated to a considerable extent by cytokines, it is important to note alterations in cytokine expression patterns found in the intestinal mucosa of IBD patients. IBD is characterized by an enhanced expression of proinflammatory cytokines such as IL-1β, IL-6, IL-8, and TNF-α in actively inflamed tissues.[61,91] These pro-inflammatory cytokines support MC progenitor and Eo recruitment by stimulating endothelial cells to express elevated levels of adhesion molecules such as VCAM-1 and ICAMs, or, as in the case of IL-8, by directly mediating chemotaxis of MC progenitors and Eos.[26,29,30] Furthermore, MAdCAM-1 was found to be more widely expressed at inflammatory foci of ulcerative colitis and Crohn's disease than in controls.[92] Distinct cytokine expression patterns related to the predominance of Th1 or Th2 polarized lymphocyte subsets during the course of IBD have been proposed also. It was hypothesized that long-standing Crohn's disease is related to a Th1-dominated immune response characterized by enhanced levels of IL-2, IL-12, and IFN-γ whereas ulcerative colitis and early phases of Crohn's disease show typical features of a Th2-type response with increased levels of IL-5 and IL-4 promoting intestinal Eo and MC survival and proliferation, respectively.[93-95] Moreover, enhanced expression of chemokines such as eotaxin, RANTES, and MCP-3 known to mediate chemotaxis of Eos and MCs was observed in tissues derived from IBD patients.[31,96,97]

TGF-β_1 known for its crucial role in mediating oral tolerance and its profound suppressive effects on intestinal MCs and Eos (ref. 28 and our own unpublished observations) was reported to be highly expressed in inflammatory infiltrates in IBD.[98,99] Most interestingly, a recent study[100] demonstrated over-expression of Smad7 in IBD mucosa and mucosal T cells leading to impaired TGF-β signaling and abrogation of immunoregulatory properties of exogenous TGF-β_1. The isolated T cells restored responsiveness towards exogenous TGF-β_1 following treatment with specific antisense oligonucleotides against Smad7.[100] Thus, it might be hypothesized that over-expression of TGF-β_1 under these conditions reflects an ineffective compensatory mechanism due to impaired signaling pathways or a counter-regulatory mechanism to suppress excessive inflammation. These findings on Smad7 may be related to the recent reports of a 3020insC variant of NOD2 gene in families with Crohn's disease.[101,102] Since the wild-type NOD2 gene product activates NF-κB to interfere with Smad7 expression,[103,104] a mutated NOD2 protein might lead to impaired NF-κB response to pro-inflammatory signals (e.g., challenge with bacterial antigens) and in turn might result in increased Smad7 expression and abrogation of the TGF-β signaling cascade. This could finally induce breakdown of oral tolerance mechanisms and subsequent chronic tissue inflammation, destruction and fibrotic transformation as seen in the course of IBD.

Structure and Function of The Enteric Nervous System

In the gastrointestinal tract, coordination and control of a wide variety of physiological processes has to be orchestrated constitutively. These include transport of luminal contents, secretion of ions and water, absorption of ions, water and nutrients, regulation of blood supply, elimination of waste and noxious substances, defense against pathogens, or tolerance induction towards nonpathogenic antigens. To fulfill these functions, the gut wall is equipped with approximately 10^8 interconnected neurons that together represent, second to the brain, the largest accumulation of nerve cells in the body. Moreover, in contrast to most other organs, this gut-residing part of the nervous system is able to exert pivotal functions independently of higher control centers in the central nervous system (CNS) even ex vivo when isolated from the body. These unique features lead to the term enteric nervous system (ENS) as a third independent part of the autonomic nervous system.[105] Unlike sympathetic or parasympathetic ganglia that mainly represent relay distribution centers for signals transmitted from the CNS to the peripheral organs, ENS ganglia are highly interconnected with each other to allow local integrative processing and storage of a library of programs for distinct patterns of gut behavior. Since this network of neurons structurally and functionally resemble higher integration centers found in the brain or in the spinal cord, the ENS is also referred to as the "local minibrain" or "enteric" or "second brain".[106,107] Influence on ENS functions from higher CNS centers as well as flux of information from the sensory apparatus of the gut to the CNS occur through efferent and afferent neurons, respectively, belonging to the sympathetic and parasympathetic nervous systems. Parasympathetic innervation of the gut via the vagus nerve and pelvic sacral nerves increases the tone of gastrointestinal smooth muscles, enhances peristalsis, relaxes intestinal sphincters, stimulates exocrine secretion from the glandular epithelium, and enhances secretion of peptide hormones from enteroendocrine cells. In contrast, sympathetic stimulation of the gut via splanchnic, hypogastric and colonic nerves mainly leads to opposing effects. Rather than directly controlling effector functions at the single cell level, these pre or post-ganglionic fibers project to ganglia within the two main plexuses of the ENS, namely the myenteric plexus located between the circular and the longitudinal muscle layers and the submucous plexus situated between the submucosa and the mucosa. Via interconnected networks of sensory neurons, interneurons, and secretomotor neurons, the submucous plexus mainly regulates

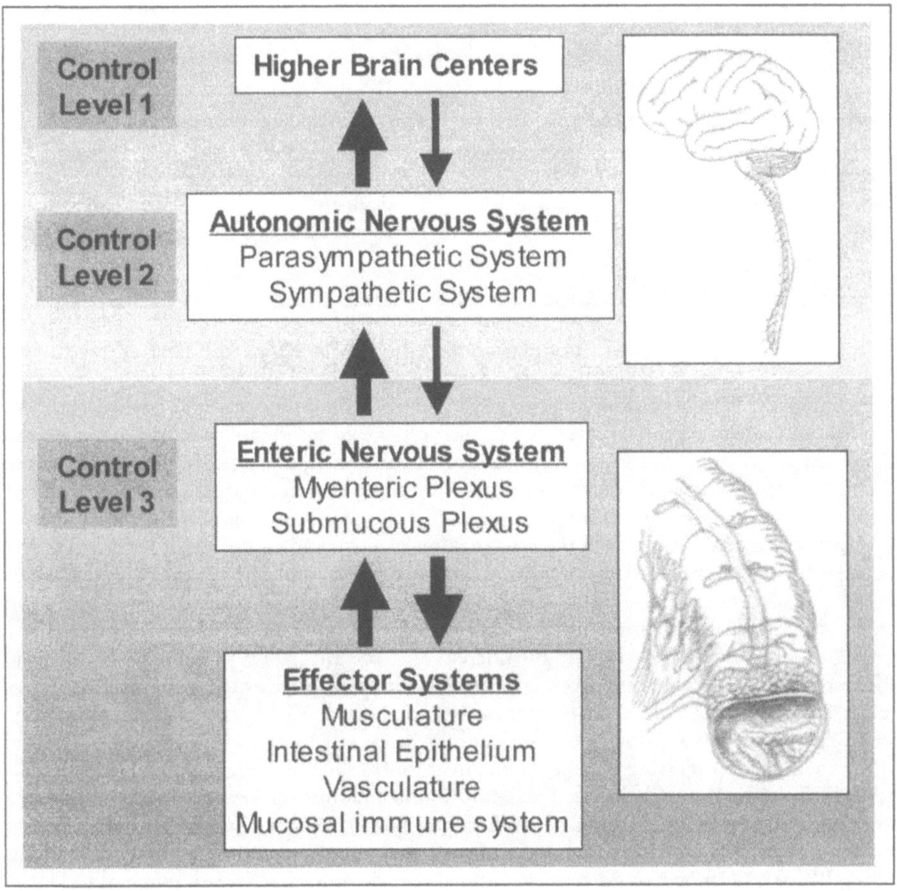

Figure 4. Hierarchy of neuronal input into the gastrointestinal (GI) tract and its effector systems. Only 10% of axons are of afferent origin projecting from higher control levels in the spinal cord or in the brain to the GI tract. In contrast, 90% of axons are of efferent origin transmitting information from the GI tract to central nervous structures. Interestingly, the enteric nervous system is autonomous in nature, capable of regulating intestinal functions independently of higher control centers. Adapted from Wood[106] and Schemann.[108]

mucosal functions and blood supply, whereas the myenteric plexus controls smooth muscle activity and coordinated peristaltic movements in cooperation with the nonneural pacemaker system comprising of the interstitial cells of Cajal.[106] This kind of circuitry explains the relative small number of autonomic efferent fibers found in the gut and why, for instance, efferent fibers of the vagus nerve represent only approximately 10% of all vagal fibers whereas the great majority of them are of afferent function. The hierarchy underlying this integrative regulation of gastrointestinal function at the distinct levels is summarized in Figure 4.

Neuron subclasses in the gut differ in features such as location, extent and polarization of projections, content of and responsiveness towards certain neurotransmitters (neurochemical code), formation of synaptic connections, and the extent to which they are modulated by nonneural substances such as inflammatory mediators. Information flux across synaptic connections between different gut neurons may be regulated by presynaptic modulation via acetylcholine, histamine, neuropeptide Y (NPY), nitric oxide (NO), norpeinephrine (NE), or

Table 1. Some intestinal diseases characterized by altered ENS function

With Obvious Impact of the ENS	With Possible Impact of the ENS
Irritable bowel syndrome	Inflammatory bowel disease
Adynamic Ileus	Food allergy
Hirschsprung's disease	Pseudomembranous colitis
Achalasia	
Intestinal pseudoobstruction	

serotonin, and by postsynaptic modulation via enterochromaffine cell-derived mediators, inflammatory mediators, exogenous substances, nutrients, or toxins.[108]

Functionally, afferent neurons, interneurons, and secretomotor neurons can be distinguished. Extrinsic and intrinsic primary afferent neurons respond to information on luminal chemical milieu, deep body temperature, and mechanical alterations. In this process, mediators such as SP and serotonin released by neighbor cells such as enterochromaffine cells upon chemical or mechanical stimulation play an important role.[108] The precise function of different types of nociceptors or mechanoceptors in the transmission of acute and chronic pain sensation is still not fully elucidated. Interestingly, so called silent nociceptors implicated in the chronic form of visceral pain become sensitized by inflammatory mediators resulting in markedly lower activation thresholds.[106] The interneurons form a wide network of interconnected neurons that allow integrative information flux through the efferent and afferent pathways required for intrinsic reflex programs. Transmitters involved in the function of different interneuron subtypes include acetylcholine (ACh), somatostatin, serotonin, NO, or VIP.[109] Intrinsic motor neurons receive signals from local primary afferent neurons, from ascending and descending pathways, and by themselves projecting in both directions to form synaptic contact within the circular and the longitudinal muscle layers. Excitatory motor neurons release ACh, while inhibitory motor neurons act via NO, adenosine triphosphate (ATP), VIP, or pituitary adenylate cyclase activating peptide (PACAP). In addition, secretomotor and vasomotor neurons containing ACh, NPY, or VIP project to the mucosa and regulate the intestinal secretion rate of fluid and ions, as well as blood supply via vasodilator mechanisms.[108]

All these different neuron types act in concert to perform meaningful gut behavior. The required coordination is regulated by the different control centers within the gut wall in the submucous and the myenteric plexus, whereas, less complex, direct axon reflex programs include only a small number of neurons. Peristalitic movement as a common motor programm, for instance, depends on aboral relaxation and simultaneous oral contraction of a high number of smooth muscle cells, and is mediated by excitatory and inhibitory interconnected motor neurons. Furthermore, the impact of the ENS is emphasized by the list of diseases characterized by on impaired neural and motility functions, respectively, of the gut (Table 1).

Mechansims of Mast Cell / Eosinophil-Neuroimmune-Crosstalk in the Gut

Anatomic Basis for Neuroimmune-Crosstalk Involving Mast Cells and Eosinophils

Peripheral nerves branch in every organ of the organism and target several types of immuno-competent tissue cells. Functional impact of such colocalization has been observed for fibroblasts, epithelial cells, bone marrow stromal cells, dendritic cells, macrophages, MCs, Eos, basophils, lymphocytes, endothelial cells, pericytes, or thymocytes.[1] Especially, the phenomenon of MC-nerve associations was addressed by several morphologic studies and evidence for dense innervation of MCs in tissues such as skin, lung, intestine, lymphoid tissue, or synovial tissue was reported in animals and humans.[5] Moreover, MCs were also frequently found in peripheral nerves, in autonomic ganglia, and within different regions of the CNS with preferential accumulation in the thalamus.[110,111]

The first morphological evidence for intimate MC-nerve association in the gastrointestinal tract arose from studies by Stach and Heine, and later, ultrastuctural membrane to membrane contact of both cell types in the rat intestine was observed by Newson et al.[111-113] Stead and coworkers quantitatively investigated the microanatomical relationship between MCs and enteric nerves in normal rat and human intestine, as well as in parasite-infected rat intestine, and reported extensive associations.[114] In normal rats, 50% of the mucosal MCs were in close proximity to nerve fibers. In nematode infected rats showing mucosal MC hyperplasia, 67% of MCs were touching subepithelial nerves, and an additional 20% were located within 2 microns of nerves. These nerves stained positive for SP, ACh and/or CGRP, respectively, and vagal origin of some afferent fibers was identified by injecting marker substances in the nodose ganglion.[5,114-116] Using electron microscopy, 8% of MCs were found to exhibit membrane to membrane contact with unmyelinated axons, and an additional 31% were situated less than 250 nm from nerves. Interestingly, some MCs appeared to even embrace nerve bundles through extended pseudopodia.[114] In the human gastrointestinal tract, similar observations with respect to frequency and type of MC-nerve associations were demonstrated. Depending on the different anatomical regions, 47-77% of all mucosal MCs were juxtaposed to nerves with highest incidence of contact in the appendix,[117] and also membrane to membrane contact formations were observed.

The extent to which Eos appose in close proximity to nerves in different tissues has been studied less extensively. However, there were some reports indicting that Eo-nerve associations occur in skin, lung and intestine in rodents and / or man, respectively.[115,118,119] In the jejunal mucosa of normal rats, 3.3% of Eos occurred within 100 nm of neural processes and an additional 23.3% between 101 and 500 nm of these structures.[115]. Accordingly, studies in human tissue obtained from patients with early gastrointestinal involvement in the course of systematic connective-tissue disease revealed intimate association of Eos and neurons within the submucous and myenteric plexus.[120]

Possible Molecular Mechanisms of Mast Cell / Eosinophil-Nerve Associations

The mechanisms underlying the close association of MCs / Eos and nerves in diverse tissues are unclear, however , several studies point towards a role of neurotrophins and cytokines in this respect. Neurotrophic growth factors such as NGF, brain-derived neurotrophic factor (BDNF), neurotrophins (NT)-3 to 7, glia cell-derived neurotrophic factor, neuroturin, persephin, ciliary neurotrophic factor (CNTF), or leukaemia inhibitory factor (LIF) play important roles during neuronal development and induce proliferation, differentiation, survival, and transmitter production in cholinergic and petidergic neurons.[121,122] Interestingly, many of these neuronal growth factors and cytokines are also produced by MCs and Eos, and MCs and Eos by themselves express neurotrophin receptors and respond to these factors (Figs. 5, 6). Expression of NGF was observed in rat peritoneal MCs and in human MCs such as the immature MC line HMC-1 or cord blood-derived MCs that, in addition, were shown to express mRNA encoding for BDNF and NT-3.[123-125] Very recently, expression, storage, and release of NGF and NT-3 was also reported for human Eos.[126] Moreover, conditioned medium from HMC-1 cells was able to induced NGF-dependent neurite outgrowth from cultured chicken embryonic sensory ganglia, and, accordingly, NGF in the supernatant of stimulated Eos was sufficient to promote neurite extension of cultured PC-12 pheochromocytoma cell line.[125,126] Furthermore, direct trophic and functional interaction between MCs and nerves could be demonstrated using a coculture system of murine superior cervical ganglia with rat peritoneal MCs or rat leukemia basophilic (RBL) cells known to resemble mucosal MCs.[127-131] Another neurotrophin produced by rat MCs and human Eos is LIF, that, apart

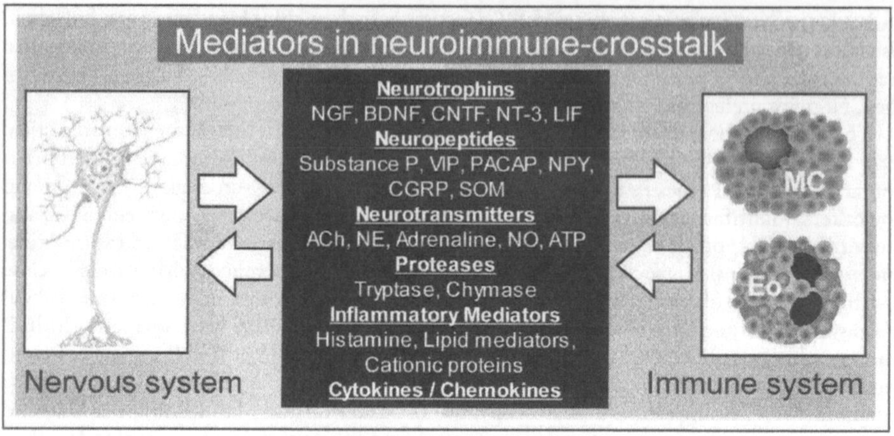

Figure 5. Mediators involved in neuroimmune-crosstalk are produced by and act on cells of both, the nervous and the immune system. Noteworthy, neuropeptides and some neurotrophins are also produced by MCs and Eos, respectively. In contrast, most of the classical neurotransmitters are of neuronal origin but act on MCs and Eos. Abbreviations: NGF, nerve growth factor; BDNF, brain-derived neurotrophic factor; CNTF, ciliary neurotrophic factor; NT-3, neurotrophin-3; LIF, leukaemia inhibitory factor; VIP, vasoactive intestinal peptide; PACAP, pituitary adenylate cyclase activating peptide; NPY, neuropeptide Y; CGRP, calcitonin-gene related peptide; SOM, Somatostation; ACh, acetylcholine; NE, norepinephrine; NO, nitric oxide; ATP, adenosine triphosphate.

from its immuno-regulatory properties, is also known as a differentiation factor for cholinergic neurons and a promoter of SP production, and, therefore, might support MC / Eo-nerve interactions.[132-134]

Apart from neurotrophins, also cytokines such as IL-1, IL-5, IL-9, βFGF, PDGF or TGF-β known to be expressed in MCs or Eos, respectively, display neurotrophic activities and support neurite outgrowth to form cell to cell contacts.[40,48,135,136] Alternatively, since neurons are capable of expressing different chemokines and respond with enhanced production following challenge with proinflammatory stimuli,[137] MCs and Eos might be specifically recruited to nerves by expression of chemokines such as eotaxin or RANTES under inflammatory conditions. Such inflammation-induced recruitment to nerves was reported, for instance, in a guinea pig asthma model, where Eos selectively migrated to airway nerves following antigen challenge.[118] In the same study, this phenomenon was also reported in human patients with fatal asthma where a dense infiltration of Eos around the airways was observed. In addition to possibly involved chemokines, several neuropeptides known to be released by afferent nerves were found to exert chemotactic effects on human Eos in vitro and, therefore, might contribute to inflammation-induced enhancement of Eo-nerve associations.[138]

Effects of Neurotrophins on Mast Cells and Eosinophils

A great number of neurotrophins, neuropeptides, and neurotransmitters has been investigated for their possible role in regulating MC and Eo growth and function in the last 25 years (Fig. 6; Tables 2, 3). Among them, NGF gathered the widest attention, since Aloe and Levi-Montalcini reported increased MC numbers in tissues of neonatal rats following NGF administration.[139] In accordance with this finding, NGF was later shown to promote in vitro growth and differentiation of murine and human MCs and to exert chemotactic effects in rat peritoneal MCs.[140-145] Similar effects were also reported for human Eos whose differentiation and survival was found to be enhanced by NGF.[146,147] Several studies identified NGF also as a MC activator or priming agent directly inducing murine and human MC degranulation in the presence of phosphatidyl serine,[148,149] and, as shown for murine MCs, enhancing production

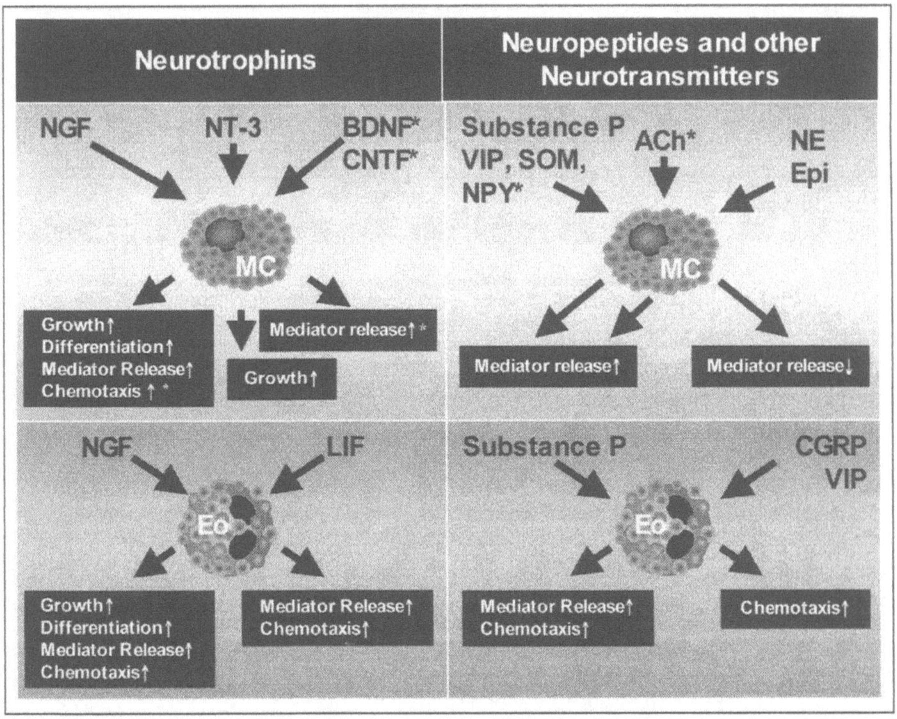

Figure 6. Effects of some neurotrophins, neuropeptides and other neurotransmitters on MCs and Eos. Note that effects of the different mediators on MCs were investigated in MCs of different species and different anatomical locations. Due to MC heterogeneity, these effects are not easily transferable to human intestinal MCs in general. All effects shown for Eos were observed with human cells. *Only shown for murine MCs. Abbreviations: See legend of Figure 5.

and release of several cytokines and proinflammatory mediators.[150-153] Moreover, NGF was shown to induce chemotaxis and support the larvicidal activity of human Eos,[147] whereas others found release of EPO in response to rather high concentrations of NGF.[154]

Apart from NGF, other neurotrophic growth factors were found to affect MC and Eo function (Fig. 6; Tables 2, 3). BDNF and CNTF, for instance, induced histamine release of isolated rat thalamic MCs,[155] whereas preincubation with LIF resulted in increased EPO release and chemotactic activity in human Eos following stimulation.[133] Furthermore, our own unpublished results show that NT-3 supports the SCF-dependent growth of in vitro cultured human intestinal MCs.

Effects of Neuropeptides and Other Neurotransmitters on Mast Cells and Eosinophils

Several neuropeptides have been investigated for potential effects on MC and Eo function (Fig. 6; Tables 2, 3). Potent secretagogue effects were observed for NPY, SP, somatostatin, VIP, and neurotensin in rat peritoneal MCs.[156,157] In accordance with this, human skin MCs resembling the MC subtype found in the rat peritoneum released histamine upon challenge with SP, VIP, somatostatin, and PACAP.[43,158] In contrast to this MC subtype, intestinal mucosal MCs in rats were only responsive towards SP, whereas other neuropeptides failed to induce mediator release.[157] Human mucosal MCs isolated from normal lung or gut tissue, however, were not affected by any of these neuropeptides indicating heterogeneous regulation of MCs

Table 2. Effect of neurotrophins, neuropeptides and other neurotransmitters on mast cells

	Effects	Human	Murine	References
Neurotrophins				
NGF	Promotion of growth and differentiation	yes	yes	139-144
	Chemotaxis	–	yes	145
	Induction of mediator production and release	yes	yes	148-153
NT-3	Promotion of growth	yes	–	*
BDNF CNTF	Induction of mediator release in thalamic MCs	–	yes	155
Neuropeptides				
SP VIP NPY SOM	Induction of mediator release in "connective tissue" (i.e. rat peritoneal, human skin) but not "mucosal" (i.e. murine and human intestine or lung) MCs	yes	yes	43,44, 156-158
Others				
ACh	Induction of mediator release and enhancement of antigen-induced mediator release	no	yes	168,169, *
	Reduced MC density after disruption of parasympathetic nerve supply in vivo	yes	yes	175-177
Epinephrine	Inhibition of mediator release	yes	yes	180-82, 185,*
Norepinephrine	Inhibition of mediator release	–	yes	183,184

* Our own unpublished observations

residing in connective tissue and mucosa (ref. 43 and our own unpublished results). The mechanism by which SP activates MCs is unclear since conflicting data have been published on NK receptor expression in MCs and on the potency of NK receptor antagonists to block the SP triggered MC activation.[130,159-164] Interestingly, we found expression of NK-1 receptor mRNA and protein in human intestinal MCs following IgE receptor crosslinking suggesting that mucosal MCs may become responsive towards neuropeptides when preactivated or primed, for instance under conditions of acute or chronic inflammation (our own unpublished results). In this respect, it is interesting to note that Raithel et al demonstrated histamine release by SP from intestinal tissue samples of IBD patients but not from that of healthy controls.[88]

Effects of the above mentioned neuropeptides on Eo function have been investigated to a much lesser extent. Direct stimulatory effects of SP, CGRP, secretoneurin, VIP and secretin on human Eo migration were reported,[138,165] and, in analogy to SP-elicited release of MC mediators, guinea pig and eotaxin-primed normal human Eos released EDN following stimulation by SP.[166,167]

Evidence for direct effects of autonomic nerves on MC and Eo function came from numerous studies employing animal models of either nerve stimulation or dissection, or from in vitro experiments investigating ACh, epinephrine, or norepinephrine effects on purified MCs and Eos.

ACh was found to elicit atropine-sensitive histamine secretion form purified rat MCs,[168,169] to enhance antigen-induced histamine release from human lung tissue,[170] and to mediate MC degranulation following electrical stimulation of parasympathetic nerves.[171-174] Moreover,

Table 3. Effect of neurotrophins and neuropeptides on human eosinophils

	Effects	References
Neurotrophins		
NGF	Promotion of survival and differentiation	146,147
	Chemotaxis	147
	Induction of mediator release	154
	Increased cytotoxic activity	147
LIF	Enhancement of mediator release and chemotaxis following activation	133
Neuropeptides		
SP	Induction of mediator release	167
	Chemotaxis	138
CGRP	Chemotaxis	138
VIP		
Others		
Adrenergic agonists	Inhibition of mediator production and release	188

disruption of parasympathetic nerve supply resulted in reduced MC density, degranulation index, and tissue histamine content in animals and humans.[175-177] In contrast, other studies failed to show any direct secretory effect of ACh on rat MCs.[178,179]and cultured human intestinal MCs (our own unpublished results).

The great majority of studies investigating effects of catecholamines or adrenergic receptor agonists on MC function revealed inhibitory effects. Inhibition of antigen-induced histamine release in lung tissue by epinephrine[180] was mediated via β-adrenergic receptors involving cAMP,[181,182] whereas other investigators observed inhibition of MC activation by NE via α-adrenergic receptors.[183,184] Cultured human intestinal MCs were also found to be sensitive towards inhibitory effects of epinephrine (our own unpublished results) or β-adrenergic agonist salbutamol[185] with respect to IgE-dependent mediator release. In addition, sympathetic nerve stimulation was shown to inhibit antigen-induced histamine secretion from lung tissue in vivo[172] and to increase MC serotonin content. Accordingly, sympathetic ganglionectomy had the opposite effect.[186] In contrast, a recent study revealed increased dural MC numbers and histamine levels 60 days after superior cervical ganglionectomy suggesting a long-term trophic effect of sympathetic nerve degeneration on MCs in the dura mater.[187]

Moreover, the inhibitory effects of β-adrenergic agonists are not restricted to MCs since activation of other inflammatory cells including Eos was also found suppressed in the presence of such substances.[188]

Neurophysiological Effects of Mast Cell and Eosinophil Activation

As stated above, both, MCs and Eos, can produce, store and release several factors known to exert neurotrophic functions in that they regulate survival, neurite outgrowth, and neurotransmitter production in enteric neurons (Fig. 5). These factors include NGF, NT-3, BDNF, LIF, and a broad spectrum of cytokines.[48,52,53,124,133] Moreover, human Eos have been identified as a source of neuropeptides such as SP, VIP, CGRP, and somatostatin.[52,53] One might hypothesize that neuropeptide-controlled events such as regulation of blood supply, mucus and electrolyte secretion, or visceral pain sensation might, therefore, be modulated by Eo activation and degranulation, however, experimental evidence for such a role is still lacking. In this respect, it is interesting to note that the MC-derived neutral proteases tryptase and chymase are

Table 4. Neural effects of mast cell and eosinophil mediators

	Effects	MC	Eos	References
Histamine	Excitation of sensory nerves via H_1 receptors	X	–	206a
	Slow synaptic excitation of enteric neurons via H_2 receptors			204
	Presynaptic inhibition of neurotransmitter release in enteric neurons via H_3 receptors			203
	Regulation of intestinal secretion and motility			201,205,208
Tryptase	Degradation of neuropeptides	X	–	34
	Regulation of protease activated receptors on enteric neurons implicated in intestinal ion secretion			190
Chymase	Degradation of SP	X	–	34
Leukotrienes	Increased excitability in sensory neurons	X	X	218
	Regulation of intestinal secretion			197,201
PGD_2	Increased excitability in sensory neurons	X	–	217
	Regulation of intestinal secretion			197,201
MBP	Allosteric antagonist at inhibitory muscarinic M2 receptors	–	X	46
	Stimulation of SP release from cultured sensory neurons			47
	Stimuation of pulmonary C-fiber afferents			193
ECP	Neurotoxicity	–	X	195
	Stimuation of pulmonary C-fiber afferents			193
EDN/EPX	Neurotoxicity	–	X	195
EPO	Stimuation of pulmonary C-fiber afferents	–	X	193
Cytokines (e.g. IL 1, IL-5, IL-6, IL-9, βFGF, TGF-β)	Regulation of neuronal development	X	X	135
	Regulation of enteric neuronal excitability and neuotransmitter content / release			220,221

able to degrade neuropeptides and, therefore, might regulate extent and duration of neuropeptide-mediated changes in enteric neurophysiology. In addition, it was recently shown that tryptase is able to regulate activation of proteinase-activated receptors (PARs) expressed on myenteric and sensory neurons and thus intestinal ion secretion in rat jejunum (Table 4).[189-191]

Eo granule constituents such as MBP, ECP, or EDN are implicated in regulating neuronal function and survival (Table 4). MBP was found to stimulate SP release from cultured sensory neurons.[47]and was identified as a key player in mediating airway hyperreactivity in a murine model of asthma directly interfering with neuronal transmission by acting as allosteric antagonist at the inhibitory muscarinic M2 receptor on parasympathetic airway nerves.[46,118,192] Other Eo-derived cationic proteins were found to simulate vagal C-afferent fibers in the rat lung resulting in irregular breathing pattern after intratracheal installation.[193,194] Involvement of vagal afferents in these experiments was demonstrated by direct recording of single-unit activities and by prevention of cationic protein effects following neonatal capsaicin treatment, known to ablate SP containing sensory C fibers.[193,194] Moreover, ECP and EDN were also shown to exert potent cytotxic activity towards neurons[195] which possibly contribute to neuronal damage observed in inflammed intestinal tissue in close proximity to activated inflammatory cells.[120,117,196]

In a series of elegant studies employing in vitro and in vivo animal models of intestinal hypersensitivity, a large amount of data on the mechanisms of MC-nerve-epithelial cell interactions in the gut mucosa was gained. In these experiments, rodents were sensitized to ovalbumin or infected with parasites causing MC hyperplasia. Effects of antigen challenge or electric field stimulation on intestinal electrolyte secretion or intestinal barrier function were assessed using classic Ussing chambers experiments, electron microscopy, or measurement of tracer molecule recovery in the peripheral blood, respectively.[197] Antigen challenge resulted in MC degranulation, decreased mucosal histamine content, and increased mucosal ion secretion measured as short-circuit current (Isc) change in Ussing chambers, thus indicating a relationship between MC activation and mucosal fluid transport.[198,199] Further evidence for the important role of MCs came from studies in which MC-stabilizing agents[199]or MC-deficient mice were used.[200] In these animals the antigen-induced Isc change was reduced by 70% and could be restored by reconstitution of MC deficiency using normal bone marrow-derived MC precursors. Pretreatment with a histamine receptor-1 antagonist, a 5-HT-2 receptor antagonist, or a cyclooxygenase inhibitor also significantly decreased antigen-induced Isc changes pointing towards an involvement of histamine, serotonine, and lipid mediators (leukotrienes, prostaglandins, platelet-activating factor). Such factors were also shown to elicit chloride secretion when added to intestinal mucosa or epithelial cell lines, respectively.[197,201] Most interestingly, in these models also the neurotoxin tetrodotoxin, the muscarinic inhibitor atropine, and treatment of neonatal rats with capsaicin to deplete SP containing nerves suppressed the mucosal response to antigen and to exogenous histamine or serotonin. Moreover, in MC deficient mice, the Isc response evoked by electrical stimulation of enteric nerves was shown to be 50% lower than in normal mice.[200-202]

Other investigators used direct single cell recordings in the colonic submucous plexus of sensitized guinea pigs. Histamine released from mucosal MCs upon antigenic challenge was shown to induce complex effects on excitatory postsynaptic potentials by modulating pre and postsynaptic activity through distinct receptor subclasses[203,204] resulting in altered intestinal motility.[205] Moreover, histamine was shown to stimulate mesenteric afferents via the H1 recepter.[206]

Taken together, these data indicate a bi-directional interaction between mucosal MCs, intrinsic nerves, and intestinal epithelial cells in the regulation of intestinal ion secretion that was later, in part, also confirmed using human intestinal tissue.[207-208]

Such MC-nerve-epithelial cell interactions were also shown to regulate the increased intestinal permeability observed after antigen challenge in sensitized rats. This was demonstrated using intestinal mucosa from MC deficient rats or tetrodotoxin pretreated mucosa that showed markedly reduced secretory responses.[209,210]

In other studies, the role of MC-nerve interactions in the intestinal response towards different types of stressors was investigated. The investigators found that acute or chronic stress resulted in MC degranulation accompanied by enhancement of intestinal permeability and mucin and fluid secretion. These effects were absent in MC deficient animals, could be blocked by the muscarinic antagonist atropine, and most interestingly, could be mimicked by intracerebral injection of corticotropin-releasing factor (CRF) in conscious, nonstressed rats. Accordingly, an inhibitor of CRF prevented intestinal effects of stress.[211-214] Moreover, application of cold pain stress in humans also resulted in increased luminal release of MC mediators and jejunal water secretion as assessed using a closed-segment perfusion technique.[215] These data clearly indicate the ability of the central nervous system to modulate intestinal MC activity by mechanisms involving central CRH release and cholinergic nerves, and extent previous findings implicating the brain in psychological or Pavlovian conditioning of peripheral MC degranulation.[216] Vice versa, antigenic challenge of hypothalamic MCs in dogs was recently found to activate the hypothalamic-pituitary-adrenal-axis resulting in increased cortisol secretion.[217]

Numerous other studies investigated peripheral neuromodulatory effects associated with MC activation recorded in extraintestinal tissue such as airways or ganglia. A large body of data was generated, and MC activation was shown to directly stimulate neurons, increase their excitability, increase synaptic efficacy, decrease neuronal activation thresholds, or change the neurotransmitter content in sensory and autonomic nerves, respectively.[218]

Alterations of Neuroimmune-Crosstalk in IBD

Major intestinal symptoms presented by IBD patients include abdominal pain, diarrhea, tenesmus, constipation, or vomiting. Noteworthy, these symptoms possibly reflect altered visceral sensitivity and disturbance of normal gut motility suggesting a role of dysregulated ENS function in these diseases (Fig. 7). Indeed, intestinal inflammation was shown to induce remodeling of ENS neurons[219] and to affect neurotransmitter content and release of enteric nerves in rats.[220,221] Therefore, it is tempting to speculate that such processes may also account for altered ENS structure and function in human IBD. Supporting this hypothesis, numerous studies investigating nerve morphology, neurotransmitter content and distribution, as well as functional properties of human intestinal tissue obtained from IBD patients demonstrated profound ENS alterations (Table 5).

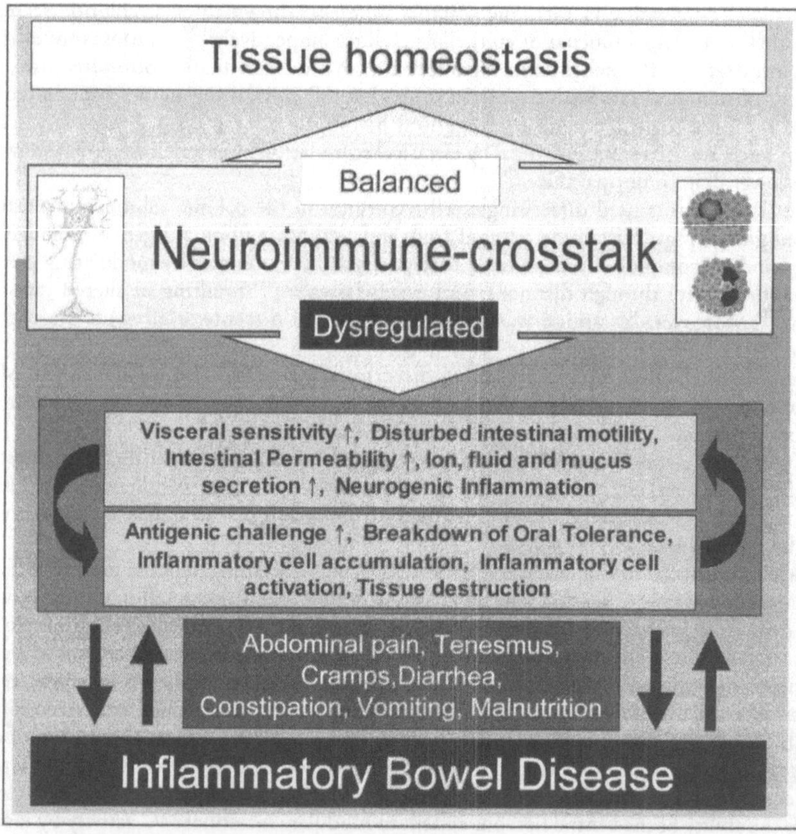

Figure 7. Impact of neuroimmune-crosstalk in IBD (Hypothesis). Neuroimmune-crosstalk regulates intestinal neurophysiology and oral tolerance mechanisms providing tissue homeostasis. Dysregulated neuroimmune-crosstalk in the course of IBD leads to disturbed gut functions and intestinal inflammation causing the classical symptoms of IBD patients.

Table 5. Alterations of the ENS in IBD

	Alterations	CD	UC	References
Morphology and structure	Increased size and frequency of varicosities	–	yes	224
	Degenerative transformation of axons	yes	no	62,222,223
	Deceased numbers and degeneration of interstitial cells of Cajal	yes	yes	225,226
SP	Increased expression of SP in neuronal and non-neuronal elements positively correlated with grade of inflammation	yes	yes	229,233
	Decreased (severe inflammation) or increased (moderate inflammation) numbers and immunoreactivity of SP containing nerves	yes	yes	234,235
	Increased tissue levels positively correlated with grade of inflammation and cellular inflammatory infiltrate	no	yes	236-238
	Decreased tissue levels	yes	no	238
	Increased expression of SP receptor NK-1	yes	yes	239
	Induction of histamine release in inflamed IBD and non-inflamed CD tissue samples, but not in controls	yes	yes	88
	Impaired responses of muscle layer preparations following stimulation with SP	–	yes	240
	Reduced intestinal secretion rate following stimulation of tissue samples with SP	yes	–	241
VIP	Increased numbers of VIP containing nerves increased VIP tissue levels	yes	no	245
	Increased numbers of VIP containing nerves	–	yes	232,240
	Increased numbers of VIP containing nerves	yes	–	228
	Decreased numbers of VIP containing nerves and decreased VIP tissue levels	yes	yes	229,234,236, 242,243
	Increased VIP plasma levels during active disease	yes	yes	246,247
CGRP	Hypertrophic transformation of CGRP containing nerves in non-inflamed tissue	yes	–	228
SOM	Decreased SOM tissue levels	–	yes	250
	Induced vascular expression of SOM receptors in inflamed tissue	yes	yes	249
Catecholamines	Decreased epinephrine tissue levels in inflamed and non-inflamed mucosa	yes	no	270
	Increased catecholamine-containing structures in the myenteric plexus	no	yes	224,271
NGF	Increased expression of NGF and its receptor TrkA	yes	yes	269

In Crohn's disease, several investigators described degenerative transformation of axons exerting a thickened, coarse, or swollen appearance,[62,222,223] and in ulcerative colitis, alterations of size and frequency of neuronal varicosities were found.[224] Interestingly, these phenomena were also observed in uninvolved sites of the gut and often occurred in axons that were located in close proximity to activated MCs or Eos, supporting the hypothesis that mediators released by these cells are responsible for this neuronal damage.[83,117,196] Moreover, decreased numbers of interstitial cells of Cajal with degenerative morphology were reported in IBD, possibly also contributing to the disturbed intestinal motility observed in these patients.[225,226]

Qualitative and quantitative alterations of neuronal distribution pattern and neurochemical composition, as well as expression of neuropeptides in nonneural elements were also frequently reported in inflamed and noninflamed tissue (Table 5).[227,228] The great majority of immunohistochemical studies revealed increased numbers and immunoreactivity of SP containing nerves in both IBD entities including regions in which SP expression was absent in normal intestine, e.g., in the longitudianal muscle layer (Table 5).[229-235] Moreover, SP concentrations in tissues derived from ulcerative colitis patients were found to be increased and positively correlated with the histological inflammation score and cellular inflammatory infiltrate comprising of mononucleated cells, neutrophils and Eos.[236-238] Other studies addressed expression profiles of SP receptors and consistently found marked up-regulation of the NK-1 receptor throughout the whole gut wall in inflamed and noninflamed tissue of IBD patients.[239] Interestingly, in these tissues, very high concentrations of SP receptors were also ectopically expressed within germinal centers of lymph nodules. These altered expression profiles of SP and its receptors observed in IBD tissue were also reflected by altered responsiveness of these tissue samples towards SP stimulation in vitro (Table 5). In tissue samples derived from inflamed IBD or uninvolved Crohn's disease gut, SP was shown to elicit histamine release, whereas normal tissue was found to be insensitive towards SP stimulation in this respect.[88] Others found impaired in vitro responses of muscle layer preparations derived from ulcerative colitis patients or reduced mucosal secretion rates in tissue from Crohn's disease patients following stimulation with SP, possibly reflecting receptor desensitization.[240,241]

VIP containing nerves were also affected in IBD (Table 5). Most of the studies revealed diminished VIP immunoreactivity, decreased numbers of VIP containing nerves, or decreased tissue VIP concentrations, respectively, depending on severity of inflammation in both diseases.[229,234,236,242-244] In contrast, others demonstrated increased numbers of VIP positive nerves in IBD.[228,232,240,245] Additionally, one study could demonstrate that decreased VIP concentrations were accompanied with diminished inhibitory neural input into circular muscle in Crohn's disease.[242] Finally, assessment of VIP plasma levels revealed an increase during active disease compared to remission phases in patients with Crohn's disease or ulcerative colitis.[246,247]

Apart from SP and VIP, other neuropeptides such as CGRP or SOM are also candidates implicated in intestinal pathophysiology and inflammation seen in IBD (Table 5).[248,249] CGRP containing nerve fibers were found to display morphologic features of hypertrophic transformation in noninflamed rectum of patients with Crohn's disease.[228] SOM content of the rectal mucosa was found to be decreased in ulcerative colitis,[250] whereas induction of SOM receptor expression was reported in tissue sections of both IBD entities but not in control tissue.[249]

Taken together, the above mentioned studies demonstrate substantial changes in intestinal neuropeptide expression and function in the course of IBD (Table 5). The data strongly point towards an important role of these neuropeptides in acute and chronic intestinal inflammation. The discrepancy between results reported in some of the above mentioned studies might be explained by distinct methodological approaches and, particularly, not readily comparable sample collections with respect to anatomic localization, inflammation grade, disease stage, or previous pharmacological therapy. Given the pivotal role of neuropeptides such as SP, VIP, CGRP or SOM in the regulation of intestinal motility and secretion of ions, fluid and mucus, these alterations obviously contribute to symptoms such as diarrhea, tenesmus, constipation, or vomiting frequently associated with IBD (Fig. 7). Moreover, it is supposed that SP regulates chronic or persistent visceral pain sensations[251] and increased visceral sensitivity by recruiting "silent" afferent fibers or nociceptors that normally do not respond to the strongest of mechanical stimuli.[252] Under inflammatory conditions, these receptors become activated, display markedly lower activation thresholds,[252,253] and are, therefore, good candidates for mediating visercal hypersensitivity and hyperalgesia associated symptoms found in IBD patients.[254,255]

Noteably, MC and Eos were found to contribute to neuropeptide homeostasis in tissue. Eos produce SP, VIP, CGRP or SOM,[52,53] whereas MC-derived proteases degrade neuropeptides.

Additionally, both cell types were also identified as targets of proinflammatory effects of these neuropeptides (Tables 2, 3). Sensory nerves may be involved in the inflammatory process through the antidromic release of SP and other neuropeptides from nociceptive nerves or C-fibers following direct stimulation or triggering by local axon reflex, the phenomenon referred to as neurogenic inflammation.[254,256] In accordance with this hypothesis, in murine models of experimental colitis, neutralization of SP or its preferential receptor NK-1 was found to reduce the severity of inflammation,[257,258] whereas impaired degradation of SP in the neutral endopeptidase knock out model worsened inflammation.[259] Direct evidence for neural involvement in human IBD arose from the early observation that surgical denervation of the pelvis or vagotomy had been used successfully to ameliorate refractory IBD.[260,261] Moreover, spinal cord stimulation for treatment of peripheral pain was reported to associate with relapse of remitting ulcerative colitis.[262,263] Furthermore, therapeutic benefit from the use of local lidocaine application in the treatment of ulcerative colitis or experimental colitis in rats clearly point towards a direct neuronal involvement in the modulation of intestinal inflammation.[264,265] However, the role of neuropeptides in intestinal inflammation are not proinflammatory in general. In contrast to their role in neurogenic inflammation where neuropeptides are released from sensory fibers in an antidromic fashion, neuropeptide-containing sensory afferents are also known to contribute to mucosal protection. This was shown in animal models of experimental colitis, where sensory denervation with the neurotoxin capsaicin or pretreatment with a CGRP receptor antagonist promoted acute and chronic inflammation.[266-268]

As outlined above, also NGF was identified as a central player in enteric neuroimmune-crosstalk exerting trophic effects on nerves and proinflammatory effects on MCs, Eos, and other inflammatory cells. An impact of this pathway in neuroimmune interaction was suggested by the finding of elevated expression of NGF and its receptor TrkA in ulcerative colitis and Crohn's disease.[269] It is therefore conceivable that NGF might promote pronounced MC/Eo-nerve association and increased activation of MCs and Eos observed in IBD mucosa.[83,196]

Other important transmitters in neuroimmune-crosstalk such as catecholamines were suggested to play anti-inflammatory roles by acting as inhibitors of a broad range of inflammatory cells including Eos and intestinal MCs. Accordingly, reduced tissue levels of epinephrine and other monoamines were found in inflamed and noninflamed tissues in Crohn's disease but not in ulcerative colitis.[270,271] In contrast, adrenergic innervation appeared to be increased in ulcerative colitis without altered tissue catecholamine levels.[271]

An overview on the hypothetical impact of dysregulated neuroimmune-crosstalk on pathogenesis of IBD and its symptoms based on the studies discussed above, is presented in Figure 7.

Conclusions

Intestinal MCs and Eos are capable of generating and releasing a broad variety of mediators implicated in diverse physiological and pathophysiological processes in the gut (Fig. 2). These processes include neurophysiologic gut functions such as intestinal secretion or motility, as well as wound healing, inflammation, or fibrotic tissue transformation (Fig. 3). Interestingly, morphologic studies revealed close associations of MCs and Eos with nerve fibers and, additionally, a great number of functional studies demonstrated bi-directional communication between these cell types (Tables 2-4; Figs. 5, 6).

In the course of IBD, MCs and Eos accumulate and become activated. In parallel, disturbance of morphology, neurochemical composition, and function of the ENS has consistently been observed in involved and uninvolved tissue samples of IBD patients (Table 5). In this respect, it is interesting to note that also immune cells such as MCs or Eos were identified as a source of several neuropeptides, neurotrophins, and other mediators directly acting on neurons (Fig. 5). Thus, it is tempting to speculate that alterations of intestinal neuroimmune-crosstalk involving MCs and Eos may have an impact in the pathogenesis of IBD.

On the one hand, these alterations may cause disturbed intestinal secretion, motility, or permeability underlying symptoms of IBD patients such as diarrhea, cramps, constipation or malnutrition. On the other hand, over-expression of neurotrophins or certain neuropeptides as seen in IBD may trigger or sustain inflammation (neurogenic inflammtion) since e.g., SP or NGF are known as proinflammatory MC and Eo agonists. Moreover, dysregulation of neuropeptides and enhanced local production of proinflammatory mediators by MCs and Eos may sensitize or excite nociceptive structures leading to visceral hyperalgesia.

One of the important questions in this respect is whether these ENS alterations precede the onset of intestinal inflammation as suggested by studies demonstrating a deranged ENS structure and function even in uninvolved sites in IBD patients. A primary dysbalance of neuronal mediators exerting proinflammatory (e.g., SP, NGF) and anti-inflammatory (e.g., CGRP, epinephrine) effects on immune cells would then support the development and maintenance of chronic intestinal inflammation caused by as yet unknown agents.

Alternatively, these changes may occur secondary in the course of intestinal inflammation associated with the release of proinflammatory and cytotoxic mediators from MCs and Eos directly affecting ENS functions. This was shown in models of experimental colitis in animals, where intestinal inflammation induced remodeling of ENS neurons.

In any case, the above mentioned studies clearly point towards a role of the ENS in chronic intestinal inflammation. To further understand the mechanisms of neural participation in IBD, more studies are necessary. Of particular interest in this respect are studies addressing the time course of ENS alterations during the development of IBD in humans or IBD-like experimental models in animals. Correlation with histologic inflammation score and clinical symptoms would possibly allow to define a more precise role of different neural mediators in this context. Moreover, targeting of neural mediators by antibody or antagonist strategies would possibly be of therapeutic benefit, as shown at least in some animal models.

References

1. Downing JE, Miyan JA. Neural immunoregulation: Emerging roles for nerves in immune homeostasis and disease. Immunol Today 2000; 21:281-9.
2. Anisman H, Baines MG, Berczi I et al. Neuroimmune mechanisms in health and disease: 2. Disease. CMAJ 1996; 155:1075-82.
3. Stead RH. Innervation of mucosal immune cells in the gastrointestinal tract. Reg Immunol 1992; 4:91-9.
4. McKay DM, Bienenstock J. The interaction between mast cells and nerves in the gastrointestinal tract. Immunol Today 1994; 15:533-8.
5. Williams RM, Bienenstock J, Stead RH. Mast cells: The neuroimmune connection. Chem Immunol 1995; 61:208-35.
6. Bradding P, Okayama Y, Howarth PH et al. Heterogeneity of human mast cells based on cytokine content. J Immunol 1995; 155:297-307.
7. Weidner N, Austen KF. Heterogenity of mast cells at multiple body sites. Path Res Pract 1993; 189:156-162.
8. Metcalfe DD, Baram D, Mekori YA. Mast cells. Physiol Rev 1997; 77:1033-1079.
9. Bischoff SC, Wedemeyer J, Herrmann A et al. Quantitative assessment of intestinal eosinophils and mast cells in inflammatory bowel disease. Histopathology 1996; 28:1-13.
10. Kato M, Kephart GM, Morikawa A et al. Eosinophil infiltration and degranulation in normal human tissues: Evidence for eosinophil degranulation in normal gastrointestinal tract. Int Arch Allergy Immunol 2001; 125(Suppl 1):55-8.
11. Weller PF. The immunobiology of eosinophils. N Engl J Med 1991; 324:1110-8.
12. Gelbmann CM, Mestermann S, Gross V et al. Strictures in Crohn's disease are characterised by an accumulation of mast cells colocalised with laminin but not with fibronectin or vitronectin. Gut 1999; 45:210-217.
13. Levi Schaffer F, Austen KF, Caulfield JP et al. Coculture of human lung-derived mast cells with mouse 3T3 fibroblasts: Morphology and IgE-mediated release of histamine, prostaglandin D2, and leukotrienes. J Immunol 1987; 139:494-500.
14. Mierke CT, Ballmeier M, Werner U et al. Human endothelial cells regulate survival and proliferation of human mast cells. J Exp Med 2000; 192:801-811.
15. Bischoff SC, Schwengberg S, Raab R et al. Functional properties of human intestinal mast cells cultured in a new culture system: Enhancement of IgE receptor-dependent mediator release and response to stem cell factor. J Immunol 1997; 159:5560-5567.

16. Kitamura Y, Go S, Hatanaka K. Decrease of mast cells in W/Wv mice and their increase by bone marrow transplantation. Blood 1978; 52:447-52.
17. Kitamura Y, Go S. Decreased production of mast cells in S1/S1d anemic mice. Blood 1979; 53:492-7.
18. Longley Jr BJ Metcalfe DD Tharp M et al. Activating and dominant inactivating c-kit catalytic domain mutations in distinct clinical forms of human mastocytosis. Proc Natl Acad Sci USA 1999; 96:1609-14.
19. Lorentz A, Schuppan D, Gebert A et al. Regulatory effects of stem cell factor and interleukin-4 on adhesion of human mast cells to extracellular matrix proteins. Blood 2002; 99:966-72.
20. Artis D, Humphreys NE, Potten CS et al. Beta7 integrin-deficient mice: Delayed leukocyte recruitment and attenuated protective immunity in the small intestine during enteric helminth infection. Eur J Immunol 2000; 30:1656-64.
21. Gurish MF, Tao H, Abonia JP et al. Intestinal mast cell progenitors require CD49dbeta7 (alpha4beta7 integrin) for tissue-specific homing. J Exp Med 2001; 194:1243-52.
22. Bischoff SC, Sellge G, Lorentz A et al. IL-4 enhances proliferation and mediator release in mature human mast cells. Proc Natl Acad Sci USA 1999; 96:8080-8085.
23. Gebhardt T, Sellge G, Lorentz A et al. Cultured human intestinal mast cells express functional IL-3 receptors and respond to IL-3 by enhancing growth and IgE receptor-dependent mediator release. Eur J Immunol. 2002; 32:2308-16
24. Hartmann K, Henz BM, Kruger-Krasagakes S et al. C3a and C5a stimulate chemotaxis of human mast cells. Blood 1997; 89:2863-70.
25. Romagnani P, De Paulis A, Beltrame C et al. Tryptase-chymase double-positive human mast cells express the eotaxin receptor CCR3 and are attracted by CCR3-binding chemokines. Am J Pathol 1999; 155:1195-204.
26. Olsson N, Piek E, ten Dijke P et al. Human mast cell migration in response to members of the transforming growth factor-beta family. J Leukoc Biol 2000; 67:350-6.
27. Cyr MM, Denburg JA. Systemic aspects of allergic disease: The role of the bone marrow. Curr Opin Immunol 2001; 13:727-32.
28. Alam R, Forsythe P, Stafford S et al. Transforming growth factor beta abrogates the effects of hematopoietins on eosinophils and induces their apoptosis. J Exp Med 1994; 179:1041-5.
29. Bochner BS, Schleimer RP. The role of adhesion molecules in human eosinophil and basophil recruitment. J Allergy Clin Immunol 1994; 94:427-38.
30. Rothenberg ME, Mishra A, Brandt EB et al. Gastrointestinal eosinophils in health and disease. Adv Immunol 2001; 78:291-328.
31. Garcia-Zepeda EA, Rothenberg ME, Ownbey RT et al. Human eotaxin is a specific chemoattractant for eosinophil cells and provides a new mechanism to explain tissue eosinophilia. Nat Med 1996; 2:449-56.
32. Mishra A, Hogan SP, Lee JJ et al. Fundamental signals that regulate eosinophil homing to the gastrointestinal tract. J Clin Invest 1999; 103:1719-27.
33. Hogan SP, Mishra A, Brandt EB et al. A pathological function for eotaxin and eosinophils in eosinophilic gastrointestinal inflammation. Nat Immunol 2001; 2:353-60.
34. Bischoff SC. Regulation and function of human intestinal mast cells. In: Marone G, Lichtenstein LM, Galli SJ, eds. Mast cells and basophils. London: Academic press, 2000:541-565.
35. Jutel MT, Watanabe S, Klunker M et al. Histamine regulates T-cell and antibody responses by differential expression of H1 and H2 receptors. Nature 2001; 413:420-425.
36. Hirai HK, Tanaka O, Yoshie K et al. Prostaglandin D2 selectively induces chemotaxis in T helper type 2 cells, eosinophils, and basophils via seven-transmembrane receptor CRTH2. J Exp Med 2001; 193:255-61.
37. Lorentz A, Schwengberg S, Mierke C et al. Human intestinal mast cells produce IL-5 in vitro upon IgE receptor cross-linking and in vivo in the course of intestinal inflammatory disease. Eur J Immunol 1999; 29:1496-503.
38. Bischoff SC, Lorentz A, Schwengberg S et al. Mast cells are an important cellular source of tumour necrosis factor alpha in human intestinal tissue. Gut 1999; 44:643-652.
39. Lorentz A, Schwengberg S, Sellge G et al. Human intestinal mast cells are capable of producing different cytokine profiles: Role of IgE receptor cross-linking and IL-4. J Immunol 2000; 164:43-48.
40. Kanbe NM, Kurosawa H, Nagata T et al. Production of fibrogenic cytokines by cord blood-derived cultured human mast cells. J Allergy Clin Immunol 2000; 106:85-90.
41. Bacci S, Faussone-Pellegrini S, Mayer B et al. Distribution of mast cells in human ileocecal region. Dig Dis Sci 1995; 40:357-65.
42. Echtenacher BD, Mannel N, Hultner L. Critical protective role of mast cells in a model of acute septic peritonitis. Nature 1996; 381:75-77.

43. Lawrence ID, Warner JA, Cohan VL et al. Purification and characterization of human skin mast cells. Evidence for human mast cell heterogeneity. J Immunol 1987; 139:3062-9.
44. Lowman MA, Benyon RC, Church MK. Characterization of neuropeptide-induced histamine release from human dispersed skin mast cells. Br J Pharmacol 1988; 95:121-30.
45. Bischoff SC, Schwengberg S, Wordelmann K et al. Effect of c-kit ligand, stem cell factor, on mediator release by human intestinal mast cells isolated from patients with inflammatory bowel disease and controls. Gut 1996; 38:104-111.
46. Jacoby DB, Gleich GJ, Fryer AD. Human eosinophil major basic protein is an endogenous allosteric antagonist at the inhibitory muscarinic M2 receptor. J Clin Invest 1993; 91:1314-8.
47. Garland A, Necheles J, White SR et al. Activated eosinophils elicit substance P release from cultured dorsal root ganglion neurons. Am J Physiol 1997; 273:L1096-102.
48. Lacy P, Moqbel R. Eosinophil cytokines. In: Marone G, ed. Human eosinophils. Biological and clinical aspects. Karger Basel, 2000:134-155.
49. Bischoff SC, Nguyen QT, Mierke C et al. Transforming growth factor β production by eosinophils in human intestinal mucosa. Imunnology Letters 1999; 69:91, (Abstract).
50. Desreumaux P, Bloget F, Seguy D et al. Interleukin 3, granulocyte-macrophage colony-stimulating factor, and interleukin 5 in eosinophilic gastroenteritis. Gastroenterology 1996; 110:768-74.
51. Seegert D, Rosenstiel P, Pfahler H et al.Increased expression of IL-16 in inflammatory bowel disease. Gut 2001; 48(3):326-32.
52. Aliakbari J, Sreedharan SP, Turck CW et al. Selective localization of vasoactive intestinal peptide and substance P in human eosinophils. Biochem Biophys Res Commun 1987; 148:1440-5.
53. Metwali A, Blum AM, Ferraris L et al. Eosinophils within the healthy or inflamed human intestine produce substance and vasoactive intestinal peptide. J Neuroimmunol 1994; 52:69-78.
54. Church MK, Lichtenstein LM, Simon Hu et al. Effector cells of allergy. In Holgate ST, Church MK, Lichtenstein LM, eds. Allergy. London: Mosby, 2001; 303-324.
55. Raible DG, Schulman ES, DiMuzio J et al. Mast cell mediators prostaglandin-D2 and histamine activate human eosinophils. J Immunol 1992; 148:3536-42.
56. Shi HZ, Humbles A, Gerard C et al. Lymph node trafficking and antigen presentation by endobronchial eosinophils. J Clin Invest 2000; 105:945-53.
57. Villa I, Skokos D, Tkaczyk C et al.Capacity of mouse mast cells to prime T cells and to induce specific antibody responses in vivo. Immunology 2001; 102:165-72.
58. Woerly G, Roger N, Loiseau S et al. Expression of CD28 and CD86 by human eosinophils and role in the secretion of type 1 cytokines (interleukin 2 and interferon gamma): Inhibition by immunoglobulin a complexes. J Exp Med 1999; 190:487-95.
59. Dimitriadou V, Mecheri S, Koutsilieris M et al. Expression of functional major histocompatibility complex class II molecules on HMC-1 human mast cells. J Leukoc Biol 1998; 64:791-9.
60. Wershil BK. Role of mast cells and basophils in gastrointestinal inflammation. Chem Immunol 1995; 62:187-203.
61. Fiocchi C. Inflammatory bowel disease: Etiology and Pathogenesis. Gastroenterology 1998; 115:182-205.
62. Dvorak AM, Monahan RA, Osage JE et al. Crohn's disease: Transmission electron microscopic studies. II. Immunologic inflammatory response. Alterations of mast cells, basophils, eosinophils, and the microvasculature. Hum Pathol 1980; 11:606-19.
63. Willoughby CP, Piris J, Truelove SC. Tissue eosinophils in ulcerative colitis. Scand J Gastroenterol 1979; 14:395-9.
64. Sarin SK, Malhotra V, Sen Gupta S et al. Significance of eosinophil and mast cell counts in rectal mucosa in ulcerative colitis. A prospective controlled study. Dig Dis Sci 1987; 32:363-7.
65. Hallgren R, Colombel JF, Dahl R et al. Neutrophil and eosinophil involvement of the small bowel in patients with celiac disease and Crohn's disease: Studies on the secretion rate and immunohistochemical localization of granulocyte granule constituents. Am J Med 1989; 86:56-64.
66. Raab Y, Fredens K, Gerdin B et al. Eosinophil activation in ulcerative colitis: Studies on mucosal release and localization of eosinophil granule constituents. Dig Dis Sci 1998; 43:1061-70.
67. Jeziorska M, Haboubi N, Schofield P et al. Distribution and activation of eosinophils in inflammatory bowel disease using an improved immunohistochemical technique. J Pathol 2001; 194:484-92.
68. Berstad A, Borkje B, Riedel B et al. Increased fecal eosinophil cationic protein in inflammatory bowel disease. Hepatogastroenterology 1993; 40:276-8.
69. Levy AM, Gleich GJ, Sandborn WJ et al. Increased eosinophil granule proteins in gut lavage fluid from patients with inflammatory bowel disease. Mayo Clin Proc 1997; 72:117-23.

70. Luck W, Becker M, Niggemann B et al. In vitro release of eosinophil cationic protein from peripheral eosinophils reflects disease activity in childhood Crohn disease and ulcerative colitis. Eur J Pediatr 1997; 156:921-4.
71. Bischoff SC, Grabowsky J, Manns MP. Quantification of inflammatory mediators in stool samples of patients with inflammatory bowel disorders and controls. Dig Dis Sci 1997; 42:394-403.
72. Troncone R, Caputo N, Esposito V et al. Increased concentrations of eosinophilic cationic protein in whole-gut lavage fluid from children with inflammatory bowel disease. J Pediatr Gastroenterol Nutr 1999; 28:164-8.
73. Bischoff SC, Mayer J, Nguyen QT et al. Immunnohistological assessment of intestinal eosinophil activation in patients with eosinophilic gastroenteritis and inflammatory bowel disease. Am J Gastroenterol 1999; 94:3396-9.
74. Carlson M, Raab Y, Peterson C et al. Increased intraluminal release of eosinophil granule proteins EPO, ECP, EPX, and cytokines in ulcerative colitis and proctitis in segmental perfusion. Am J Gastroenterol 1999; 94:1876-83.
75. Saitoh O, Kojima K, Sugi K et al. Fecal eosinophil granule-derived proteins reflect disease activity in inflammatory bowel disease. Am J Gastroenterol 1999; 94:3513-20.
76. Pronk-Admiraal CJ, Linskens RK, Van Bodegraven AA et al. Serum eosinophil cationic protein in active and quiescent ulcerative colitis. Clin Chem Lab Med 2000; 38:619-22.
77. Winterkamp S, Raithel M, Hahn EG. Secretion and tissue content of eosinophil cationic protein in Crohn's disease. J Clin Gastroenterol 2000; 30:170-5.
78. Lloyd G, Green FH, Fox H et al. Mast cells and immunoglobulin E in inflammatory bowel disease. Gut 1975; 16:861-66.
79. Sanderson IR, Leung KB, Pearce FL et al. Lamina propria mast cells in biopsies from children with Crohn's disease. J Clin Pathol 1986; 39:279-83.
80. Balazs M, Illyes G, Vadasz G. Mast cells in ulcerative colitis. Quantitative and ultrastructural studies. Virchows Arch B Cell Pathol Incl Mol Pathol 1989; 57:353-60.
81. Nolte H, Spjeldnaes N, Kruse A et al. Histamine release from gut mast cells from patients with inflammatory bowel diseases. Gut 1990; 31:791-4.
82. King T, Biddle W, Bhatia P et al. Colonic mucosal mast cell distribution at line of demarcation of active ulcerative colitis. Dig Dis Sci 1992; 37:490-5.
83. Dvorak AM, McLeod RS, Onderdonk A et al. Ultrastructural evidence for piecemeal and anaphylactic degranulation of human gut mucosal mast cells in vivo. Int Arch Allergy Immunol 1992; 99:74-83.
84. Araki Y, Kakegawa T, Stadil F. Mast cells and histamine release in Crohn's disease. Kurume Med J 1993; 40:93-9.
85. Knutson L, Ahrenstedt O, Odlind B et al. The jejunal secretion of histamine is increased in active Crohn's disease. Gastroenterology 1990; 98:849-54.
86. Raithel M, Matek M, Baenkler HW et al. Mucosal histamine content and histamine secretion in Crohn's disease, ulcerative colitis and allergic enteropathy. Int Arch Allergy Immunol 1995; 108:127-33.
87. Weidenhiller M, Raithel M, Winterkamp S et al. Methylhistamine in Crohn's disease (CD): Increased production and elevated urine excretion correlates with disease activity. Inflamm Res 2000; 49:S35-6.
88. Raithel M, Schneider HT, Hahn EG. Effect of substance P on histamine secretion from gut mucosa in inflammatory bowel disease. Scand J Gastroenterol 1999; 34:496-503.
89. Raithel M, Winterkamp S, Pacurar A et al. Release of mast cell tryptase from human colorectal mucosa in inflammatory bowel disease. Scand J Gastroenterol 2001; 36:174-9.
90. Fox CC, Lazenby AJ, Moore WC et al. Enhancement of human intestinal mast cell mediator release in active ulcerative colitis. Gastroenterology 1990; 99:119-24.
91. Funakoshi K, Sugimura K, Anezaki K et al. Spectrum of cytokine gene expression in intestinal mucosal lesions of Crohn's disease and ulcerative colitis. Digestion 1998; 59:73-8.
92. Briskin M, Winsor-Hines D, Shyjan A et al. Human mucosal addressin cell adhesion molecule-1 is preferentially expressed in intestinal tract and associated lymphoid tissue. Am J Pathol 1997; 151:97-110.
93. Dubucquoi S, Janin A, Klein O et al. Activated eosinophils and interleukin 5 expression in early recurrence of Crohn's disease. Gut 1995; 37:242-6.
94. Fuss IJ, Neurath M, Boirivant M et al. Disparate CD4+ lamina propria (LP) lymphokine secretion profiles in inflammatory bowel disease. Crohn's disease LP cells manifest increased secretion of IFN-gamma, whereas ulcerative colitis LP cells manifest increased secretion of IL-5. J Immunol 1996; 157:1261-70.

95. Desreumaux P, Brandt E, Gambiez L et al. Distinct cytokine patterns in early and chronic ileal lesions of Crohn's disease. Gastroenterology 1997; 113:118-26.
96. Mazzucchelli L, Hauser C, Zgraggen K et al. Differential in situ expression of the genes encoding the chemokines MCP-1 and RANTES in human inflammatory bowel disease. J Pathol 1996; 178:201-6.
97. Wedemeyer J, Lorentz A, Goke M et al. Enhanced production of monocyte chemotactic protein 3 in inflammatory bowel disease mucosa. Gut 1999; 44:629-35.
98. Babyatsky MW, Rossiter G, Podolsky DK. Expression of transforming growth factors alpha and beta in colonic mucosa in inflammatory bowel disease. Gastroenterology 1996; 110:975-84.
99. Xian CJ, Xu X, Mardell CE et al. Site-specific changes in transforming growth factor-alpha and -beta1 expression in colonic mucosa of adolescents with inflammatory bowel disease. Scand J Gastroenterol 1999; 34:591-600.
100. Monteleone G, Kumberova A, Croft NM et al. Blocking Smad7 restores TGF-beta1 signaling in chronic inflammatory bowel disease. J Clin Invest 2001; 108:601-9.
101. Ogura Y, Bonen DK, Inohara N et al. A frameshift mutation in NOD2 associated with susceptibility to Crohn's disease. Nature 2001; 411:603-6.
102. Hugot JP, Chamaillard M, Zouali H et al. Association of NOD2 leucine-rich repeat variants with susceptibility to Crohn's disease. Nature 2001; 411:599-603.
103. Nagarajan RP, Chen F, Li W et al. Repression of transforming-growth-factor-beta-mediated transcription by nuclear factor kappaB. Biochem J 2000; 348(Pt 3):591-6.
104. Ogura Y, Inohara N, Benito A et al. Nod2, a Nod1/Apaf-1 family member that is restricted to monocytes and activates NF-kappaB. J Biol Chem 2001; 276:4812-8.
105. Langley JN. The autonomic nervous system, Part 1. Cambridge: Heffner and Sons, 1921.
106. Wood JD, Alpers DH, Andrews PL. Fundamentals of neurogastroenterology. Gut 1999; 45:II6-II16.
107. Gershon MD. In The Second Brain. New York: Harper Collins, 1998:1-312.
108. Schemann M. Regulation of gastrointestinal functions by the enteric nervous system. Proc Soc Nutr Physiol 2001; 10:17-25.
109. Costa M, Brookes SJ, Hennig GW. Anatomy and physiology of the enteric nervous system. Gut 2000; 47:iv15-9.
110. Florenzano F, Bentivoglio M. Degranulation, density, and distribution of mast cells in the rat thalamus: A light and electron microscopic study in basal conditions and after intracerebroventricular administration of nerve growth factor. J Comp Neurol 2000; 424:651-69.
111. Heine H, Forster FJ. Relationships between mast cells and preterminal nerve fibers. Z Mikrosk Anat Forsch 1975; 89:934-7.
112. Stach W. [Nerve plexuses of the duodenal villi. Light and electron microscopic studies]. Acta Anat 1973; 85:216-31.
113. Newson B, Dahlstrom A, Enerback L et al. Suggestive evidence for a direct innervation of mucosal mast cells. Neuroscience 1983; 10:565-70.
114. Stead RH, Tomioka M, Quinonez G et al. Intestinal mucosal mast cells in normal and nematode-infected rat intestines are in intimate contact with peptidergic nerves. Proc Natl Acad Sci USA 1987; 84:2975-9.
115. Arizono N, Matsuda S, Hattori T et al. Anatomical variation in mast cell nerve associations in the rat small intestine, heart, lung, and skin. Similarities of distances between neural processes and mast cells, eosinophils, or plasma cells in the jejunal lamina propria. Lab Invest 1990; 62:626-34.
116. Williams RM, Berthoud HR, Stead RH. Vagal afferent nerve fibres contact mast cells in rat small intestinal mucosa. Neuroimmunomodulation 1997; 4:266-70.
117. Stead RH, Dixon MF, Bramwell NH et al. Mast cells are closely apposed to nerves in the human gastrointestinal mucosa. Gastroenterology 1989; 97:575-85.
118. Costello RW, Schofield BH, Kephart GM et al. Localization of eosinophils to airway nerves and effect on neuronal M2 muscarinic receptor function. Am J Physiol 1997; 273:L93-103.
119. Johansson O, Liang Y, Marcusson JA et al. Eosinophil cationic protein- and eosinophil-derived neurotoxin/eosinophil protein X-immunoreactive eosinophils in prurigo nodularis. Arch Dermatol Res 2000; 292:371-8.
120. DeSchryver-Kecskemeti K, Clouse RE. Perineural and intraneural inflammatory infiltrates in the intestines of patients with systemic connective-tissue disease. Arch Pathol Lab Med 1989; 113:394-8.
121. Thoenen H, Edgar D. Neurotrophic factors. Science 1985; 229:238-242.
122. Chalazonitis A, Pham TD, Rothman TP et al. Neurotrophin-3 is required for the survival-differentiation of subsets of developing enteric neurons. J Neurosci 2001; 21:5620-36
123. Leon A, Buriani A, Dal Toso R et al. Mast cells synthesize, store, and release nerve growth factor. Proc Natl Acad Sci USA 1994; 91:3739-43.

124. Tam SY, Tsai M, Yamaguchi M et al. Expression of functional TrkA receptor tyrosine kinase in the HMC-1 human mast cell line and in human mast cells. Blood 1997; 90:1807-20.
125. Nilsson G, Forsberg-Nilsson K, Xiang Z et al. Human mast cells express functional TrkA and are a source of nerve growth factor. Eur J Immunol 1997; 27:2295-301.
126. Kobayashi H, Gleich GJ, Butterfield JH et al. Human eosinophils produce neurotrophins and secrete nerve growth factor on immunologic stimuli. Blood 2002; 99:2214-20.
127. Blennerhassett MG, Tomioka M, Bienenstock J. Formation of contacts between mast cells and sympathetic neurons in vitro. Cell Tissue Res 1991; 265:121-8.
128. Blennerhassett MG, Bienenstock J. Sympathetic nerve contact causes maturation of mast cells in vitro. J Neurobiol 1998; 35:173-82.
129. Blennerhassett MG, Janiszewski J, Bienenstock J. Sympathetic nerve contact alters membrane resistance of cells of the RBL-2H3 mucosal mast cell line. Am J Respir Cell Mol Biol 1992; 6:504-9.
130. Suzuki R, Furuno T, McKay DM et al. Direct neurite-mast cell communication in vitro occurs via the neuropeptide substance P. J Immunol 1999; 163:2410-5.
131. Suzuki R, Furuno T, Teshima R et al. Bi-directional relationship of in vitro mast cell-nerve communication observed by confocal laser scanning microscopy. Biol Pharm Bull 2001; 24:291-4.
132. Marshall JS, Gauldie J, Nielsen L et al. Leukemia inhibitory factor production by rat mast cells. Eur J Immunol 1993; 23:2116-20.
133. Zheng X, Knight DA, Zhou D et al. Leukemia inhibitory factor is synthesized and released by human eosinophils and modulates activation state and chemotaxis. J Allergy Clin Immunol 1999; 104:136-44.
134. Yamamori T, Fukada K, Aebersold R et al. The cholinergic neuronal differentiation factor from heart cells is identical to leukemia inhibitory factor. Science 1989; 246:1412-6.
135. Mehler MF, Rozental R, Dougherty M et al. Cytokine regulation of neuronal differentiation of hippocampal progenitor cells. Nature 1993; 362:62-5.
136. Henz B, Hermes B, Welker P. Interactions between neurotrophins and mast cells. In: Marone G, Lichtenstein LM, Galli SJ, eds. Mast cells and basophils. London: Academic press, 2000:341-354.
137. Bajetto A, Bonavia R, Barbero S et al. Chemokines and their receptors in the central nervous system. Front Neuroendocrinol 2001; 22:147-84.
138. Dunzendorfer S, Meierhofer C, Wiedermann CJ. Signaling in neuropeptide-induced migration of human eosinophils. J Leukoc Biol 1998; 64:828-34.
139. Aloe L, Levi-Montalcini R. Mast cells increase in tissues of neonatal rats injected with the nerve growth factor. Brain Res 1977; 133:358-66.
140. Matsuda H, Kannan Y, Ushio H et al. Nerve growth factor induces development of connective tissue-type mast cells in vitro from murine bone marrow cells. J Exp Med 1991; 174:7-14.
141. Horigome K, Bullock ED, Johnson Jr EM. Effects of nerve growth factor on rat peritoneal mast cells. Survival promotion and immediate-early gene induction. J Biol Chem 1994; 269:2695-702.
142. Kawamoto K, Okada T, Kannan Y et al. Nerve growth factor prevents apoptosis of rat peritoneal mast cells through the trk proto-oncogene receptor. Blood 1995; 86:4638-44.
143. Welker P, Grabbe J, Gibbs B et al. Nerve growth factor-beta induces mast-cell marker expression during in vitro culture of human umbilical cord blood cells. Immunology 2000; 99:418-26.
144. Kanbe N, Kurosawa M, Miyachi Y et al. Nerve growth factor prevents apoptosis of cord blood-derived human cultured mast cells synergistically with stem cell factor. Clin Exp Allergy 2000; 30:1113-20.
145. Sawada J, Itakura A, Tanaka A et al. Nerve growth factor functions as a chemoattractant for mast cells through both mitogen-activated protein kinase and phosphatidylinositol 3-kinase signalling pathways. Blood 2000; 95:2052-8.
146. Matsuda H, Coughlin MD, Bienenstock J et al. Nerve growth factor promotes human hemopoietic colony growth and differentiation. Proc Natl Acad Sci USA 1988; 85:6508-12.
147. Hamada A, Watanabe N, Ohtomo H et al. Nerve growth factor enhances survival and cytotoxic activity of human eosinophils. Br J Haematol 1996; 93:299-302.
148. Bruni A, Bigon E, Boarato E et al. Interaction between nerve growth factor and lysophosphatidylserine on rat peritoneal mast cells. FEBS Lett 1982; 138:190-2.
149. Purcell WM, Atterwill CK. Human placental mast cells as an in vitro model system in aspects of neuro-immunotoxicity testing. Hum Exp Toxicol 1994; 13:429-33.
150. Bullock ED, Johnson Jr EM. Nerve growth factor induces the expression of certain cytokine genes and bcl-2 in mast cells. Potential role in survival promotion. J Biol Chem 1996; 271:27500-8.
151. Marshall JS, Gomi K, Blennerhassett MG et al. Nerve growth factor modifies the expression of inflammatory cytokines by mast cells via a prostanoid-dependent mechanism. J Immunol 1999; 162:4271-6.

152. Tomioka M, Stead RH, Nielsen L et al. Nerve growth factor enhances antigen and other secretagogue-induced histamine release from rat peritoneal mast cells in the absence of phosphatidylserine. J Allergy Clin Immunol 1988; 82:599-607.
153. Tada K, Murakami M, Kambe T et al. Induction of cyclooxygenase-2 by secretory phospholipases A2 in nerve growth factor-stimulated rat serosal mast cells is facilitated by interaction with fibroblasts and mediated by a mechanism independent of their enzymatic functions. J Immunol 1998; 161:5008-15.
154. Solomon A, Aloe L, Pe'er J et al. Nerve growth factor is preformed in and activates human peripheral blood eosinophils. J Allergy Clin Immunol 1998; 102:454-60.
155. Purcell WM, Westgate C, Atterwill CK. Rat brain mast cells: An in vitro paradigm for assessing the toxic effects of neurotropic therapeutics. Neurotoxicology 1996; 17:845-50.
156. Arzubiaga C, Morrow J, Roberts IInd LJ et al. Neuropeptide Y, a putative cotransmitter in noradrenergic neurons, induces mast cell degranulation but not prostaglandin D2 release. J Allergy Clin Immunol 1991; 87:88-93.
157. Shanahan F, Denburg JA, Fox J et al. Mast cell heterogeneity: Effects of neuroenteric peptides on histamine release. J Immunol 1985; 135:1331-7.
158. Odum L, Petersen LJ, Skov PS et al. Pituitary adenylate cyclase activating polypeptide (PACAP) is localized in human dermal neurons and causes histamine release from skin mast cells. Inflamm Res 1998; 47:488-92.
159. Repke H, Bienert M. Mast cell activation-a receptor-independent mode of substance P action? FEBS Lett 1987; 221:236-40.
160. Church MK, Lowman MA, Robinson C et al. Interaction of neuropeptides with human mast cells. Int Arch Allergy Appl Immunol 1989; 88:70-8.
161. Cocchiara R, Lampiasi N, Albeggiani G et al. Mast cell production of TNF-alpha induced by substance P evidence for a modulatory role of substance P-antagonists. J Neuroimmunol 1999; 101:128-36.
162. Krumins SA, Broomfield CA. Evidence of NK1 and NK2 tachykinin receptors and their involvement in histamine release in a murine mast cell line. Neuropeptides 1992; 21:65-72.
163. Brooks AC, Whelan CJ. Reactive oxygen species generation by mast cells in response to substance P: A NK1-receptor-mediated event. Inflamm Res 1999; 48:S121.
164. Okada T, Hirayama Y, Kishi S et al. Functional neurokinin NK-1 receptor expression in rat peritoneal mast cells. Inflamm Res 1999; 48:274-9.
165. Manley HC, Haynes LW. Eosinophil chemotactic response to rat CGRP-1 is increased after exposure to trypsin or guinea-pig lung particulate fraction. Neuropeptides 1989; 13:29-34.
166. Kroegel C, Giembycz MA, Barnes PJ. Characterization of eosinophil cell activation by peptides. Differential effects of substance P, melittin, and FMET-Leu-Phe. J Immunol 1990; 145:2581-7.
167. El-Shazly A, Ishikawa T. Novel cooperation between eotaxin and substance-P in inducing eosinophil-derived neurotoxin release. Mediators Inflamm 1999; 8:177-9.
168. Fantozzi R, Masini E, Blandina P et al. Release of histamine from rat mast cells by acetylcholine. Nature 1978; 273:473-4.
169. Blandina P, Fantozzi R, Mannaioni PF et al. Characteristics of histamine release evoked by acetylcholine in isolated rat mast cells. J Physiol 1980; 301:281-93.
170. Kaliner M, Orange RP, Austen KF. Immunological release of histamine and slow reacting substance of anaphylaxis from human lung. J Exp Med 1972; 136:556-67.
171. Bani-Sacchi T, Barattini M, Bianchi S et al. The release of histamine by parasympathetic stimulation in guinea-pig auricle and rat ileum. J Physiol 1986; 371:29-43.
172. Leff AR, Stimler NP, Munoz NM et al. Augmentation of respiratory mast cell secretion of histamine caused by vagus nerve stimulation during antigen challenge. J Immunol 1986; 136:1066-73.
173. Dimitriadou V, Buzzi MG, Moskowitz MA et al. Trigeminal sensory fiber stimulation induces morphological changes reflecting secretion in rat dura mater mast cells. Neuroscience 1991; 44:97-112.
174. Masini E, Rucci L, Cirri-Borghi MB et al. Stimulation and resection of Vidian nerve in patients with chronic hypertrophic nonallergic rhinitis: Effect on histamine content in nasal mucosa. Agents Actions 1986; 18:251-3.
175. Rucci L, Masini E, Arbi Riccardi R et al. Vidian nerve resection, histamine turnover and mucosal mast cell function in patients with chronic hypertrophic nonallergic rhinitis. Agents Actions 1989; 28:224-30.
176. Ganguly AK, Sathiamoorthy SS, Bhatnagar OP. Effect of sub-diaphragmatic vagotomy on gastric mucosal mast cell population in pylorus ligated rats. Q J Exp Physiol Cogn Med Sci 1978; 63:89-92.
177. Gottwald T, Lhotak S, Stead RH. Effect of truncal vagotomy and capsaicin on mast cells and IgA-positive plasma cells in rat jejunal mucosa. Neurogastroenterol Motil 1997; 9:25-32.

178. Kiernan JA. Effects of known and suspected neurotransmitter substances and of some nucleotides on isolated mast cells. Experientia 1972; 28:653-655.
179. Kazimmeierizak W, Diamant B. Mechanisms of histamine release in anaphylactic and anaphylactoid reactions. Prog Allergy 1978; 24:295-365.
180. Assem ES, Schild HO. Inhibition by sympathomimetic amines of histamine release by antigen in passively sensitized human lung. Nature 1969; 224:1028-9.
181. Assem ES. Adrenergic mechanisms and immediate-type allergy. Clin Allergy 1974; 4:185-94.
182. Soll AH, Toomey M. Beta-adrenergic and prostanoid inhibition of canine fundic mucosal mast cells. Am J Physiol 1989; 256:G727-32.
183. Alm PE, Bloom GD. Effect of norepinephrine on in vitro histamine release from rat mast cells. Int Arch Allergy Appl Immunol 1979; 60:60-7.
184. Alm PE, Bloom GD. What - if any - is the role of adrenergic mechanisms in histamine release from mast cells? Agents Actions 1981; 11:60-6.
185. Befus AD, Dyck N, Goodacre R et al. Mast cells from the human intestinal lamina propria. Isolation, histochemical subtypes, and functional characterization. J Immunol 1987; 138:2604-10.
186. Ferrante F, Ricci A, Felici L et al. Suggestive evidence for a functional association between mast cells and sympathetic nerves in menigeal membranes. Acta Histochem Cytochem 1990; 23:637-46.
187. Bergerot A, Reynier-Rebuffel AM, Callebert J et al. Long-term superior cervical sympathectomy induces mast cell hyperplasia and increases histamine and serotonin content in the rat dura mater. Neuroscience 2000; 96:205-13.
188. Barnes PJ. Effect of beta-agonists on inflammatory cells. J Allergy Clin Immunol 1999; 104:S10-7.
189. Vergnolle N, Macnaughton WK, Al-Ani B et al. Proteinase-activated receptor 2 (PAR2)-activating peptides: Identification of a receptor distinct from PAR2 that regulates intestinal transport. Proc Natl Acad Sci USA 1998; 95:7766-71.
190. Corvera CU, Dery O, McConalogue K et al. Thrombin and mast cell tryptase regulate guinea-pig myenteric neurons through proteinase-activated receptors-1 and -2. J Physiol 1999; 517:741-56.
191. de Garavilla L, Vergnolle N, Young SH et al. Agonists of proteinase-activated receptor 1 induce plasma extravasation by a neurogenic mechanism. Br J Pharmacol 2001; 133:975-87.
192. Evans CM, Fryer AD, Jacoby DB et al. Pretreatment with antibody to eosinophil major basic protein prevents hyperresponsiveness by protecting neuronal M2 muscarinic receptors in antigen-challenged guinea pigs. J Clin Invest 1997; 100:2254-62.
193. Lee LY, Gu Q, Gleich GJ. Effects of human eosinophil granule-derived cationic proteins on C-fiber afferents in the rat lung. J Appl Physiol 2001; 91:1318-26.
194. Coyle AJ, Perretti F, Manzini S et al. Cationic protein-induced sensory nerve activation: Role of substance P in airway hyperresponsiveness and plasma protein extravasation. J Clin Invest 1994; 94:2301-6.
195. Fredens K, Dahl R, Venge P. The Gordon phenomenon induced by the eosinophil cationic protein and eosinophil protein X. J Allergy Clin Immunol 1982; 70:361-6.
196. Dvorak AM. Ultrastructural evidence for release of major basic protein-containing crystalline cores of eosinophil granules in vivo: Cytotoxic potential in Crohn's disease. J Immunol 1980; 125:460-2.
197. Yu LC, Perdue MH. Role of mast cells in intestinal mucosal function: Studies in models of hypersensitivity and stress. Immunol Rev 2001; 179:61-73.
198. Perdue MH, Forstner JF, Roomi NW et al. Epithelial response to intestinal anaphylaxis in rats: Goblet cell secretion and enterocyte damage. Am J Physiol 1984; 247:G632-7.
199. Perdue MH, Gall DG. Intestinal anaphylaxis in the rat: Jejunal response to in vitro antigen exposure. Am J Physiol 1986; 250:G427-31.
200. Perdue MH, Masson S, Wershil BK et al. Role of mast cells in ion transport abnormalities associated with intestinal anaphylaxis. Correction of the diminished secretory response in genetically mast cell-deficient W/Wv mice by bone marrow transplantation. J Clin Invest 1991; 87:687-93.
201. Crowe SE, Sestini P, Perdue MH. Allergic reactions of rat jejunal mucosa. Ion transport responses to luminal antigen and inflammatory mediators. Gastroenterology 1990; 99:74-82.
202. Castro GA, Harari Y, Russell D. Mediators of anaphylaxis-induced ion transport changes in small intestine. Am J Physiol 1987; 253:G540-8.
203. Frieling T, Cooke HJ, Wood JD. Neuroimmune communication in the submucous plexus of guinea pig colon after sensitization to milk antigen. Am J Physiol 1994; 267:G1087-93.
204. Wood JD. Histamine signals in enteric neuroimmune interactions. Ann NY Acad Sci 1992; 664:275-83.
205. Scott RB, Tan DT, Miampamba M et al. Anaphylaxis-induced alterations in intestinal motility: Role of extrinsic neural pathways. Am J Physiol 1998; 275:G812-21.
206. Crowe SE, Perdue MH. Anti-immunoglobulin E-stimulated ion transport in human large and small intestine. Gastroenterology 1993; 105:764-72.

206a. Jiang W, Kreis ME, Eastwood C et al. 5-HT(3) and histamine H(1) receptors mediate afferent nerve sensitivity to intestinal anaphylaxis in rats. Gastroenterology 2000; 119:1267-75.
207. Riegler M, Castagliuolo I, So PT et al. Effects of substance P on human colonic mucosa in vitro. Am J Physiol 1999; 276:G1473-83.
208. Crowe SE, Luthra GK, Perdue MH. Mast cell mediated ion transport in intestine from patients with and without inflammatory bowel disease. Gut 1997; 41:785-92.
209. Crowe SE, Soda K, Stanisz AM et al. Intestinal permeability in allergic rats: Nerve involvement in antigen-induced changes. Am J Physiol 1993; 264:G617-23.
210. Berin MC, Kiliaan AJ, Yang PC et al. The influence of mast cells on pathways of transepithelial antigen transport in rat intestine. J Immunol 1998; 161:2561-6.
211. Santos J, Yang PC, Soderholm JD et al. Role of mast cells in chronic stress induced colonic epithelial barrier dysfunction in the rat. Gut 2001; 48:630-6.
212. Castagliuolo I, Wershil BK, Karalis K et al. Colonic mucin release in response to immobilization stress is mast cell dependent. Am J Physiol 1998; 274:G1094-1100.
213. Pothoulakis C, Castagliuolo I, Leeman SE. Neuroimmune mechanisms of intestinal responses to stress. Role of corticotropin-releasing factor and neurotensin. Ann NY Acad Sci 1998; 840:635-48.
214. Kiliaan AJ, Saunders PR, Bijlsma PB et al. Stress stimulates transepithelial macromolecular uptake in rat jejunum. Am J Physiol 1998; 275:G1037-44.
215. Santos J, Saperas E, Nogueiras C et al. Release of mast cell mediators into the jejunum by cold pain stress in humans. Gastroenterology 1998; 114:640-8.
216. MacQueen G, Marshall J, Perdue M et al. Pavlovian conditioning of rat mucosal mast cells to secrete rat mast cell protease II. Science 1989; 243:83-5.
217. Matsumoto I, Inoue Y, Shimada T et al. Brain mast cells act as an immune gate to the hypothalamic-pituitary-adrenal axis in dogs. J Exp Med 2001; 194:71-8.
218. Undem BJ, Kajekar R, Myers AC. Modulation of peripheral neurotransmission associated with mast cell activation. In: Marone G, Lichtenstein LM, Galli SJ, eds. Mast cells and basophils. London: Academic press, 2000:355-366.
219. Stead RH. Nerve remodelling during intestinal inflammation. Ann NY Acad Sci 1992; 664:443-55.
220. Collins SM, Hurst SM, Main C et al. Effect of inflammation of enteric nerves. Cytokine-induced changes in neurotransmitter content and release. Ann NY Acad Sci 1992; 664:415-24.
221. Xia Y, Hu HZ, Liu S et al. IL-1beta and IL-6 excite neurons and suppress nicotinic and noradrenergic neurotransmission in guinea pig enteric nervous system. J Clin Invest 1999; 103:1309-16.
222. Dvorak AM, Silen W. Differentiation between Crohn's disease and other inflammatory conditions by electron microscopy. Ann Surg 1985; 201:53-63.
223. Steinhoff MM, Kodner IJ, DeSchryver-Kecskemeti K. Axonal degeneration/necrosis: A possible ultrastructural marker for Crohn's disease. Mod Pathol 1988; 1:182-7.
224. Kyosola K, Penttila O, Salaspuro M. Rectal mucosal adrenergic innervation and enterochromaffin cells in ulcerative colitis and irritable colon. Scand J Gastroenterol 1977; 12:363-7.
225. Rumessen JJ. Ultrastructure of interstitial cells of Cajal at the colonic submuscular border in patients with ulcerative colitis. Gastroenterology 1996; 111:1447-55.
226. Porcher C, Baldo M, Henry M et al. Deficiency of interstitial cells of Cajal in the small intestine of patients with Crohn's disease. Am J Gastroenterol 2002; 97:118-25.
227. Belai A, Boulos PB, Robson T et al. Neurochemical coding in the small intestine of patients with Crohn's disease. Gut 1997; 40:767-74.
228. Schneider J, Jehle EC, Starlinger MJ et al. Neurotransmitter coding of enteric neurones in the submucous plexus is changed in noninflamed rectum of patients with Crohn's disease. Neurogastroenterol Motil 2001; 13:255-64.
229. Mazumdar S, Das KM. Immunocytochemical localization of vasoactive intestinal peptide and substance P in the colon from normal subjects and patients with inflammatory bowel disease. Am J Gastroenterol 1992; 87:176-81.
230. Keranen U, Kiviluoto T, Jarvinen H et al. Changes in substance P-immunoreactive innervation of human colon associated with ulcerative colitis. Dig Dis Sci 1995; 40:2250-8.
231. Keranen U, Jarvinen H, Karkkainen P et al. Substance P-an underlying factor for pouchitis? Prospective study of substance P- and vasoactive intestinal polypeptide-immunoreactive innervation and mast cells. Dig Dis Sci 1996; 41:1665-71.
232. Keranen U, Jarvinen H, Kiviluoto T et al. Substance P and vasoactive intestinal polypeptide-immunoreactive innervation in normal and inflamed pouches after restorative proctocolectomy for ulcerative colitis. Dig Dis Sci 1996; 41:1658-64.
233. Watanabe T, Kubota Y, Muto T. Substance P containing nerve fibers in rectal mucosa of ulcerative colitis. Dis Colon Rectum 1997; 40:718-25.

234. Kimura M, Masuda T, Hiwatashi N et al. Changes in neuropeptide-containing nerves in human colonic mucosa with inflammatory bowel disease. Pathol Int 1994; 44:624-34.
235. Watanabe T, Kubota Y, Muto T. Substance P containing nerve fibers in ulcerative colitis. Int J Colorectal Dis 1998; 13:61-7.
236. Koch TR, Carney JA, Go VL. Distribution and quantitation of gut neuropeptides in normal intestine and inflammatory bowel diseases. Dig Dis Sci 1987; 32:369-76.
237. Goldin E, Karmeli F, Selinger Z et al. Colonic substance P levels are increased in ulcerative colitis and decreased in chronic severe constipation. Dig Dis Sci 1989; 34:754-7.
238. Bernstein CN, Robert ME, Eysselein VE. Rectal substance P concentrations are increased in ulcerative colitis but not in Crohn's disease. Am J Gastroenterol 1993; 88:908-13.
239. Mantyh CR, Gates TS, Zimmerman RP et al. Receptor binding sites for substance P, but not substance K or neuromedin K, are expressed in high concentrations by arterioles, venules, and lymph nodules in surgical specimens obtained from patients with ulcerative colitis and Crohn disease. Proc Natl Acad Sci USA 1988; 85:3235-9.
240. Tomita R, Tanjoh K, Fujisaki S et al. Peptidergic nerves in the colon of patients with ulcerative colitis. Hepatogastroenterology 2000; 47:400-4.
241. Moriarty D, Goldhill J, Selve N et al. Human colonic anti-secretory activity of the potent NK(1) antagonist, SR140333: Assessment of potential anti-diarrhoeal activity in food allergy and inflammatory bowel disease. Br J Pharmacol 2001; 133:1346-54.
242. Koch TR, Carney JA, Go VL et al. Altered inhibitory innervation of circular smooth muscle in Crohn's colitis. Association with decreased vasoactive intestinal polypeptide levels. Gastroenterology 1990; 98:1437-44.
243. Kubota Y, Petras RE, Ottaway CA et al. Colonic vasoactive intestinal peptide nerves in inflammatory bowel disease. Gastroenterology 1992; 102:1242-51.
244. Schulte-Bockholt A, Fink JG, Meier DA et al. Expression of mRNA for vasoactive intestinal peptide in normal human colon and during inflammation. Mol Cell Biochem 1995; 142:1-7.
245. O'Morain C, Bishop AE, McGregor GP et al. Vasoactive intestinal peptide concentrations and immunocytochemical studies in rectal biopsies from patients with inflammatory bowel disease. Gut 1984; 25:57-61.
246. Duffy LC, Zielezny MA, Riepenhoff-Talty M et al. Vasoactive intestinal peptide as a laboratory supplement to clinical activity index in inflammatory bowel disease. Dig Dis Sci 1989; 34:1528-35.
247. Todorovic V, Janic B, Koko V et al. Colonic vasoactive intestinal polypeptide (VIP) in ulcerative colitis a radioimmunoassay and immunohistochemical study. Hepatogastroenterology 1996; 43:483-8.
248. Holzer P. Implications of tachykinins and calcitonin gene-related peptide in inflammatory bowel disease. Digestion 1998; 59:269-83.
249. Reubi JC, Mazzucchelli L, Laissue JA. Intestinal vessels express a high density of somatostatin receptors in human inflammatory bowel disease. Gastroenterology 1994; 106:951-9.
250. Yamamoto H, Morise K, Kusugami K et al. Abnormal neuropeptide concentration in rectal mucosa of patients with inflammatory bowel disease. J Gastroenterol 1996; 31:525-32.
251. Cervero F. Visceral pain-central sensitisation. Gut 2000; 47:iv56-7.
252. Bueno L, Fioramonti J, Delvaux M et al. Mediators and pharmacology of visceral sensitivity: From basic to clinical investigations. Gastroenterology 1997; 112:1714-43.
253. Gebhart GF. Visceral pain-peripheral sensitisation. Gut 2000; 47:iv54-5.
254. Collins SM. The immunomodulation of enteric neuromuscular function: Implications for motility and inflammatory disorders. Gastroenterology 1996; 111:1683-99.
255. Farthing MJ, Lennard-jones JE. Sensibility of the rectum to distension and the anorectal distension reflex in ulcerative colitis. Gut 1978; 19:64-9.
256. Mayer EA, Raybould H, Koelbel C. Neuropeptides, inflammation, and motility. Dig Dis Sci 1988; 33:71S-77S.
257. Agro A, Stanisz AM. Inhibition of murine intestinal inflammation by anti-substance P antibody. Reg Immunol 1993; 5:120-6.
258. Sonea IM, Palmer MV, Akili D et al. Treatment with neurokinin-1 receptor antagonist reduces severity of inflammatory bowel disease induced by cryptosporidium parvum. Clin Diagn Lab Immunol 2002; 9:333-40.
259. Sturiale S, Barbara G, Qiu B et al. Neutral endopeptidase (EC 3.4.24.11) terminates colitis by degrading substance P. Proc Natl Acad Sci USA 1999; 96:11653-8.
260. Shafiroff GP, Hinton J. Denervation of the pelvic colon for ulcerative colitis. Surg Forum 1950; 134-39.
261. Thorek P. Vagotomy for idiopathic ulcerative colitis and regional enteritis. JAMA 1951; 145:140-46.
262. Kemler MA, Barendse GA, Van Kleef M. Relapsing ulcerative colitis associated with spinal cord stimulation. Gastroenterology 1999; 117:215-7.

263. Barbara G, De Giorgio R, Stanghellini V et al. Relapsing ulcerative colitis after spinal cord stimulation: A case of intestinal neurogenic inflammation? (letter) Gastroenterology 1999; 117:1256-1257.
264. Bjorck S, Dahlstrom A, Johansson L et al. Treatment of the mucosa with local anaesthetics in ulcerative colitis. Agents Actions 1992; C60-72.
265. McCafferty DM, Sharkey KA, Wallace JL. Beneficial effects of local or systemic lidocaine in experimental colitis. Am J Physiol 1994; 266:G560-7.
266. Evangelista S, Meli A. Influence of capsaicin-sensitive fibres on experimentally-induced colitis in rats. J Pharm Pharmacol 1989; 41:574-5.
267. Eysselein VE, Reinshagen M, Patel A et al. Calcitonin gene-related peptide in inflammatory bowel disease and experimentally induced colitis. Ann NY Acad Sci 1992; 657:319-27.
268. Reinshagen M, Flamig G, Ernst S et al. Calcitonin gene-related peptide mediates the protective effect of sensory nerves in a model of colonic injury. J Pharmacol Exp Ther 1998; 286:657-61.
269. di Mola FF, Friess H, Zhu ZW et al. Nerve growth factor and Trk high affinity receptor (TrkA) gene expression in inflammatory bowel disease. Gut 2000; 46:670-9.
270. Magro F, Vieira-Coelho MA, Fraga S et al. Impaired synthesis or cellular storage of norepinephrine, dopamine, and 5-hydroxytryptamine in human inflammatory bowel disease. Dig Dis Sci 2002; 47:216-24.
271. Koch TR, Cave DR, Ford H et al. Histofluorescent and radioenzymatic analysis of colonic catecholamines in man. J Auton Nerv Syst 1984; 11:383-91.

Chapter 13

Multiparameter Analysis of Immunogenetic Mechanisms in Clinical Diagnosis and Management of Inflammatory Bowel Disease

Jonathan Braun and Stephan R. Targan

Abstract

The integrity of the intestinal mucosa depends on a functional coordination of the epithelium, lumenal microorganisms, and the local immune system. The mammalian immune system is superbly organized for innate and adaptive recognition of microbial antigens,[1,2] a defensive capacity that must be balanced against the tissue damage produced by immune activity to preserve normal intestinal function.[3] Inflammatory bowel disease (IBD) is generally thought to reflect an impairment in this balance, due to a combination of host genetic traits that shift the balance of immune and epithelial function to commensal microbiota, and perhaps the composition or activity of certain microbial elements as well.

There has been much progress defining the fundamental disorders of these host traits, immunologic processes, and microbial targets in inflammatory bowel disease.[4] Other fields of clinical and geologic microbiology are teaching us about the dynamic interaction of commensal bacteria with their host environment.[5,6] These lines of investigation have revealed not only important insights about inflammatory bowel disease (IBD) pathogenesis, but also defined technologies and tools useful for its diagnosis and clinical management. This review focuses on these advances at the translational interface. We will first consider the innate anti-microbial response, centering on the utility of NOD2 genotyping for predicting disease susceptibility, prognosis, and therapeutic response profile. We will then turn to the adaptive anti-microbial response, focusing on the application of antibodies to fungal and bacterial species and products for Crohn's disease (CD) diagnosis and prognosis, and immunogenetics of T cell immunosuppression management. Finally, we will describe autoimmune mechanisms in IBD, with particular attention to autoantibodies in IBD diagnosis and infliximab responsiveness. We will conclude with the concept of multiparameter analysis of patients, to refine patient characterization and stratification in diagnosis and clinical management.

NOD2 and Innate Anti-Microbial Processes in IBD Diagnosis

During the past decade, a major effort by several laboratories in genome-wide analysis of at-risk human populations has culminated in the identification of several loci associated with CD and/or ulcerative (UC) susceptibility.[7] The greatest progress has occurred with the IBD1 locus on chromosome 16, which proved to reflect mutations in the NOD2/CARD15 gene.[8,9]

Immune Mechanisms in Inflammatory Bowel Disease, edited by Richard S. Blumberg and Markus F. Neurath.

This protein encodes an important innate immune microbial recogntion molecule—an intracellular peptidoglycan receptor which activates CARD and NF-κB dependent pathways of cellular activation.[10] NOD2 mutations primarily if not exclusively affect CD susceptibility, indicating that this protein pertains to the pathobiology of CD rather than UC. Several NOD2 alleles are associated with CD disease risk, and relate to loss-of-function missense mutations of the NOD2 protein.[11]

Several studies have shed light on the utility of NOD2 mutations as adjuncts to disease diagnosis. First, it should be noted that the overall genetic contribution to CD is estimated to be only a minority of disease risk (about 36%),[12] and NOD2 is only one of several loci contributing to this facet of disease pathobiology. Second, even among individuals with CD, most individuals are heterozygote for loss-of-function NOD2 alleles, which would be expected to confer lesser risk than the homozygote state. A practical issue is the number of pertinent disease-susceptibility alleles, since a large number of such alleles, each at low frequency, might additively be a large percent of genetic disease risk in the population. Fortunately, it so far appears that only three haplotypes account for most pertinent NOD2 mutations.[11,13]

With these considerations in mind, studies to this point indicate that the heterozygote state is associated with about 2-fold excess disease risk, and about 24-fold in the homozygote state.[13-15] Since the prevalence of CD is rather low in the population (about 0.1%), neither the heterozygote nor homozygote states of these NOD alleles are suitable as screening diagnostic criteria for CD. However, the allelic state may be useful in validating diagnosis in patients in combination with other independent parameters (clinical, immunologic, or other susceptibility genes) permitting disease stratification (see below).

A more immediate use of NOD2 allelism is for disease phenotype and prognosis. Patterns of CD clinical phenotype are substantially hereditary,[16] and several recent studies concur that NOD2 mutations are an important predictor of fibrostenosing ileal disease.[13-15] Such information is important in prognosis, and for stratifying and selecting patients for therapies targeting these facets of CD disease activity. Finally, cellular NOD2 protein levels may add further refinement to this assessment, since there is evidence for NOD2 expression in epithelial cells, whose level and function may contribute to mucosal-microbial homeostasis.[17]

Like NOD2, a number of additional genes or loci are promising for CD or UC diagnosis. The IBD2 (a chromosome 12 haplotype) and IBD5 (a 250-kb haplotype the 5q31 cytokine gene cluster) loci confer independent disease susceptibility with NOD2 in CD patients.[18,19] IBD2 is of additional interest because it is the strongest haplotype associated with UC disease risk, a form of IBD with historically less evidence for genetic susceptibility. This suggests that the pertinent gene will represent a common molecule regulating mucosal inflammation in both disease settings.[20] Even before the pertinent genes are identified, the IBD2 and IBD5 haplotypes are probably analytically suitable and merit assessment for their contribution to disease phenotype and prognosis.

Monitoring of certain candidate genes in innate immune function may find an eventual place in IBD clinical management, and we highlight four as examples of future prospects. First, the association of IBD with the major histocompatibility locus has been under investigation for decades.[21-23] While attention has previously focused on the HLA class 1 and 2 genes, the locus also includes TNF-α and lymphotoxin-α genes. A recent study by Taylor and colleagues indicates that certain haplotypes in this region predict responsiveness to anti-TNF (infliximab) therapy in CD.[24] This finding opens up the exciting possibility that such genotyping may be valuable in stratifying patients for biologicals or small molecules pertinent to the TNF facet of immune regulation or effector function. Second, troglitazone, a nominal agonist of PPAR-γ (or the closely related LXR family) have potent anti-inflammatory actions due to the regulatory action of these transcription factors on the macrophage activation state.[25,26] PPAR-γ epithelial expression is reported to be deficient in most UC patients, and also in some CD patients.[27] Since this class of small molecules is emerging as new therapy in IBD (particularly CD), it would seem wise to assess the relationship and basis for impaired expression with the outcome of clinical response.

Third, defensins- a family of anti-microbial peptides produced by hemopoeitic and epithelial cells—have emerged as important players in epithelial-commensal bacterial interaction in the gut.[28] Fellerman and colleagues recently have reviewed evidence that impaired epithelial production of certain beta-defensins may contribute to disordered mucosal-microbiologic homeostasis in IBD, particularly UC.[29] Since changes in beta-defensin isoforms are heterogeneous, assessment of beta-defensin status may be a useful discriminator of biologic distinct patient subsets, whose prognosis or response to treatment would be more homogeneous.

Fourth, epithelial integrity forms an innate barrier to bacteria and their products, and impairments of this barrier result in chronic colitis in transgenic mouse systems.[30] P-glycoprotein (encoded by the multidrug resistance-1 gene, MDR1) is widely expressed in intestinal epithelial cells and forms a barrier to bacteria-dependent intestinal inflammation bacterial invasion and incursion of microbial products.[31] A recent study of a human MDR1 single nucleotide polymorphism C3435T, associated with lower intestinal P-glycoprotein expression, demonstrated that this polymorphism selectively predisposes to development of UC.[32] These observations suggest that assessment of polymorphic MDR1 expression, either in biopsy specimens or in blood lymphocytes may permit this trait to be integrated into a patient stratification approach.[33]

Anti-Microbial Antibodies and Adaptive Immune Components in IBD Diagnosis

Classic pathologic diagnostic parameters. The classic diagnostic parameters for UC and CD include pathologic and laboratory measurements, which in effect constitute applied cell biology and biochemistry. In the case of CD, this includes mucosal accumulation and activation of monocyte-macrophages and lymphocytes, with features of activation evocative of TH1-like responses, including granuloma formation (which experimentally is largely IFN-γ dependent).[34-36] Distinctly, UC is manifested not only by mononuclear cell infiltration, but also trafficking of granulocytes to form crypt abscesses, and cytokine and chemokine profiles mediating this effector response.[36-39] Disease activity is assessed in part by laboratory tests for inflammatory products of these immune-mediated processes. For example, serum acute phase reactant proteins (e.g., C-reactive protein) are elicited by the hepatocyte response to systemic levels of IL-1, IL-6, and TNF-α generated from the intestinal immune response.[25,40]

Anti-Microbial Antibodies

IgG seroreactivity to a variety of microbial species has been associated with IBD, particularly Crohn's disease.[41-43] These findings agree with the view that adaptive anti-microbial immunity is a component of CD pathogenesis, although the diversity of microorganisms raise several issues regarding the nature of the microbial encounter and immunologic response.[44] Nonetheless, such antibodies have emerged as useful analytes for IBD diagnosis.

The most widely used of these analytes is ASCA (antibodies to the cell wall polysaccharide of *Saccharomyces cerevisiae).* IgG and IgA antibodies are highly specific for CD, and moderately sensitive for this patient group (about 60% of patients).[45,46] The antigenic stimulus for these antibodies is uncertain, since the core epitope is found among diverse fungal and plant cell walls.[47] It is possible that the response relates to dietary antigens,[48,49] or a previously unappreciated existence of a fungal component in the enteric microbial community.[49] Finally, it is notable that studies of CD patients and first-degree family members indicate that ASCA seroreactivity (negative or positive) is a familial trait of both patients and unaffected family members.[50,51]

Table 1. Genetic loci in IBD disease stratification

Parameter	Description	Phenotype	Assay	Reference
IBD1	nod2	CD; fibrostenosing ileal disease	microsatellite	13-15
IBD2	12q14	nod2-independent CD; ulcerative colitis	microsatellite	18
IBD5	5q31	nod2-independent CD	microsatel-lite	19,20
MHC	LT-a	Infliximab response	microsatellite	24
β-defensins	Anti-microbial peptides	CD	Endoscopic biopsy immunohistochemistry	29
Mdr1	P-glycoprotein	UC	Single-nucleotide polymorphism; lymphocyte flow cytometry; sucrose absorption permeability	32,33
Thiopurine methlytrans-ferase	Polymorphic enzyme controlling thiopurine bioavailability	6-mercaptopurine and azathioprine dosing	Erythrocyte enzyme activity	72

Mycobacterial species are associated with CD-like disease in human and other species, and several groups have reported elevated levels of anti-mycobacterial antibodies in CD patients.[52-56] Analytically, immunoassay detection has focused on recombinant protein antigens from *M. paratuberculosis* with impressive reports of specificity and sensitivity to CD (~75% and ~90%) compared to normals and UC patients.[57,58] While the contribution of *M. paratuberculosis* as a proinflammatory species in CD remains controversial, anti-mycobacterial antibiotics make the search for such a patient population appealing. Multiparameter assessment with serology and tissue based microbial detection may provide an avenue to identify such patients.[59]

TonB-linked outer membrane proteins of certain human intestinal commensals (*Bacteroides caccae, B. thetaiotamicron; E. coli*) have also emerged as useful antigenic targets for CD serodiagnosis.[60-62] These proteins (OmpW, SusC, and OmpC) are highly homologous to RagA of *Porphyromonas gingivalis*, an immunologic virulence factor for this periodontal disease pathogen. Similarly, the product of certain pseudomonads (the *P. fluorescens* protein PfiT, and the embedded I2 peptide) encode an antigen with ~60% IgG and IgA seroreactivity in CD patients.[62-64] Pseudomonads are rare commensals, and may be present in the gut lumen in part as a dietary component. However, *P. fluorescens* is molecularly detectable by PCR in the majority of CD lesions,[63] and PfiT itself has T cell superantigen bioactivity.[65] This implies that seropositive patients may display biologically distinct disease behavior useful in patient stratification (see below).[62]

IBD-Related T Cell Function

Disordered features of T cell microbial recognition and effector function anti-microbial T cell response is likely to be a central aspect of IBD disease biology.[66,67] To this point, we have lacked technologies for clinically useful assessment of the T cell in IBD. Minimal cells are available from biopsy or resection specimens, and require elaborate handling unsuitable for the clinical setting; and, blood lymphocytes do not generally reflect the repertoire and differentiation state of tissue-based immune responses. An interesting solution may be now emerging from advances in the understanding of chemokine-based regulation of mucosal populations. CCR9, the receptor for a small-intestine specific chemokine TECK (CCL25), is highly expressed on blood lymphocytes recirculating to the small intestine.[68] Initial work indicates that this marker allows sampling of mucosal T cells "in transit", so they can be characterized with regard to antigenic specificities, activation state, or effector mechanisms reflective of the mucosal immune state. It appears that additional chemokine receptors may similarly permit assessment of other lymphocyte subpopulations and mucosal sites.[69]

T cell immunosuppression is a major strategy in IBD therapy, including the use of 6-mercaptopurine congeners acting through small G-protein inhibition of T cell activation.[70] An impediment to optimal clinical use of these agents is their substantial hematologic toxicity, which is not adequately predictable under conventional dosing. An important host parameter affecting bioavailability and metabolism is thiopurine methlytransferase, whose level of activity is genetically polymorphic.[71] A recent study suggested that measurement of this enzyme activity in surrogate cells (erythrocytes) can be useful in predicting exceptional resistance or toxicity with 6-MP and azathioprine.[72] This type of investigation highlights the opportunities for pharmacogenetic diagnostics as an early opportunity in facets of IBD clinical management.

Autoantibodies in IBD Diagnosis

Antibodies to a perinuclear neutrophil antigen, pANCA, are a sensitive and specific criterion for diagnosis of UC, and a clinically distinct subset of CD.[73,74] Antigenically, they are distinguished from other cANCAs and pANCAs by sensitivity of their antigen to DNAse I treatment and localization to the inner nuclear membrane leaflet.[75,76] At least a component of these antigens appear to include histone H1 and HMG family members.[77-80] It remains unknown whether these antibodies are a disease marker or a pathogenic factor. Recent advances in experimental autoimmune vasculitis have defined a comprehensive pathophysiologic mechanism relating cANCA (anti-MPO antibodies) to vascular and tissue damage.[81] This casts a fresh light on the previously reported occurrence of anti-endothelial antibodies in IBD,[82,83] and might serve as a guide to pathophysiologic assessment of pANCA in UC.

pANCA expression is a familial trait, since it is concordant in monozygotic twins, present in unaffected family members, and associated with a certain MHC II haplotype.[23,24,50,77,84] In genetic models of colitis in mouse, pANCA antibodies are also detected in UC-like phenotypes.[85] These observations indicate that pANCA expression is a trait associated with immunogenetic susceptibility for ulcerative colitis. Moreover, multiparameter assessment of this antibody trait and genetic polymorphisms (e.g., the TNFA/LTA locus) is useful not only in diagnosis, but in discrimination of a CD patient subset with a distinct distribution of disease and responsiveness to infliximab.[24,62,86,87]

Anti-epithelial antibodies, recently focusing on tropomyosin, have emerged as a disease-related seroreactivity in UC.[88-91] These antibodies are also observed in a number of mouse model systems preceding onset of clinical disease, and mediate antibody-dependent cytotoxicity in vivo.[85,92] Anti-tropomyosin antibodies thus would merit incorporation as part of a multi-parameter assessment (using genetic or other antibodies), to assess whether this trait would enhance current capabilities disease prognosis or management.

Multiparameter Analysis of Immunogenetic Traits in IBD Diagnosis

In this chapter, we have summarized different classes of immunogenetic parameters (genetic allelisms, and antibody levels to microbial and autoantigens) pertinent to IBD diagnosis, disease pattern, or response to treatment. Since these parameters are in most cases biologically divergent, it is reasonable to imagine that an integrated assessment of these traits in

Table 2. Serum antibodies in IBD disease stratification

Parameter	Description	Phenotype	Assay	Reference
ASCA	*Saccharomyces cerevisciae* cell wall polysaccharide	CD	ELISA	50,51
p35, IS900	*Mycobacterium paratuberculosis*	CD	ELISA; tissue PCR	59
OmpC OmpW	*E. coli, Bacteroides caccae*	CD	ELISA	60-62
PfiT (I2)	*Pseudomonas fluorescens*	CD	ELISA	62-64
CCR9 T cells	Homing receptor for small-intestine	CD disease activity	Flow cytometry	69
pANCA	Neutrophil perinuclear antigen	UC; infliximab-resistant CD	ELISA immunohistochemistry;	24,62,86,87
Tropomyosin	Anti-epithelial antibody	UC	ELISA	88-91
Antiendothelial		CD	Immunohistochemistry	82,83

patients would permit better resolution of clinically meaningful disease subsets. If so, such patient stratification could be useful for both IBD research, and to augment clinical assessment of patients for diagnosis and treatment.

Several studies have now provided evidence in support of this approach. As noted above, antibodies to certain microbial products and autoantigens (ASCA and pANCA) identify subsets of individuals with UC and CD with distinct clinica l courses.[50,87] Combined host genetic (MHC) and anti-microbial analysis have identified patient subsets distinguished by disease prognosis[23] or infliximab responsiveness.[24] Recently, our group has assessed a large CD patient population to assess their concordance with multiple anti-microbial antibodies (ASCA, PfiT, and outer membrane porins) and the pANCA autoantibody.[62] This study demonstrated several points about patient heterogeneity in CD. First, although each of the anti-microbial antibodies was present at similar frequencies in the CD population, the expression of these antibodies was nonconcordant, and in each individual was a stable phenotype. In particular, patients were divergent with respect to anti-fungal and anti-bacterial antibodies, indicating that these responses reflect different biologic groups of CD patients. Second, pANCA autoantibody levels were discordant with the two sets of anti-microbial antibody traits. Notably, these parameters were each stable over time in individual patients, suggesting that they are intrinsic traits of these patients. Taken together, this multiparameter analysis reflects the predicted biologic heterogeneity of the CD population (Fig. 1). Since NOD2 polymorphisms are already known to convey clinical heterogeneity distinct from some of these immunologic markers,[14] inclusion of NOD2 genotyping would be expected to further refine the definition of patients with regard to their clinical phenotype.

In summary, progress in basic IBD immunogenetics is now beginning to provide tools useful for clinically meaningful stratification of patients with this set of diseases. In some cases, available parameters may be useful as an adjunct to disease diagnosis. More important, these parameters, due to their reflection of distinct aspects of the disease biology, are likely to provide tools to identify patients with differing patterns of disease, prognosis, and response to treatment. Integrated multiparameter assessment is expected to be useful in stratification of more

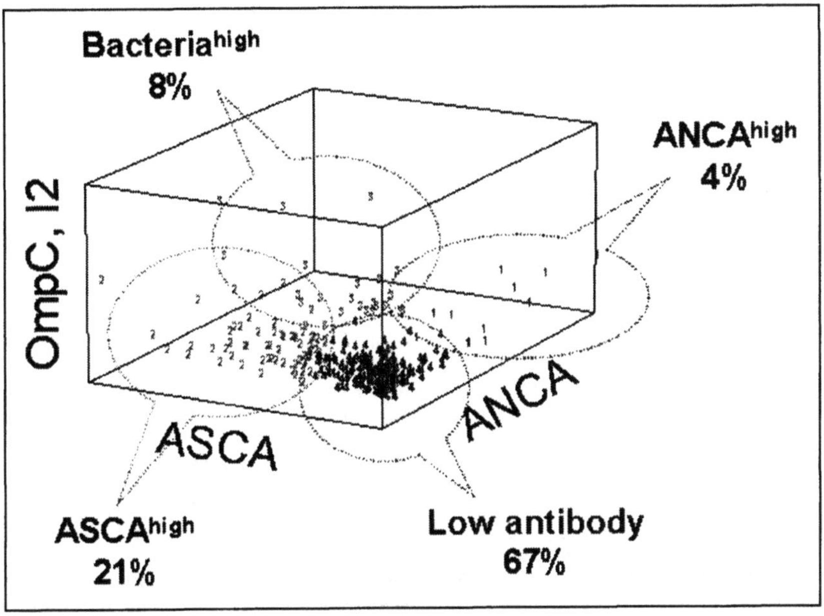

Figure 1. IBD patient stratification by multi parameter antibody analysis. A three-dimensional scatterplot was constructed from antibody levels using 5 specificities in 307 patients. Clusters of patients defined by antibody patterns are indicated by balloons.

biologically homogeneous patient subsets for functional and genetic pathogenesis research, trials of therapy, and guidance at the bedside in clinical management.

References

1. Starzl TE, Zinkernagel RM. Mechanisms of Disease: Antigen localization and migration in immunity and tolerance. N Engl J Med 1998; 339(26):1905-1913.
2. Janeway Jr CA, Medzhitov R. Innate immune recognition. Annu Rev Immunol 2002; 20:197-216.
3. Sartor RB. The influence of normal microbial flora on the development of chronic mucosal inflammation. Res Immunol 1997; 148:567-576.
4. Strober W, Fuss IJ, Blumberg RS. The immunology of mucosal models of inflammation. Annu Rev Immunol 2002; 20:495-549.
5. Covacci A, Telford JL, Del Giudice G et al. Helicobacter pylori virulence and genetic geography. Science 1999; 284:1328-1333.
6. Torsvik V, Ovreas L. Microbial diversity and function in soil: From genes to ecosystems. Curr Opin Microbiol 2002; 5(3):240-245.
7. Bonen DK, Cho JH. The genetics of inflammatory bowel disease. Gastroenterology 2003; 124(2):521-536.
8. Hugot JP, Chamaillard M, Zouali H et al. Association of NOD2 leucine-rich repeat variants with susceptibility to Crohn's disease. Nature 2001; 411(6837):599-603.
9. Ogura Y, Bonen DK, Inohara N et al. A frameshift mutation in NOD2 associated with susceptibility to Crohn's disease. Nature 2001; 411(6837):603-606.
10. Inohara N, Nunez G. NODs: Intracellular proteins involved in inflammation and apoptosis. Nat Rev Immunol 2003; 3(5):371-382.
11. Bonen DK, Ogura Y, Nicolae DL et al. Crohn's disease-associated NOD2 variants share a signaling defect in response to lipopolysaccharide and peptidoglycan. Gastroenterology 2003; 124(1):140-146.
12. Satsangi J, Parkes M, Jewell DP et al. Genetics of inflammatory bowel disease. Clin Sci 1998; 94(5):473-478.
13. Cuthbert AP, Fisher SA, Mirza MM et al. The contribution of NOD2 gene mutations to the risk and site of disease in inflammatory bowel disease. Gastroenterology 2002; 122(4):867-874.

14. Abreu MT, Taylor KD, Lin YC et al. Mutations in NOD2 are associated with fibrostenosing disease in patients with Crohn's disease. Gastroenterology 2002; 123(3):679-688.
15. Ahmad T, Armuzzi A, Bunce M et al. The molecular classification of the clinical manifestations of Crohn's disease. Gastroenterology 2002; 122(4):854-866.
16. Halfvarson J, Bodin L, Tysk C et al. Inflammatory bowel disease in a Swedish twin cohort: A long-term follow-up of concordance and clinical characteristics. Gastroenterology 2003; 124(7):1767-1773.
17. Hisamatsu T, Suzuki M, Reinecker HC et al. CARD15/NOD2 functions as an antibacterial factor in human intestinal epithelial cells. Gastroenterology 2003; 124(4):993-1000.
18. Giallourakis C, Stoll M, Miller K et al. IBD5 is a general risk factor for inflammatory bowel disease: Replication of association with Crohn disease and identification of a novel association with ulcerative colitis. Am J Hum Genet 2003; 73(1):205-211.
19. Curran ME, Lau KF, Hampe J et al. Genetic analysis of inflammatory bowel disease in a large European cohort supports linkage to chromosomes 12 and 16. Gastroenterology 1998; 115(5):1066-1071.
20. Parkes M, Satsangi J, Jewell DP et al. Ulcerative colitis is more strongly linked to chromosome 12 than Crohn's disease. Gut 2001; 49(2):311.
21. Bouma G, Oudkerk P, Crusius JB et al. Evidence for genetic heterogeneity in inflammatory bowel disease (IBD); HLA genes in the predisposition to suffer from ulcerative colitis (UC) and Crohn's disease (CD). Clinical and Experimental Immunology 1997; 109(1):175-179.
22. Naom I, Lee J, Ford D et al. Analysis of the contribution of HLA genes to genetic predisposition in inflammatory bowel disease. American Journal of Human Genetics 1996; 59(1):226-233.
23. Yang H-Y, Rotter JI, Toyoda H et al. Ulcerative colitis: A genetically heterogeneous disorder defined by genetic (HLA class II) and subclinical (antineutrophil cytoplasmic antibodies) markers. J Clin Invest 1993; 92(2):1080-1084.
24. Taylor KD, Plevy SE, Yang H et al. ANCA pattern and LTA haplotype relationship to clinical responses to anti-TNF antibody treatment in Crohn's disease. Gastroenterology 2001; 120(6):1347-1355.
25. Ghanim H, Garg R, Aljada A et al. Suppression of nuclear factor-kappaB and stimulation of inhibitor kappaB by troglitazone: Evidence for an anti-inflammatory effect and a potential antiatherosclerotic effect in the obese. J Clin Endocrinol Metab 2001; 86(3):1306-1312.
26. Joseph SB, Castrillo A, Laffitte BA et al. Reciprocal regulation of inflammation and lipid metabolism by liver X receptors. Nat Med 2003; 9(2):213-219.
27. Dubuquoy L, Jansson EA, Deeb S et al. Impaired expression of peroxisome proliferator-activated receptor gamma in ulcerative colitis. Gastroenterology 2003; 124(5):1265-1276.
28. Stappenbeck TS, Hooper LV, Gordon JI. Developmental regulation of intestinal angiogenesis by indigenous microbes via Paneth cells. Proc Natl Acad Sci USA 2002; 99(24):15451-15455.
29. Fellermann K, Wehkamp J, Herrlinger KR et al. Crohn's disease: A defensin deficiency syndrome? Eur J Gastroenterol Hepatol 2003; 15(6):627-634.
30. Hermiston ML, Gordon JI. Inflammatory bowel disease and adenomas in mice expressing a dominant negative N-cadherin. Science 1995; 270:1203-1207.
31. Panwala CM, Jones JC, Viney JL. A novel model of inflammatory bowel disease: Mice deficient for the multiple drug resistance gene, mdr1a, spontaneously develop colitis. J Immunol 1998; 161(10):5733-5744.
32. Schwab M, Schaeffeler E, Marx C et al. Association between the C3435T MDR1 gene polymorphism and susceptibility for ulcerative colitis. Gastroenterology 2003; 124(1):26-33.
33. Yacyshyn B, Maksymowych W, Bowen-Yacyshyn MB. Differences in P-glycoprotein-170 expression and activity between Crohn's disease and ulcerative colitis. Hum Immunol 1999; 60(8):677-687.
34. Kakazu T, Hara J, Matsumoto T et al. Type 1 T-helper cell predominance in granulomas of Crohn's disease. Am J Gastroenterol 1999; 94(8):2149-2155.
35. Neurath MF, Weigmann B, Finotto S et al. The transcription factor T-bet regulates mucosal T cell activation in experimental colitis and Crohn's disease. J Exp Med 2002; 195(9):1129-1143.
36. Dohi T, Fujihashi K, Kiyono H et al. Mice deficient in Th1- and Th2-type cytokines develop distinct forms of hapten-induced colitis. Gastroenterology 2000; 119(3):724-733.
37. Fuss IJ, Neurath M, Boirivant M et al. Disparate CD4+ lamina propria (LP) lymphokine secretion profiles in inflammatory bowel disease. Crohn's disease LP cells manifest increased secretion of IFN-gamma, whereas ulcerative colitis LP cells manifest increased secretion of IL-5. J Immunol 1996; 157:1261-1270.
38. Uguccioni M, Gionchetti P, Robbiani DF et al. Increased expression of IP-10, IL-8, MCP-1, and MCP-3 in ulcerative colitis. American Journal of Pathology 1999; 155(2):331-336.
39. Boirivant M, Fuss IJ, Chu A et al. Oxazolone colitis: A murine model of T helper cell type 2 colitis treatable with antibodies to interleukin 4. J Exp Med 1998; 188(10):1929-1939.

40. Ghosh S, Goldin E, Gordon FH et al. Natalizumab for active Crohn's disease. N Engl J Med 2003; 348(1):24-32.
41. Blaser MJ, Miller RA, Lacher J et al. Patients with active Crohn's disease have elevated serum antibodies to antigens of seven enteric bacterial pathogens. Gastroenterology 1984; 87:888-894.
42. Auer IO, Roder A, Wensinck F et al. Selected bacterial antibodies in Crohn's disease and ulcerative colitis. Scandinavian Journal of Gastroenterology 1983; 18(2):217-223.
43. O'Mahony S, Anderson N, Nuki G et al. Systemic and mucosal antibodies to Klebsiella in patients with ankylosing spondylitis and Crohn's disease. Ann Rheum Dis 1992; 51:1296-1300.
44. Braun J. Unsettling facts of life: Bacterial commensalism, epithelial adherence, and inflammatory bowel disease. Gastroenterology 2002; 122(1):228-230.
45. McKenzie H, Main J, Pennington CR et al. Antibody to selected strains of Saccharomyces cerevisiae (baker's and brewer's yeast) and Candida albicans in Crohn's disease. Gut 1990; 31:536-538.
46. Barnes RMR, Allan S, Taylor-robinson CH et al. Serum antibodies reactive with Saccharomyces cerevisiae in inflammatory bowel disease: Is IgA antibody a marker for Crohn's disease. Int Arch Allergy Appl Immunol 1990; 92:9-15.
47. Sendid B, Colombel JF, Jacquinot PM et al. Specific antibody response to oligomannosidic epitopes in Crohn's disease. Clin Diag Lab Immunol 1996; 3:219-226.
48. Davidson IW, Lloyd RS, Whorewell PJ et al. Antibodies to maize in patients with Crohn's disease, ulcerative colitis, and coeliac disease. Clin Exp Immunol 1979; 35:147-148.
49. Sonnenberg A. Occupational distribution of inflammatory bowel disease among German employees. Gut 1990; 31:1037-1040.
50. Quinton JF, Sendid B, Reumaux D et al. Anti-Saccharomyces cerevisiae mannan antibodies combined with antineutrophil cytoplasmic autoantibodies in inflammatory bowel disease: Prevalence and diagnostic role. Gut 1998; 42(6):788-791.
51. Sutton C, Yang H-Y, Rotter JI et al. Familial expression of anti-Saccharomyces cerevisiae mannan antibodies (ASCA) in affected and unaffected relatives of Crohn's disease patients. Gut 2000; 46:58-63.
52. Wayne LG, Hollander D, Anderson B et al. Immunoglobulin A (IgA) and IgG serum antibodies to mycobacterial antigens in Crohn's disease patients and their relatives. J Clin Microbiol 1992; 30:2013-2018.
53. Vannuffel P, Dieterich C, Naerhuyzen B et al. Occurrence, in Crohn's disease, of antibodies directed against a species-specific recombinant polypeptide of Mycobacterium paratuberculosis. Clin Diag Lab Immunol 1994; 1:241-243.
54. Elsaghier A, Prantera C, Moreno C et al. Antibodies to Mycobacterium paratuberculosis-specific protein antigens in Crohn's disease. Clin Exp Immunol 1992; 90:503-508.
55. Suenaga K, Yokoyama Y, Nishimori I et al. Serum antibodies to Mycobacterium paratuberculosis in patients with Crohn's disease. Dig Dis Sci 1999; 44(6):1202-1207.
56. Cohavy O, Harth G, Horwitz MA et al. Identification of a novel mycobacterial histone H1 homologue (HupB) as an antigenic target of pANCA monoclonal antibody and serum IgA from patients with Crohn's disease. Infect Immun 1999; 67:6510-6517.
57. El Zaatari FA, Naser SA, Hulten K et al. Characterization of Mycobacterium paratuberculosis p36 antigen and its seroreactivities in Crohn's disease. Curr Microbiol 1999; 39(2):115-119.
58. Naser SA, Hulten K, Shafran I et al. Specific seroreactivity of Crohn's disease patients against p35 and p36 antigens of M. avium subsp. paratuberculosis. Vet Microbiol 2000; 77(3-4):497-504.
59. Collins MT, Lisby G, Moser C et al. Results of multiple diagnostic tests for Mycobacterium avium subsp. Paratuberculosis in patients with inflammatory bowel disease and in controls. J Clin Microbiol 2000; 38(12):4373-4381.
60. Cohavy O, Bruckner D, Eggena ME et al. Colonic bacteria express an ulcerative colitis pANCA-related protein epitope. Infect Immun 2000; 68:1542-1548.
61. Wei B, Dalwadi H, Gordon LK et al. Molecular cloning of a Bacteroides caccae TonB-linked outer membrane protein associated with inflammatory bowel disease. Infect Immun 2001; 69(10):6044-6054.
62. Landers CJ, Cohavy O, Misra R et al. Selected loss of tolerance evidenced by Crohn's disease-associated immune responses to auto- and microbial antigens. Gastroenterology 2002; 123(3):689-699.
63. Sutton CL, Kim J, Yamane A et al. Identification of a novel bacterial sequence associated with Crohn's disease. Gastroenterology 2000; 119:23-28.
64. Wei B, Huang T, Dalwadi H et al. Pseudomonas fluorescens Encodes the Crohn's Disease-Associated I2 Sequence and T-Cell Superantigen. Infect Immun 2002; 70(12):6567-6575.
65. Dalwadi H, Wei B, Kronenberg M et al. The Crohn's disease-associated bacterial protein I2 is a novel enteric T cell superantigen. Immunity 2001; 15(1):149-158.

66. Duchmann R, May E, Heike M et al. T cell specificity and cross reactivity towards enterobacteria, Bacteroides, Bifidobacterium, and antigens from resident intestinal flora in humans. Gut 1999; 44(6):812-818.
67. Podolsky DK. Inflammatory bowel disease. N Engl J Med 2002; 347(6):417-429.
68. Papadakis KA, Prehn J, Moreno ST et al. CCR9-positive lymphocytes and thymus-expressed chemokine distinguish small bowel from colonic Crohn's disease. Gastroenterology 2001; 121(2):246-254.
69. Kunkel EJ, Kim CH, Lazarus NH et al. CCR10 expression is a common feature of circulating and mucosal epithelial tissue IgA Ab-secreting cells. J Clin Invest 2003; 111(7):1001-1010.
70. Tiede I, Fritz G, Strand S et al. CD28-dependent Rac1 activation is the molecular target of azathioprine in primary human CD4+ T lymphocytes. J Clin Invest 2003; 111(8):1133-1145.
71. Hanauer SB, Present DH. The state of the art in the management of inflammatory bowel disease. Rev Gastroenterol Disord 2003; 3(2):81-92.
72. Dubinsky MC, Yang H, Hassard PV et al. 6-MP metabolite profiles provide a biochemical explanation for 6-MP resistance in patients with inflammatory bowel disease. Gastroenterology 2002; 122(4):904-915.
73. Rump JA, Scholmerich J, Gross V et al. A new type of perinuclear anti-neutrophil cytoplasmic antibody (p-ANCA) in active ulcerative colitis but not in Crohn's disease. Immunobiology 1990; 181(4-5):406-413.
74. Duerr RH, Targan SR, Landers CJ et al. Anti-neutrophil cytoplasmic antibodies in ulcerative colitis. Comparison with other colitides/diarrheal illnesses. Gastroenterology 1991; 100:1590-1596.
75. Billing P, Tahir S, Calfin B et al. Nuclear localization of the antigen detected by ulcerative colitis-associated perinuclear antineutrophil cytoplasmic antibodies. Am J Pathol 1995; 147:979-987.
76. Vidrich A, Lee J, James E et al. Segregation of pANCA antigenic recognition by DNase treatment of neutrophils: Ulcerative colitis, type 1 autoimmune hepatitis, and primary sclerosing cholangitis. J Clin Immunol 1995; 15:293-299.
77. Folwaczny C, Noehl N, Endres SP et al. Antinuclear autoantibodies in patients with inflammatory bowel disease - High prevalence in first-degree relatives. Dig Dis Sci 1997; 42(8):1593-1597.
78. Eggena M, Cohavy O, Parseghian M et al. Identification of histone H1 as a cognate antigen of the ulcerative colitis-associated marker antibody pANCA. J Autoimmunity 2000; 14:83-97.
79. Reumaux D, Meziere C, Colombel J-F et al. Distinct production of autoantibodies to nuclear components in ulcerative colitis and in Crohn's disease. Clin Immunol Immunopathol 1995; 77:349-357.
80. Sobajima J, Ozaki S, Uesugi H et al. High mobility group (HMG) nonhistone chromosomal proteins HMG1 and HMG2 are significant target antigens of perinuclear anti-neutrophil cytoplasmic antibodies in autoimmune hepatitis. Gut 1999; 44(6):867-873.
81. Xiao H, Heeringa P, Hu P et al. Antineutrophil cytoplasmic autoantibodies specific for myeloperoxidase cause glomerulonephritis and vasculitis in mice. J Clin Invest 2002; 110(7):955-963.
82. Aldebert D, Masy E, Reumaux D et al. Immunoglobulin G subclass distribution of anti-endothelial cell antibodies (AECA) in patients with ulcerative colitis and Crohn's disease. Dig Dis Sci 1997; 42:2350-2355.
83. Stevens TR, Harley SL, Groom JS et al. Anti-endothelial cell antibodies in inflammatory bowel disease. Dig Dis Sci 1993; 38:426-432.
84. Yang P, Jarnerot G, Danielsson D et al. P-ANCA in monozygotic twins with inflammatory bowel disease. Gut 1995; 36(6):887-890.
85. Bhan AK, Mizoguchi E, Smith RN et al. Colitis in transgenic and knockout animals as models of human inflammatory bowel disease. Immunol Rev 1999; 169:195-207.
86. Satsangi J, Landers CJ, Welsh KI et al. The presence of anti-neutrophil antibodies reflects clinical and genetic heterogeneity within inflammatory bowel disease. Inflamm Bowel Dis 1998; 4(1):18-26.
87. Vasiliauskas EA, Kam LY, Karp LC et al. Marker antibody expression stratifies Crohn's disease into immunologically homogeneous subgroups with distinct clinical characteristics. Gut 2000; 47(4):487-496.
88. Fiocchi C, Roche JK, Michener WM. High prevalence of antibodies to intestinal epithelial antigens in patients with inflammatory bowel disease and their relatives. Ann Int Med 1989; 110:786-794.
89. Brandtzaeg P, Halstensen TS, Kett K et al. Immunobiology and immunopathology of human gut mucosa: Humoral immunity and intraepithelial lymphocytes. Gastroenterology 1989; 97:1562-1584.
90. Geng X, Biancone L, Dai HH et al. Tropomyosin isoforms in intestinal mucosa: Production of autoantibodies to tropomyosin isoforms in ulcerative colitis. Gastroenterology 1998; 114(5):912-922.
91. Sakamaki S, Takayanagi N, Yoshizaki N et al. Autoantibodies against the specific epitope of human tropomyosin(s) detected by a peptide based enzyme immunoassay in sera of patients with ulcerative colitis show antibody dependent cell mediated cytotoxicity against HLA-DPw9 transfected L cells. Gut 2000; 47(2):236-241.
92. Ohman L, Franzen L, Rudolph U et al. Immune activation in the intestinal mucosa before the onset of colitis in Galphai2-deficient mice. Scand J Immunol 2000; 52(1):80-90.

Chapter 14

The Role of Probiotics and Antibiotics in Regulating Mucosal Inflammation

Rainer Duchmann

Abstract

Antibiotic and probiotic agents have increasingly moved in the focus of basic and clinical research as well as clinical trials for IBD therapy. Both approaches modulate the intestinal flora, the former through eradication or reduction, the latter through establishment or increase of luminal bacteria. Although clinical trials provide proof of principle that both approaches can be therapeutically successfull, we just start to understand the mechanims and may get a first feeling for the potential and limitations of these "microbial" therapies. As basic research sets out to dissect the field using extensive efforts and new technologies, a more detailed exploration of the genetic, immune and microbial factors that govern the life-long crosstalk between host and intestinal flora is already opening new insight into general aspects of human immunology, immune regulation, IBD pathogenesis and therapy.

Probiotics: Animal Studies and Clinical Trials

In a widely accepted definition, Fuller defined probiotics as a live microbial feed supplement which beneficially affects the host animal by improving its microbial balance.[1] In addition to this approach of changing the intestinal flora by the addition of life microbes, modulation of the flora can also be achieved by addition of "nondigestible food ingredients that beneficially affect the host by selectively stimulating the growth, activity or both of one or a limited number of bacterial species already resident in the colon," i.e., prebiotics or a mixture of probiotics and prebiotics, i.e., synbiotics.[2]

The list of microbes that are being used as probiotics is long, and lactic acid producers such as lactobacilli and bifidobacteria are being used most often. Criteria for selection of a safe and effective probiotic at present do not include a detailed knowledge of their immunological effects but rely to large part on their pathogenicity and other more general microbial and practical aspects.[3,4]

Clinical effects of probiotics in medicine and in inflammatory bowel disease have recently been reviewed.[5-7] *Lactobacillus GG,* e.g., decreases the duration of diarrhea in infants with rotavirus infection or hospital acquired diarrhea[8,9] and decreases the incidence of atopic disease by 50% when given regularly to pregnant women and their children postpartum.[10]

In animal models of chronic intestinal inflammation, lactobacilli ameliorate acetic acid induced colitis[11] or methotrexate induced enterocolitis in rats.[12] In IL-10 deficient mice, which develop a Crohn's like disease, it was recently demonstrated that changes in the intestinal flora preceed colonic inflammation. Normalization of Lactobacillus species levels prevents these changes and attenuates colitis.[13]

In a pilot study in patients with colonic Crohn' disease, the probiotic strain E. coli Nissle 1917 was effective in remission maintenance and showed a steroid-sparing effect compared to

Immune Mechanisms in Inflammatory Bowel Disease, edited by Richard S. Blumberg and Markus F. Neurath.

placebo.[14] In ulcerative colitis, Kruis et al demonstrated that E. coli Nissle 1917 and mesalamine had equivalent effectiveness in remission maintenance[15] and similar data were reported from Leeds.[16] Kruis later confirmed his initial results in a larger study of longer duration.[17] In patients with ulcerative colitis who had undergone proctocolektomie and suffered from pouchitis, a cocktail of eight probiotic bacteria was most effective for remission maintenance.[18] In addition, beneficial effects have been reported in Crohn's disease for the yeast S. Boulardii.[19,20] S. boulardii also has beneficial effects in antibiotic-associated diarrhoe[21] and decreases the risk of recurrent C. dificile infection.[22] In summary, current data demonstrate that oral probiotic bacteriotherapy is effective in remission maintenance in patients with ulcerative colitis and pouchitis and that it may have some effect in active disease.

Probiotics: Mechanisms

As probiotics colonize the intestinal tract for a longer or shorter period of time, they can act via a variety of different mechanisms. One possibility is that they produce antimicrobial compounds and that they antagonize pathogenic bacteria, e.g., via competition for ecological niches or substrates.[22,23] In addition to this indirect way of interacting with the host via modulation of the intestinal flora, probiotics can directly interact with the host, e.g., through competition with pathogenic bacteria for binding sites on epithelial cells, by enhancing mucosal barrier function[24] or through direct immunologic effects on epithelial cells[25] and the mucosal immune system.

Regarding the question how luminal bacteria can interact with deeper immunological host structures it has been shown that although the mucosa is covered by a single cell layer of cylindric epithelium and sealed with tight junctions to exclude peptides and macromolecules, viable bacteria and bacterial components translocate through the mucosal barrier and can be found within the lamina propria where they come in contact with immune cells. This translocation occurs by active transport across epithelial cells,[26] specialized dendritic cells (DC)[27] or accidental breaches in the epithelial layer.[28] Contacting the immune system, bacterial lipopolysaccharide (LPS), peptido-glycan-polysaccharides (PG-PS)[29] and formylated chemotactic oligopeptides (FMLP)[30] can thus stimulate macrophages (Mϕ), dendritic cells (DC) and neutrophiles via specific receptors. In addition, nonself peptides, polysaccharides and glycolipids are recognized by T-cells, NK cells and NKT-cells in the mucosa. Probiotics are thus not expected to be immunologically inert but to use their immunostimulatory antigens and products to initiate a crosstalk with the host that ultimately results in the generation of protective and regulatory immune respones. On that line, probiotics should be able to exploit the full and unique capacity of the intestinal immune system to generate such responses[31] including, as is already shown for bacterial stimuli, modulation of the cytokine response[32] and induction of regulatory cells.[33,34]

Furthermore, the differential capacity of nonpathogenic vs pathogenic bacteria to interact with host structures has recently been highlighted in elegant studies which demonstrated that human epithelia show a decreased production of proinflammatory cytokines when incubated with nonpathogenic bacteria. This effect was shown to be mediated through interference with the NFκB system, more precisely through inhibition of IκB-α ubiquitination. Ubiquitination of IκB-α is necessary to translocate the transcription factor NFκB and to induce the transcription of inflammatory genes.[35] This downregulatory effect of nonpathogenic bacteria on the host NFκB system is in contrast to the effective activation of NF-κB by pathogenic bacteria via activation of extracellular and intracellular receptors of the innate immune system.[36,37] In that context it is increasingly appreciated that immune sensing of luminal bacteria through receptors of the innate immune system which recognize conserved bacterial products is a critical determinator of immune outcome. Signals provided by these receptors modulate the differentiation of antigen presenting cells and in consequence profoundly influence the quality of antigen-specific T cell responses.[38] Tight control of IL-12, which is produced by antigen presenting cells in response to pathogenic bacteria but not to normal flora, was essential to maintain tolerance to normal flora in humans and mice.[39-41]

Further studies investigating the mechanisms by which probiotics exert their beneficial effects under various circumstances are needed. Parallel with our increasing knowledge from clinical studies, this information may help to select the probiotics, the clinical situations and may be even individual patients most suitable for probiotic therapy. In addition, it remains to be seen whether early changes within the intestinal flora are present before clinical manifestation of the disease or before relapse and whether they really contribute to disease pathogenesis. If so, probiotics could be used as preventive bacteriotherapy to protect these individuals from developing IBD.

Antibiotics: Animal Studies and Clinical Trials

Reduction or eradication of luminal bacteria with antibiotics is a traditional option for therapeutic modification of the intestinal flora. In animal models of chronic intestinal inflammation, antibiotic therapy prevents disease or ameliorates established disease, depending on the animal model and the antibiotics chosen.[42-44] In IBD, metronidazole and ciprofloxacin are the most studied antibiotics. They ameliorate active and fistulizing Crohn's disease[45-48] and they are standard therapy for treatment of active pouchitis.[49]

Antibiotics: Mechanisms

Antibiotics address a certain spectrum of bacteria, reduce their number or totally eradicate them. In addition, as has been described for metronidazol and ciprofloxacin[50,51] and others, antibiotics can have direct immunologic effects.

In contrast to infectious diseases, where antibiotic therapy adresses a specific microbial target, the jury is still out to see whether there is any specific bacterium or group of bacteria that is more potent than others in inducing and perpetuating IBD.

It is known from the well studied HLA-B27/β2 microglobuline transgenic rat model, that severity of colitis and gastritis correlates with the concentration of luminal bacteria[52] and that *Bacteroides vulgatus* is a critical bacterium.[53] In the same model, selective antibiotics prevented disease but only broad spectrum antibiotics treated established disease, suggesting that the spectrum of bacteria involved in the inflammatory response broadens as the disease progresses.[43] Making things complicated, *Bacteroides* is important in the HLA-B27/β2 microglobuline transgenic rat but it is not in the IL-10 -/- model. On the same line, treatment studies with broad spectrum vs. selective antibiotics support the concept, that the role of single bacteria or groups of bacteria is dependant on the type of the abnormal host immune response and therefore varies between models.[42-44]

A variety of studies suggest that *E. coli* may be important for Crohn's disease pathogenesis. Enteroadherent *E. coli* with similar genotypes are recovered more frequently from ileal Crohn's disease lesions than from controls.[54,55] In addition, *E. coli* isolated from Crohn's disease patients were shown to be invasive and to replicate within macrophages without inducing cell death of the infected cell.[56-58] However, a combined approach using microbial cultures and state of the art molecular techniques did not reveal principle differences in the composition of the mucosal flora in IBD patients compared to controls. In the same study, there also were no indications for a special role of a single *E. coli* pathovar or bacterial translocation to lamina propria macrophages.[59] Using intestinal T cell clones, studies from our laboratory indicate that the CD4 T cell response to bacteria from the intestinal flora in IBD and controls is characterized by a large degree of crossreactivity and might be directed against conserved bacterial antigens.[60,61] In fact, if it turned out that the inflammatory potential resides more or less in the entire flora, antibiotic therapy could be expected to have only limited efficacy, according to its limited ability to subtantially reduce overall luminal bacterial content.

In that regard, identification of critical bacteria will certainly be crucial to better comprehend IBD pathogenesis and devise new and more specific antibiotic treatment strategies. Since results from different animal models suggest that certain bacteria are variably involved in the disease, it seems likely that in IBD there may also be variability between groups of individuals or even individuals. The factors which govern the selection of the intestinal flora in health and disease are still largely unknown but are likely microbiologic, immunologic and genetic.[62] The facilitated acceptance of certain bacterial species through crosstolerisation with autologous antigens[63,64] and the finding that the bacterial flora of monozygotic twins is more similar than that of dizygotic twins or unrelated individuals,[65,66] at least indicates that the intestinal flora may be genetically tailored not only for a given animal species but for each individual.

General Problems

Interventions that aim to therapeutically modulate the intestinal flora are challenged by the extreme complexity of the systems they are intervening with.

First of all, the intestinal flora itself is a very complex ecosystem with an impressive metabolic capacity and estimated to comprise 10 times more bacteria than the body contains cells. Most of our knowledge of the composition of the resident intestinal flora still comes from very laborious microbiologic and phenotypic analyses of fecal samples or luminal aspirates. These older studies employed different culture techniques to identify intestinal bacteria (***culture and phenotype***) and through technical limitations favored the growth of certain species and underestimated the importance of others. Fortunately, new types of molecular analyses with high reliability and resolving power for the study of complex bacterial samples are being developed and applied at increasing speed and allow high-throughput analysis of large quantities of bacteria without the need to culture them.[67] These techniques are much more suitable to determine and monitor the complexity and diversity of the intestinal flora and will certainly very much facilitate the investigation of factors controlling or influencing the composition of the intestinal flora in health, disease and under experimental/therapeutic intervention.

In addition, the life-long crosstalk between the intestinal flora and the host occurs at least at three different topoi, i.e., in the lumen, at the luminal/host border involving the epithelial cells and all other components of the intestinal barrier, and the mucosal immune system. Thus, effects on the mucosal immune system, which form the focus of this review, are just one possible outcome among many and may be either direct (bacteria-immune) or indirect (e.g., bacteria-bacteria-immune; metabolic-intestinal barrier-immune, etc). Thus, the effect of even a single bacterial species on the host can be quite diverse and has been assessed in a recent elegant study using monocontamination of germ free mice and global analysis of gene expression.[68]

Furthermore, we know that in the normal host, the intestinal flora is required for the establishment and maintenance of the normal intestinal integrity and function. In inflammatory bowel disease, in contrast, the intestinal flora becomes a necessary endogenous stimulus to induce or perpetuate chronic intestinal inflammation. This important role in IBD pathogenesis is well exemplified by the observation that the development of intestinal inflammation in different animal models of IBD is dependant on the presence of luminal bacteria, that inflammation can be modified by administration of antibiotics or probiotics and that CD4+ T cells stimulated by bacterial antigens can transfer disease. Variability of disease pathogenesis, e.g., between patients with different clinical IBD phenotypes or even between individuals, may add further levels of complexity. As discussed above, this makes it but more difficult to predict on immunological grounds which probiotics or antibiotics could be therapeutically more suitable than others in IBD, or why certain probiotics or antibiotics are beneficial in some clinical situations, but not or probably less in others.

Perspectives

Identification of critical bacteria or bacterial antigens would be a major step forward in order to devise more specific probiotic or antibiotic treatment strategies and requires continued research. It would also be helpful to know, whether the role of the intestinal flora, the spectrum of potential disease-relevant bacterial species and the type of immune response against them changes as a patient progresses during active phases and the course of his disease. In theory, a healthy invidual who is genetically susceptible to develop intestinal inflammation in response to a definable small group of bacteria or antigens could be amenable to preventive antibiotic or probiotic treatment. Patients with "early disease", e.g., induced by an immune response to one or several bacteria might respond well to probiotic or antibiotic therapy, whereas treatment of "late disease" with immune responses to a very diverse spectrum of bacteria or antigens might become increasingly difficult. The potential of probiotics to modulate the various aspects of human and intestinal immunology are just beginning to unravel. To name just one of many different treatment options, it could certainly be envisioned to in vitro generate autologous regulatory T cells with reactivity against probiotic antigens. These regulatory T cells could then be readministered to the patient and following probiotic therapy and intestinal probiotic colonization, be specifically stimulated to produce protective mediators and inhibit the local inflammatory process in bystander fashion. Furthermore, as has already been undertaken,[69] bacteria, either from the normal intestinal flora or attenuated mutants, can be used as DNA vectors for the local delivery of immune modifiers and/or antigens. Immune modifiers might be naturally occuring bacterial immune inhibitory factors, as they have recently been identified and sequenced in enteropathogenic E.coli or salmonella.[70,71] Alternatively, DNA coding for immune modifiers which has been artificially introduced in carrier bacteria could be efficiently transfered to host intestinal cells[72,73] and even selected cell populations.[74] Bacterial vectors might also deliver antigen in the presence of appropriate immune modifier in an attempt to modify or redirect antigen specific immune responses.

References

1. Fuller R. Probiotics in man and animals. J Appl Bacteriol 1989; 66(5):365-78.
2. Gibson G, Roberfroid M. Dietary modulation of the human colonic microbiota: Introducing the concept of prebiotcs. J Nutr 1995; 125(6):1401-1412.
3. Collins MD, Gibson GR. Probiotics, prebiotics, and synbiotics: Approaches for modulating the microbial ecology of the gut. Am J Clin Nutr 1999; 69(5):1052S-1057S.
4. Gibson GR, Fuller R. Aspects of in vitro and in vivo research approaches directed toward identifying probiotics and prebiotics for human use. J Nutr 2000; 130(2S Suppl):391S-395S.
5. Hart AL, Stagg AJ, Frame M et al. The role of the gut flora in health and disease, and its modification as therapy. Aliment Pharmacol Ther 2002; 16(8):1383-93.
6. Hart AL, Stagg AJ, Kamm MA. Use of probiotics in the treatment of inflammatory bowel disease. J Clin Gastroenterol 2003; 36(2):111-9.
7. Schultz M, Sartor RB. Probiotics and inflammatory bowel diseases. Am J Gastroenterol 2000; 95(1 Suppl):S19-21.
8. Szajewska H, Kotowska M, Mrukowicz J et al. Efficacy of lactobacillus GG in prevention of nosocomial diarrhea in infants. J Pediatr 2001; 138(3):361-365.
9. Isolauri E, Juntunen M, Rautanen T et al. A human lactobacillus strain promotes recovery from acute diarrhea in children. Pediatrics 1991; 88(1):90-97.
10. Kalliomaeki M, Salminen S, Arvilommi H et al. Probiotcs in primary prevention of atopic disease: A randomised placebo-controlled trail. Lancet 2001; 357:1076-1079.
11. Fabia R, Ar'Rajab A, Johansson ML et al. The effect of exogenous administration of lactobacillus reuteri R2LC and oat fiber on acetic acid-induced colitis in the rat. Scand J Gastroenterol 1993; 28(2):155-62.
12. Mao Y, Nobaek S, Kasravi B et al. The effects of lactobacillus strains and oat fiber on methotrexate- induced enterocolitis in rats. Gastroenterology 1996; 111(2):334-44.
13. Madsen K, Doyle J, Jewell L et al. Lactobacillus species prevents colitis in interleukin 10 gene-deficient mice. Gastroenterology 1999; 116(5):1107-1114.
14. Malchow HA. Crohn's disease and Escherichia coli. A new approach in therapy to maintain remission of colonic Crohn's disease? J Clin Gastroenterol 1997; 25(4):653-8.
15. Kruis W, Schutz E, Fric P et al. Double-blind comparison of an oral Escherichia coli preparation and mesalazine in maintaining remission of ulcerative colitis. Aliment Pharmacol Ther 1997; 11(5):853-8.
16. Rembacken BJ, Snelling AM, Hawkey PM et al. Nonpathogenic Escherichia coli versus mesalazine for the treatment of ulcerative colitis: A randomised trial. Lancet 1999; 354(9179):635-9.
17. Kruis W, Fric P, Stolte M. Maintenence of remission in ulcerative colitis is equally effective with Escherichia coli nissle 1917 and with standard mesalamine. Gastroenterology 2001; 120(5):A127.

18. Gionchetti P, Rizzello F, Venturi A et al. Oral bacteriotherapy as maintenance treatment in patients with chronic pouchitis: A double blind, placebo-conroled trial. Gastroenterology 2000; 119(2):305-9.
19. Plein K, Hotz J. Therapeutic effects of Saccharomyces boulardii on mild residual symptoms in a stable phase of Crohn's disease with special respect to chronic diarrhea—a pilot study. Z Gastroenterol 1993; 31(2):129-34.
20. Guslandi M, Mezzi G, Sorghi M et al. Saccharomyces boulardii in maintenance treatment of Crohn's disease. Dig Dis Sci 2000; 45(7):1462-4.
21. McFarland LV, Surawicz CM, Greenberg RN et al. Prevention of beta-lactam-associated diarrhea by Saccharomyces boulardii compared with placebo. Am J Gastroenterol 1995; 90(3):439-48.
22. McFarland LV, Surawicz CM, Greenberg RN et al. A randomized placebo-controlled trial of Saccharomyces boulardii in combination with standard antibiotics for Clostridium difficile disease. Jama 1994; 271(24):1913-8.
23. Flynn S, van Sinderen D, Thornton GM et al. Characterization of the genetic locus responsible for the production of ABP-118, a novel bacteriocin produced by the probiotic bacterium Lactobacillus salivarius subsp. salivarius UCC118. Microbiology 2002; 148(Pt 4):973-84.
24. Madsen K, Cornish A, Soper P et al. Probiotic bacteria enhance murine and human intestinal epithelial barrier function. Gastroenterology 2001; 121(3):580-91.
25. Cario E, Rosenberg IM, Brandwein SL et al. Lipopolysaccharide activates distinct signaling pathways in intestinal epithelial cell lines expressing Toll-like receptors. J Immunol 2000; 164(2):966-72.
26. Neutra M, Mantis N, Kraehenbuhl J. Collaboration of epithelial cells with organized mucosal lymphoid tissues. Nat Immunol 2001; 2(11):1004-1009.
27. Rescigno, Urbano M, Valzasina B et al. Dendritic cells express tight junction proteins and penetrate gut epithelial monolayers to sample bacteria. Nat Immunol 2001; 2(4):288-290.
28. O'Boyle C, MacFie J, Mitchell C et al. Microbiology of bacterial translocation in humans. Gut 1998; 42(1):29-35.
29. McCall R, Haskill J, Zimmermann E et al. Tissue interleukin-1 receptor antagonist expression in enterocolitis in resistant and susceptible rats. Gastroenterology 1994; 106:960-972.
30. Schwab J. Phlogisic properties of peptidoglycan-polysaccharide polymers from cell walls of pathogenic and normal-flora bacteria which colonize humans. Infect Immun 1993; 61:4535-4539.
31. Mowat A, Weiner H. Oral tolerance. Physiological Basis and Clinical Applications. Mucosal Immunology. 2nd ed. In: Orgra PL, Mestecky J, Lamm ME et al, eds. Academic Press, 1999:587-618.
32. Haller D, Bode C, Hammes WP et al. Nonpathogenic bacteria elicit a differential cytokine response by intestinal epithelial cell/leucocyte cocultures. Gut 2000; 47(1):79-87.
33. Khoo UY, Proctor IE, Macpherson AJ. CD4+ T cell down-regulation in human intestinal mucosa: Evidence for intestinal tolerance to luminal bacterial antigens. J Immunol 1997; 158(8):3626-34.
34. Kullberg MC, Jankovic D, Gorelick PL et al. Bacteria-triggered CD4(+) T regulatory cells suppress Helicobacter hepaticus-induced colitis. J Exp Med 2002; 196(4):505-15.
35. Neish A, Gewirtz A, Zeng H et al. Prokaryotic regulation of epithelial responses by inhibition of IkB-alpha ubiquitination. Science 2000; 289:1560-1563.
36. Zhang G, Ghosh S. Toll-like receptor-mediated NF-kappaB activation: A phylogenetically conserved paradigm in innate immunity. J Clin Invest 2001; 107(1):13-9.
37. Girardin SE, Tournebize R, Mavris M et al. CARD4/Nod1 mediates NF-kappaB and JNK activation by invasive shigella flexneri. EMBO Rep 2001; 2(8):736-42.
38. Medzhitov R, Janeway Jr C. Innate immunity. N Engl J Med 2000; 343(5):338-44.
39. Duchmann R, Neurath MF, Meyer zum Buschenfelde KH. Responses to self and nonself intestinal microflora in health and inflammatory bowel disease. Res Immunol 1997; 148(8-9):589-94.
40. Duchmann R, Kaiser I, Hermann E et al. Tolerance exists towards resident intestinal flora but is broken in active inflammatory bowel disease. Clin Exp Immunol 1995; 102(3):448-455.
41. Duchmann R, Schmitt E, Knolle P et al. Tolerance towards resident intestinal flora in mice is abrogated in experimental colitis and restored by treatment with interleukin-10 or antibodies to interleukin-12. Eur J Immunol 1996; 26:934-938.
42. Videla S, Vilaseca J, Guarner F et al. Role of intestinal microflora in chronic inflammation and ulceration of the rat colon. Gut 1994; 35(8):1090-7.
43. Rath HC, Schultz M, Freitag R et al. Different subsets of enteric bacteria induce and perpetuate experimental colitis in rats and mice. Infect Immun 2001; 69(4):2277-85.
44. Madsen KL, Doyle JS, Tavernini MM et al. Antibiotic therapy attenuates colitis in interleukin 10 gene-deficient mice. Gastroenterology 2000; 118(6):1094-105.
45. Moss A, Carbone J, Kressel H. Radiologic and clinical assessment of broad-spectrum antibiotic therapy in Crohn's disease. Am J Roentgenol 1978; 131(5):787-790.

46. Sutherland L, Singleton J, Sessions J et al. Double blind, placebo controlled trial of metronidazole in Crohn's disease. Gut 1991; 32(9):1071-5.
47. Ursing B, Alm T, Barany F et al. A comparative study of metronidazole and sulfasalazine for active Crohn's disease: The cooperative Crohn's disease study in Sweden. II. Result. Gastroenterology 1982; 83(3):550-62.
48. Sands BE. Therapy of inflammatory bowel disease. Gastroenterology 2000; 118(2 Suppl 1):S68-82.
49. Zeitz M. Guidelines of the DGVS. Pouchitis. German Society of Digestive and Metabolic Diseases. Z Gastroenterol 2001; 39(1):59-62.
50. Roche Y, Gougerot-Pocidalo MA, Fay M et al. Comparative effects of quinolones on human mononuclear leucocyte functions. J Antimicrob Chemother 1987; 19(6):781-90.
51. Grove DI, Mahmoud AA, Warren KS. Suppression of cell-mediated immunity by metronidazole. Int Arch Allergy Appl Immunol 1977; 54(5):422-7.
52. Rath H, Ikeda J, Wilson K et al. Cecal bacterial loads influence colitis and gastritis in HLA B27 transgenic rats. Gastroenterology 1997; 112:A1068.
53. Rath HC, Herfarth HH, Ikeda JS et al. Normal luminal bacteria, especially bacteroides species, mediate chronic colitis, gastritis, and arthritis in HLA-B27/human beta2 microglobulin transgenic rats. J Clin Invest 1996; 98(4):945-53.
54. Darfeuille-Michaud A, Neut C, Barnich N et al. Presence of adherent Escherichia coli strains in ileal mucosa of patients with Crohn's disease. Gastroenterology 1998; 115(6):1405-13.
55. Masseret E, Boudeau J, Colombel JF et al. Genetically related Escherichia coli strains associated with Crohn's disease. Gut 2001; 48(3):320-5.
56. Boudeau J, Glasser AL, Masseret E et al. Invasive ability of an Escherichia coli strain isolated from the ileal mucosa of a patient with Crohn's disease. Infect Immun 1999; 67(9):4499-509.
57. Boudeau J, Barnich N, Darfeuille-Michaud A. Type 1 pili-mediated adherence of Escherichia coli strain LF82 isolated from Crohn's disease is involved in bacterial invasion of intestinal epithelial cells. Mol Microbiol 2001; 39(5):1272-84.
58. Glasser AL, Boudeau J, Barnich N et al. Adherent invasive Escherichia coli strains from patients with Crohn's disease survive and replicate within macrophages without inducing host cell death. Infect Immun 2001; 69(9):5529-37.
59. Swidsinski A, Ladhoff A, Pernthaler A et al. Mucosal flora in inflammatory bowel disease. Gastroenterology 2002; 122(1):44-54.
60. Duchmann R, Marker-Hermann E, Meyer zum Buschenfelde KH. Bacteria-specific T-cell clones are selective in their reactivity towards different enterobacteria or H. pylori and increased in inflammatory bowel disease. Scand J Immunol 1996; 44(1):71-9.
61. Duchmann R, May E, Heike M et al. T cell specificity and cross reactivity towards enterobacteria, bacteroides, bifidobacterium, and antigens from resident intestinal flora in humans. Gut 1999; 44(6):812-8.
62. Duchmann R. The role of resident intestinal bacteria. Chronic inflammatory bowel diseases. Progress and Controversies. In: Holoman J, Glasa J, eds. Progress in Hepato-Pharmacology, 2000; 5:17-31.
63. Foo MC, Lee A. Antigenic cross-reaction between mouse intestine and a member of the autochthonous microflora. Infect Immun 1974; 9(6):1066-9.
64. Foo MC, Lee A. Immunological response of mice to members of the autochthonous intestinal microflora. Infect Immun 1972; 6(4):525-32.
65. Moore WE, Burmeister JA, Brooks CN et al. Investigation of the influences of puberty, genetics, and environment on the composition of subgingival periodontal floras. Infect Immun 1993; 61(7):2891-8.
66. Van de Merwe JP, Stegeman JH, Hazenberg MP. The resident faecal flora is determined by genetic characteristics of the host. Implications for Crohn's disease? Antonie Van Leeuwenhoek 1983; 49(2):119-24.
67. Wilson K, Blitchington R. Human colonic biota studied by ribosoal DNA sequence analysis. Appl Environ Microbiol 1996; 62(7):2273-227.
68. Hooper LV, Wong MH, Thelin A et al. Molecular analysis of commensal host-microbial relationships in the intestine. Science 2001; 291(5505):881-4.
69. Steidler L, Hans W, Schotte L et al. Treatment of murine colitis by Lactococcus lactis secreting interleukin- 10. Science 2000; 289(5483):1352-5.
70. Matsui K, Nagano K, Arai T et al. DNA sequencing of the gene encoding Salmonella typhimurium-derived T- cell inhibitor (STI) and characterization of the gene product, cloned STI. FEMS Immunol Med Microbiol 1998; 22(4):341-9.
71. Klapproth JM, Scaletsky IC, McNamara BP et al. A large toxin from pathogenic Escherichia coli strains that inhibits lymphocyte activation. Infect Immun 2000; 68(4):2148-55.

72. Darji A, Guzman CA, Gerstel B et al. Oral somatic transgene vaccination using attenuated S. typhimurium. Cell 1997; 91(6):765-75.
73. Fennelly GJ, Khan SA, Abadi MA et al. Mucosal DNA vaccine immunization against measles with a highly attenuated Shigella flexneri vector. J Immunol 1999; 162(3):1603-10.
74. Paglia P, Medina E, Arioli I et al. Gene transfer in dendritic cells, induced by oral DNA vaccination with Salmonella typhimurium, results in protective immunity against a murine fibrosarcoma. Blood 1998; 92(9):3172-6.

Chapter 15

From Immunogenic Mechanisms to Novel Therapeutic Approaches in Inflammatory Bowel Disease

Martin H. Holtmann and Markus F. Neurath

Abstract

Crohn's disease (CD) and ulcerative colitis (UC) are the two most common forms of chronic inflammatory bowel disease (IBD). The etiology of IBD is still unclear and should be considered as multi-factorial according to recent studies.[1] Genetic factors seem to play a pathogenetic role as well as environmental, infectious and immulogical factors. Substantial progress, however, has been made in the understanding of the pathogenesis of IBD during the past years persuing the view, that IBD could result from disturbances of the intestinal barrier and a pathologic activation of the intestinal immune response towards luminal, bacterial antigens. This paradigm has led to the identification of key players of the intestinal immune system, which represent promising targets for novel therapeutic approaches. The objective of this chapter is to provide an overview over recent advances in the elucidation of the intestinal immune system in IBD and novel therapeutic approaches that have been derived from these results. Molecular biological techniques have revealed, that many of the established conventional antiinflammatory drugs such as salicylic acids, steroids or immunuosuppressants act at the same molecules that are the target for modern biologicals, i.e., the cytokine TNF or the transcription factor NFκB. This chapter, however, focusses on novel experimental approaches such as recombinant antiinflammatory cytokines, neutralizing antibodies or antisense oligonucleotides.

Introduction

With a surface of over 300 m^2 the gut can be regarded as the largest immune organ of the human body. In the gut the organism is physiologically exposed to millions of antigens from the natural flora and the food. The intestinal immune system comprises cellular components such as T-lymphocytes, monocytes/macrophages and dendritic cells, which can be organised, i.e., the Peyer plaques in the small bowel or nonorganised, such as lamina propria mononuclear cells, and noncellular components such as cytokines and antibodies. The highly organised complex structure facilitates an effective immune response towards selected, potentially hazardous antigens, while the majority of antigens from the food and the physiological bacterial flora are tolerated.

There is increasing evidence that in IBD this balance is disturbed and that a perpetuated activation of the intestinal immune system by bacterial antigens plays a pivotal role in the pathogenesis of IBD. IBD seems to develop in patients with a genetic suseptibility who are exposed to certain environmental factors. Favored by an insufficient activation of antiinflammatory

Immune Mechanisms in Inflammatory Bowel Disease, edited by Richard S. Blumberg and Markus F. Neurath.

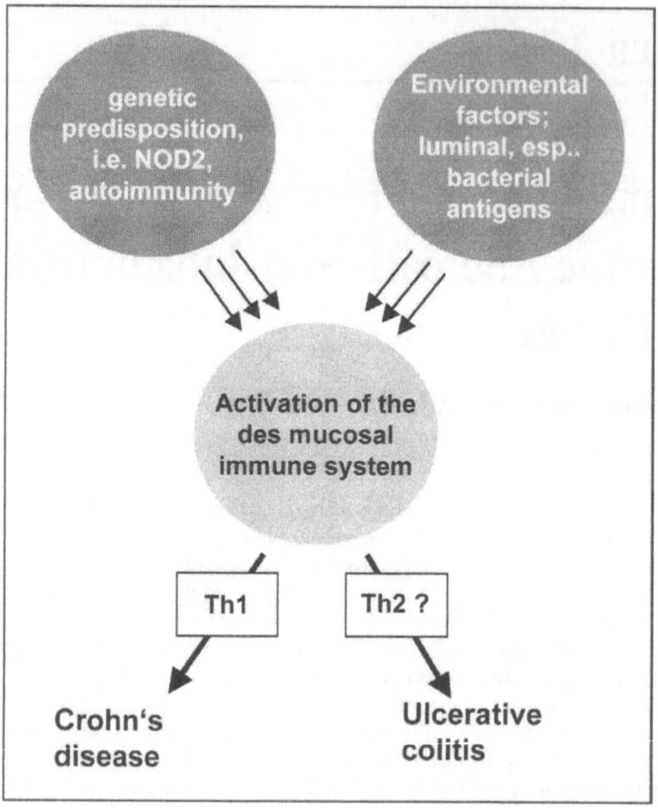

Figure 1. Patogenesis of IBD: A genetic predisposition in combination with environmental factors/ bacterial antigens leads to an activation of the mucosal immune system. For Crohn's disease this leads to a Th1 mediated intestinal inflammation, which is probably a common final pathway of a heterogeneous disease. The role of a Th2 mediated inflammatory response in the pathogenesis of ulcerative colitis is less well established.

mechanisms or a predominance of proinflammatory mechanisms this leads to chronic inflammation with the clinical manifestations of CD and UC. While the genetic susceptibility factors and bacterial antigens that contribute to the development of CD seem to be heterogenous, the activation of the intestinal immune systems appears as a common final pathway in the pathogenesis (Fig. 1).

Genetic Factors

In 1998 a large european cohort study in 359 sibling couples with IBD showed an association of CD and UC with certain loci on the chromosomes 12 and 16 by linkage analysis using microsatellite markers. The locus for CD in the pericentromeric region of chromosome 16 was termed "IBD1". In 2001 a gene on this locus could be identified by two independent research groups.[2,3] In the work from Hugot et al approximately 30 different mutations in a structurally characteristic domain in the C-terminus of the NOD2 gen could be found in a subgroup of patients with CD that might account for 5 to 10% of the cases of CD. Ogura et al identified 3 different mutations. The NOD2 protein consists of 1040 amino acids and is expressed only in monocytes. At the N-terminal end there are 2 CARD-motifs ("caspase recruitment domain"), which are structurally related to the death domains of the TNF-receptor 1 and the apoptosis receptor Fas. The C-terminal end is composed of 10 LRR ("leucine-rich repeats") which possess binding affinity for bacterial lipopolysaccharides. The physiological role of NOD2 seems to be the activation of the proinflammatory transcription factor NFκB, that plays a central role in the pathogenesis of IBD, upon bacterial antigens or bacterial products.[3] The mutations in NOD2 in patients with CD are likely to cause a loss of function that could impair the early activation of NFκB to eliminate the bacterial infections. The defective early immune response might then promote a perpetuated activation of the intestinal immune system. However, further studies are necessary to confirm this hypothesis.

In UC there is a remarkable association with certain MHC (major histocompatibility complex) class II molecules.[4] Additional evidence for a potential autoimmune genesis of UC are the organ specificity, the association with primary sclerosing cholangitis and the high prevalence of various autoantibodies such as pANCA.[5] Crossreactive structures for pANCA could be identified such as antigens from the colonic mucosa, antigens from the ciliar body of the eye and retinal ganglionic cells.[6,7] Some of the autoantibodies found in UC display cross reactivity with bacterial antigens.[8,9]

Physiological "Hyporeactivity" of the Intestinal Immune System

Physiologically the gut contains a large amount of immunocompetent cells, organized in Peyer plaques or lymhoid follicles or nonorganized in the lamina propria such as intraepithelial T-cells, epithelial cells, IgA producing B-cells, T-cells and antigen presenting cells (dendritic cells and macrophages). CD4+ T-cells can produce a variety of pro- and antiinflammatory cytokines upon antigen presentation and costimulation. According to their cytokine profile, CD4+-T-cells can be subdivided into Type 1 helper cells (Th1: IL-2, IFN-γ, TNF), Type 2 helper cells (IL-4, IL-5, IL-13), Type 3 helper cells (Th3: TGF-β) and regulatory T-cells (Tr: IL-10).

The abilitiy of lamina propria T-cells to be activated via the T-cell receptor is reduced, while stimulation via the accessory CD2/CD28 pathway is increased.[10] Under physiological conditions Th2 cytokines and the antiinflammatory cytokines IL-10 and TGF-β are predominant.[11] This is orchestrated mainly by immature dendritic cells which are able to take up luminal bacterial antigens, process them and present them to T-cells in a manner that favors a Th2 and antiinflammatory milieu.[12,13] For the controlled stimulation of T-cells by APC a second costimulatory signal is required. The most important costimulatory molecules are 2 members of the B7 family, CD80 and CD86 that interact with the receptor molecule CD28 thus eliciting a stimulatory signal.[14] Other costimulatory molecules include CD40 on APC interacting with the T-cell-ligand CD154, CD40/CD154,[15] OX40/OX40L[16] und ICOS/B7H.[17,18]

Another molecular mechanism of hyporesponsiveness in the lamina propria is the increased apoptotic rate of lamina propria T-cells compared to peripheral blood T-cells.[19] All these mechanisms contribute to the physiological hyporesponsiveness of the intestinal immune system and suggest a strong selection of antigens, against which an immune response is elecited under physiological conditions.

Activation of the Mucosal Immune System in IBD

In patients with IBD, the tightly controlled balance between responsiveness and nonresponsiveness is disturbed towards a more reactive state. For many of the cellular and humoral components if the intestinal immune system alterations and dysregulations have been identified that can be found in both CD and UC, some however are specific for either CD or UC.

Epithelial Cells

There is increasing evidence that intestinal epithelial cells do not only represent a mechanical barrier against antigens and pathogens but play an active role in the intestinal immune system. (Fig. 2). In IBD the surface marker CD14, that represents the receptor for bacterial lipopolysaccharides, is upregulated. This facilitates the elucidation of a strong signal via the

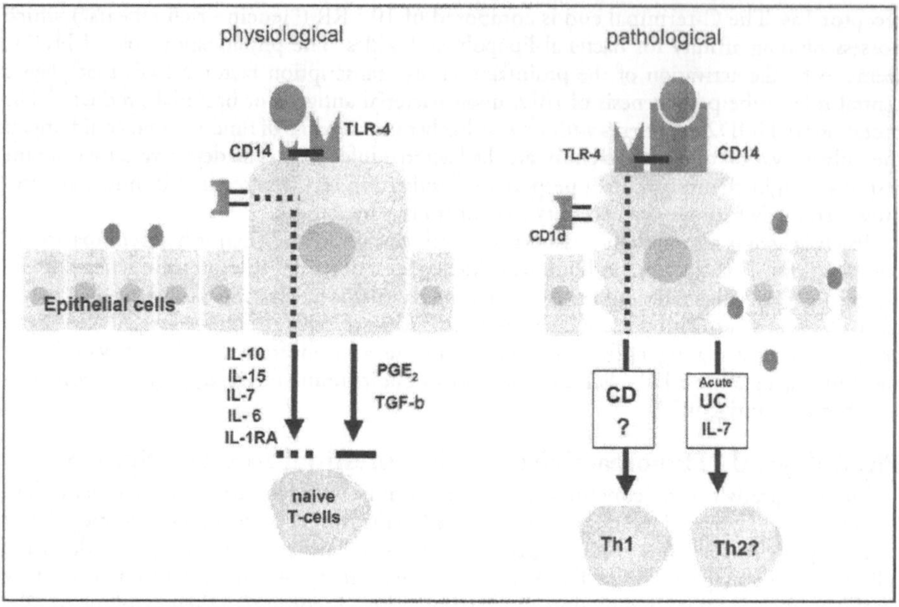

Figure 2. Role of epithelial cells. The intestinal epithelial cells do not only represent a mechanical barrier against luminal pathogenetic factors, but play an active part in the regulation of the intestinal immune system. Epithelial cells produce regulatory cytokines such as IL-10, IL-6, IL-7, IL-1RA, IL-15 and suppressive factors such as TGF-β and PGE2. In vitro, epithelial cells are capable of presenting microbial lipid antigens via the MHC (major histocompatibility complex)-class 1b molecule CD1d which leads to the production of IL-10. This controlled antigen stimulation contributes to a state of active inhibition. In IBD, CD14 on epithelial cells is upregulated. This facilitates a strong signaling activity of bacterial LPS via the Toll-like-receptor 4 (TLR-4). In active ulcerative colitis, epithelial cells produce IL-7. In the inflamed bowel, permeability is increased through metalloproteases. This promotes the invasion of luminal antigens of the bacterial flora and the food and probably activates the immune systeme.

Toll-like receptor 4 (TLR-4)[20]. Interestingly, these receptors are regulated differentially on epithelial cells from patients wiht CD and UC.[21] Epithelial cells are at least in vitro able to present microbial lipidantigen on their surface via the MHC class Ib molecule CD1d.[22] This induces the production of IL-10 by epithelial cells thus contributing to the antiinflammatory counterregulation.[23] In active UC, epithelial cells produce large amounts of IL-7.[24] This could contribute to the activation of CD4+-T-cells in the lamina propria. Interestingly, IL-7 transgenic mice develop chronic colitis, which emphasizes the potentially pathogenetic role of this cytokine.

Adhesion Molecules

The massive inflammatory infiltration of the lamina propria results from cell proliferation, diminished cell turnover by inhibition of apoptosis and increased migration of inflammatory cells from the circulation into the tissue. Migration is facilitated by adhesion molecules expressed on endothelial cells and on inflammatory cells which recruit the cells to the inflamed area. Adhesion molecules include the immunoglobulin superfamily, i.e., ICAM-1 (intercellular adhesion molecule-1) on monocytes and endothelial cells, integrins, i.e., MAdCAM (mucosal addressin cel adhesion molecule 1) and selectins. For a number of adhesion molecules upregulation in IBD has been shown.[25] The integrin MAdCAM is mucosa specific. It mediates the interaction between leukocytes and endothelial cells and is upregulated in IBD. In animal models blockade of MAdCAM by specific antibodies has shown therapeutic effects (see below). For ICAM-1 an antisense DNA approach has been utilized in a controlled study in CD.

Macrophages

Macrophages function as antigen presenting cells and stimulate T-cell activity by producing proinflammatory cytokines such as IL-1, IL-6, IL-8, IL-12 IL-18 and TNF.

These cytokines produced by macrophages exhibit specific pathologically altered expression patterns at the mRNA and protein level.[26] In patients with UC, increased levels of IL-1, IL-6 and IL-8 could be detected in inflamed areas of the colon compared to noninflamed areas. In patients with CD these cytokines were increased in both inflamed and noninflamed areas supporting the view that UC is a locally progressing inflammation while CD is rather a panenteritis with partial manifestation.[27] The synthesis of proteases, oxygen radicals and leukotriens can also be attributes to macrophage function, which is increased in the lamina propria of patients with IBD and contributes to tissue damage.

T-Lymphocytes

In CD the physiologically Th2, Th3 and Tr dominated cytokine milieu of the lamina propria shifts towards a strong Th1 profile with increased levels of TNF and INF-γ.[28,29] These proinflammatory cytokines contribute to the epithelial damage and tissue destruction via activation of matrix metalloproteinases. Production of the Th2 cytokines IL-4 and IL-5 is decreased. In UC, on the contrary, IL-5 production is increased. This could promote inflammation via activation of B-cells and production of autoantibodies. While CD has been established as a Th1 mediated disease, the association of UC with a Th2 T-cell response is less clear. While production of IL-5 is increased in UC, IL-4 levels are diminished.

Another T-cell associated pathomechanism in CD is the increased resistance of lamina propria T-cells against apoptosis.[30,31] This could be due to an increased IL-6 signaling in CD as well as an increased TNF-signaling via TNF-Receptor 2 (see below) amd is likely to contribute to the accumulation of T-cells in the gut and the perpetuation of inflammation.

Transcriptional Regulation of Th1/Th2-Differentiation in IBD

In the recent years experimental studies, especially in animal models, have contributed to the elucidation of regulatory mechanisms of Th1/Th2-differentiation at the transcriptional level and have thus dramatically extended our understanding of the pathogenesis of IBD at the molecular level (Fig. 3).

The IL-12/p35-p40 heterodimer, produced by CD8+ dendritic cells (DC's) and macrophages, is the critical cytokine for the induction of Th1-helper cell differentiation depending on the activation of the transcription factor STAT4 (signal transducer and activator of transcription 4) through phosphorylation.[13,32,33] The role of IL-12 and STAT4 activation for the Th1-mediated intestinal inflammation is well documented.[34] Conversely, STAT4 deficient T-cells do not induce a Th1-mediated colitis, while STAT4 transgenic mice have severe colitis.[35] There is data, however, suggesting that STAT4 is also activated by IL-23, a heterodimer of p19 and p40, furthermore, p19 transgenic mice develop a severe inflammatory syndrome including the bowel.[36,37]

IL-18 is also important for the mucosal Th1-response and activates the transcription factors AP-1 (c-fos/c-jun) and NFκB in T-cells. The functional relevance of IL-18 is further supported by the observation that blockade of IL-18 expression or function suppresses experimental Th1 mediated colitis.[38,39]

The cloning of a novel transcription factor of the T-box family, named T-bet, has further contributed to the elucidation of Th1-differentiation of T-helper cells.[40] T-bet can be detected in IFN-γ producing Th1-T-cells, but not in Th2-T-cells. When stimulated under conditions that lead to an Th1-differentiation, T-bet is upregulated in T-cells. T-bet deficient mice show a dramatic suppression of IFN-γ production in CD4+ T-helper cells, but not in CD8+ cytotoxic T-cells.[41] The retroviral transduction of immature T-cells or differentiated Th2 polarized T-cells with T-bet induces the production of high IFN-γ levels. T-bet thus seems to be a critical factor for Th-1 differentiation and regulation of T-cell effector function by induction of Th1 cytokine transcription. These findings are also relevant for the mucosal immune system, since T-bet deficient T-cells fail to induce Th1- mediated experimental colitis.[42]

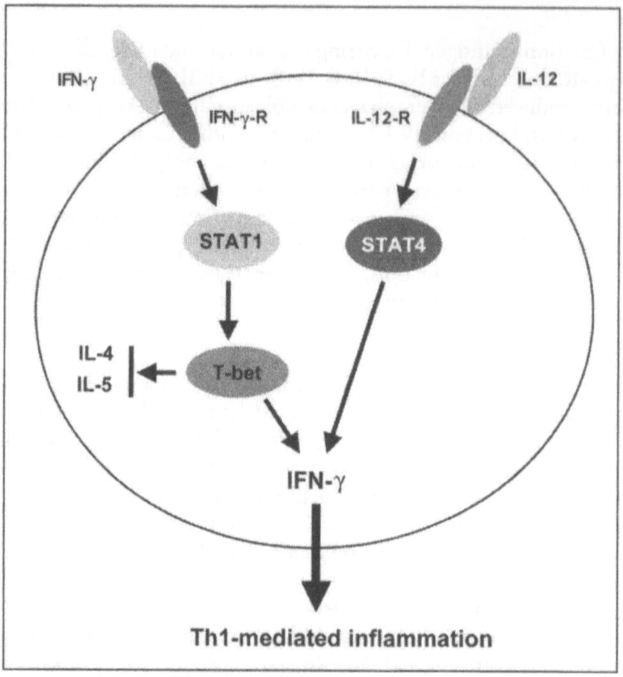

Figure 3. Transcriptional regulation of Th1/Th2 differentiation. The IL-12/p35p40-heterodimer, which is produced by CD8α+-DCs or macrophages, activates the transcription factor STAT4 (signal transducer and activator of transcription 4) via a specific cell surface receptor. IL-12 is a potent inducer of a Th1-differentiation. IFN-γ induces the transcription factor STAT1, which upregulates the transcription factor T-bet. T-bet is one of the strongest inducers of a Th1-differentiation and inhibits the synthesis of Th2 cytokines.

Cytokines

Alterations in expression at the mRNA or protein levels in IBD have been demonstrated in numerous cytokines. Since the cytokines IL-12, IL-6 and TNF are targets of established or experimental immunomodulatory strategies, they are discussed more in detail.

IL-12

IL-12 is produced by macrophages and dendritic cells. While CD goes along with an increased IL-12 production, the production of the IL-12 antagonistic cytokine EBI3 seems to be activated in UC.[43,44] These findings are functionally relevant, since IL-12 is able to induce a Th1 differentiation of naive T-cells, as it can be found in CD. The pathogenetic key role of IL-12 has been shown in various different Th1 mediated animal models of experimental colitis, in which treatment with an antagonistic anti-IL-12 antibody led to a reduction or even remission of colitis.[45] The effect of anti-IL-12 treatment was mediated by impeding Th1 T-cell differentiation and by induction of Fas-mediated T-cell apoptosis.[28]

IL-6

Recent data on the role of IL-6, which has been considered a Th2 cytokine, in experimental colitis illustrate the limitations of the Th1/Th2 paradigm in IBD. Lamina propria T-lymphocytes from patients with CD and UC produce increased amounts of IL-6 as compared to T-lymphocytes from controls.[31] In both diseases increased IL-6 production goes along with a strong activation of the signal transduction factor STAT-3. STAT-3 activates various antiapoptotic genes such as bcl-2 and bcl-xl. For patients with CD it could be shown, that these genes are indeed upregulated in lamina propria T-cells. Apoptotic resistance in these cells can effectively be suppressed in vitro by blocking IL-6 signaling with an antibody against the IL-6 receptor. This effect of IL-6 blockade could be confirmed in several Th1 models of experimental colitis in vivo.

Tumor Necrosis Factor (TNF)

TNF belongs to the best characterized proinflammatory cytokines in IBD. TNF exists in a membrane bound (mTNF) and a soluble form. There exists a large amount of direct and indirect clinical evidence as well as experimental data from murine colitis models demonstrating a pivotal role of TNF in the pathogenesis of IBD, especially in CD.[46,47] TNF is produced by lamina propria macrophages and CD4+-T-cells. In vitro CD4+-T-cells from patients with CD produce more TNF upon stimulation with TNF than CD4+-T-cells from controls.[29] Stimulation of TNF secretion by TNF itself suggests a positive feedback mechanism that could contribute to the chronification of inflammation (Fig. 4). Another potential pathomechanism could be that TNF activates endogenous matrix metalloproteases (MMP), which damage the epithelium.[48] In patients with CD, TNF expressing macrophages in the submucosa tend to aggregate along vessels with infiltration of the endothelium. The known prothrombotic effect of TNF on epithelial cells could thus be another pathomechanism of TNF in CD.[49]

TNF exerts its effects at the cellular level by means of 2 specific cell surface receptors, TNF-R1 (p60) and TNF-R2 (p80) (Fig. 5).[50] Until recently, p60 was considered the principal mediator of TNF signal transduction. This view was mainly based on the fact, that p80 binds soluble TNF (sTNF) with a 20-fold lower binding affinity than p60.[51] However, it was then found that p80 is preferentially activated by the transmembrane form of TNF (mTNF) with high affinity in a paracrine fashion.[51] The mTNF/p80 system could thus play an important

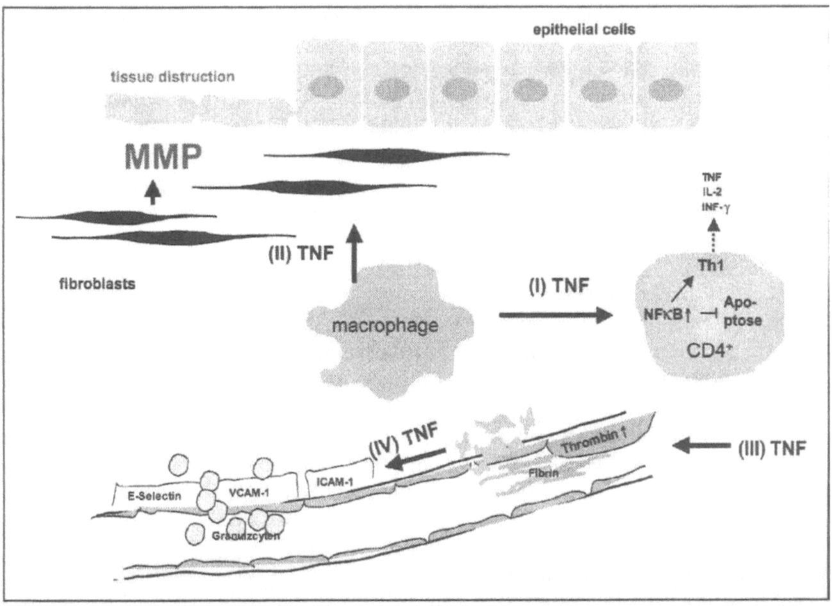

Figure 4. Pathomechanisms of TNF. Potential pathomechanisms of TNF include the activation of endogeneous Matrix-Metalloproteases, a prothrombotic effect on endothelial cells and the upregulation of adhesion molecules. These mechanism are potentiated by autostimulation of TNF production by TNF itself. (Der Internist 2002; 43:1343-1353).

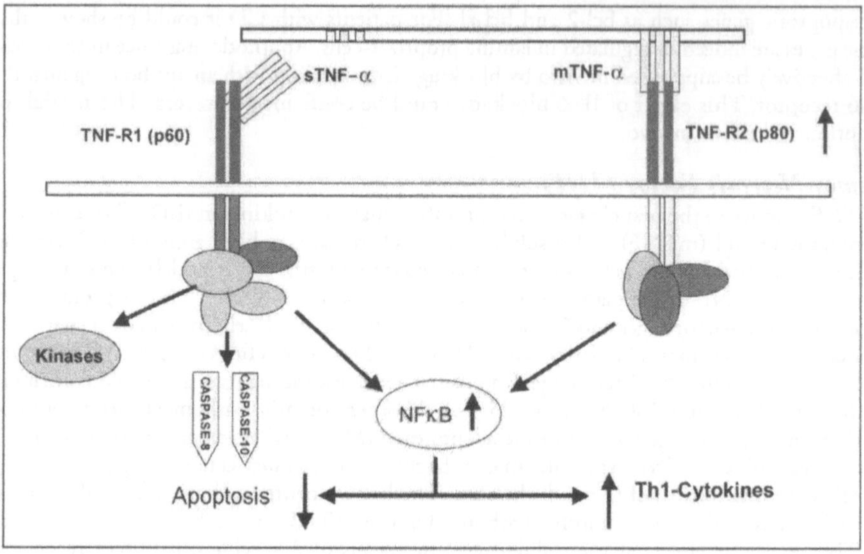

Figure 5. TNF-signaling. TNF exerts its effects via 2 specific cell surface receptors, TNF-R1 (p60) and TNF-R2 (p80). For a long time, TNF-R1 was considered as the principal mediator of TNF-effects. Recent data, however, suggest, that TNF-R2 and the TNF-R2 associated signal transduction play an important role in the pathogenesis of colitis. TNF-R1 mediated cellular responses include activation of kinases, induction of apoptosis and activation of NFκB. The main cellular response of TNF-R2 is the activation of NFκB. NFκB mediates its proinflammatory effects not only through transcriptional upregulation of proinflammatory cytokines, but also through inhibition of apoptosis. (Der Internist 2002; 43:1343-1353.)

immunoregulatory role at the local level. Unlike TNF-R1, TNF-R2 is unable to induce apoptosis, but activates NFκB that in turn has an antiapoptotic effect. An increased recruitment of TNF-R2 could alter the balance between induction and inhibition of apoptosis in CD. Clinical studies revealed that TNF-R2 is upregulated on peripheral blood and lamina propria mononuclear cells in patients with acute CD.[52] In an experimental colitis model overexpression of TNF-R2 on T-cells led to an aggravation of colitis. Underlying mechanisms were a pronounced Th1 response and a decreased apoptotic rate of T cells like in CD.

Nuclear Factor κ B (NFκB)

NFκB has been identified as a pivotal proinflammatory factor in the pathogenesis of both CD and CU and is the target of many established therapeutic strategies such as salicylic acid and corticosteroids (Fig. 6).[53,54,55] NFκB is a transcription factor, that can be found in non activated macrophages and T-cells in the cytosol as an inactive heterodimer with a p50 and p65 subunit, bound to the inhibitory molecule IκB (inhibitor of κ B).[56] Stimulation of the cell leads to the degradation of IκB by IκB kinase thus releasing NFκB. NFκB then activates the transcription of various proinflammatory genes such as TNF, IL-1, IL-6 and IL-12. Another pathomechanism could be that NFκB inhibits the apoptosis of T-cells.

Novel Therapeutic Approaches

The study of the intestinal immune system in IBD at the molecular level has led to the identification of many components of the intestinal immune system that are dysregulated. While many of these dysregulations are likely to represent secondary effects, some of them may be key steps in the pathogenesis of IBD and thus be potential targets for novel therapeutic strategies. The purpose of this chapter is to give an overview about such novel therapeutic

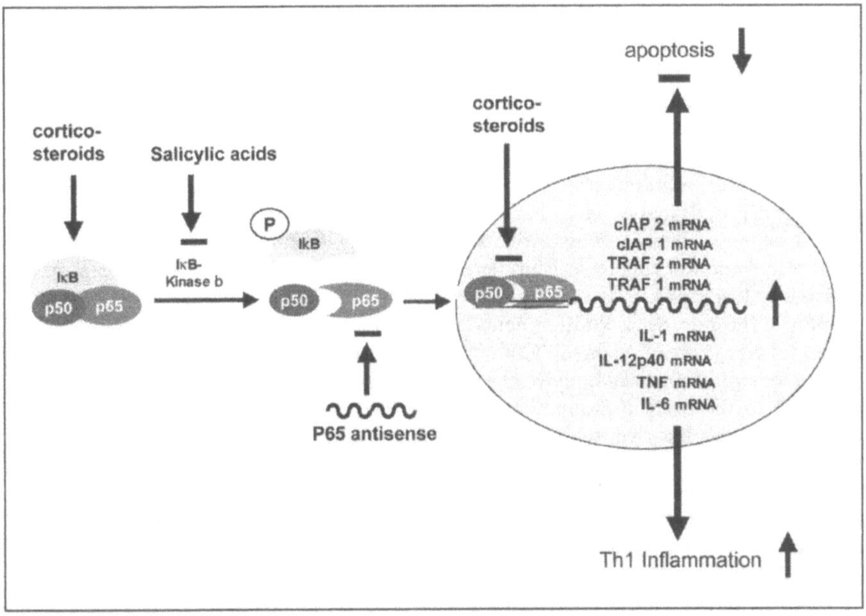

Figure 6. Activation and pharmacological modulation of NFκB. In non activated macrophages and T-cells, the transcription factor NFκB is located as an inactive heterodimer consisting of a p50 and p65 subunit, bound to the inhibitory IκB . Cellular stimulation leads to degradation of IκB by IκB-Kinase and release of NFκB. The NFκB-p50/p65-heterodimer migrates into the nucleus and activates the transcription of various proinflammatory genes such as TNF, IL-1, IL-6 and IL-12. Inhibition of T-cell apoptosis probably represents another proinflammatory mechanism. This effect is mediated by upregulation of anti-apoptotic factors of TNF-R2 signal transduction such as TRAF1, TRAF2, cIAP1 and cIAP2. NFκB is therapeutic target of many established, conventional treatment stragegies. While salicylic acids inhibit IκB-Kinase, corticosteroids induce IκB synthesis and inhibit the p50/p65-heterodimer in the nucleus through complexation. A phase 1 study on NFκB-antisense-oligonucleotides is currently being performed.

strategies focussing on recombinant peptides or proteins , that act at the level of the effector molecules of the immune systems (anti-TNF, anti-IL-12, anti-IL-6 receptor, anti-CD40L, IL-10), anti-sense constructs that act at the level of gene regulation and transcription (anti-ICAM-1, anti-NFκB p65) and small molecules that inhibit intracellular signaling pathways (MAP-Kinase-Inhibitors). The largest number of novel therapeutic strategies has evolved around the cytokine TNF, that plays a central role in the pathogenesis of IBD.

Anti-TNF Strategies

TNF blockade with the chimeric anti-TNF antibody cA2 (Infliximab) has been established as a standard approach in severe refractory cases of CD. Infliximab is a genetically engineered IgG1 murine-human monoclonal antibody with a constant region of human IgG1κ-immunoglobulin and a variable region of a monoclonal mouse anti-human antibody. cA2 binds both soluble and membrane bound TNF and most likely blocks the interaction of TNF with the TNF-receptors this way.[57,58,59] cA2 has a long half life of 9.5 days. With a single dose of 3 – 10 mg/kg bodyweight, therapeutic levels are achieved lasting for several weeks. In severe cases of refractory or steroid dependent CD a single intravenous administration of cA2 (5 mg/kg body weight) often leads to a significant clinical, histopathological and biochemical improvement.[60] Patients with fistulas usually receive a total of three infusions within six weeks, but often the fistulas already begin to close after the first treatment.[61] In a recent trial Infliximab was administered repeatedly every 8 weeks with 5 mg/kg or 10 mg/kg bodyweight respectively versus placebo for 1 year in those patients who had responded to an initial dose.[62] After 54 weeks 28,3% of the group that received 5mg/kg and 38,4% of the group that received 10 mg/kg was in remission versus 13,6% in the placebo group. The probable molecular mechanism of action of Infliximab is the binding of membrane bound TNF thus inducing apoptosis of activated lymphocytes and monocytes.[63,64]

However, the anti-TNF-strategy persued with anti-TNF-antibodies bears a number of problems. Short term side effects of Infliximab are rare and include headache, nausea, fever, urticaria, pruritus, dyspnoea and others. The long term efficacy and long term safety, however, are still a matter of debate. Most patients treated with Infliximab relapsed. Several hundred cases of reactivation of tuberculosis under treatment with Infliximab have been reported and some cases of Listeriosis.[65,66]

CDP571 (Humicade) is another genetically engineered human antibody to TNF, which has been tested in several studies in CD[67,68] and 1 uncontrolled pilot study in UC.[69]

Treatment with 10mg/kg bodyweight led to a response rate of 54 % after 12 doses weekly versus 27% in the placebo group. The treatment was well tolerated and the generation of autoantibodies was less than with Infliximab. The results of maintenance treatment remain controversial. Overall, CDP571 is moderately effective and has less side effects than Infliximab.

In another approach the human soluble tumor necrosis factor receptor fusion protein Etanercept was utilized. Etanercept inactivates soluble TNF, but not membrane bound TNF. It is effective in rheumatoid arthritis, but in CD this anti-TNF approach showed no clinical effect beyond placebo.[70]

The limitations of antibody based anti-TNF strategies have led to alternative approaches to interfere with TNF, all of them focussing on diminishing the levels of available biologically active ligand. These include the drug Thalidomide that inhibits TNF by enhancing messenger RNA degradation[71] and the phosphodiesterase inhibitor pentoxifylline that inhibits transcription of TNF by increasing intracellular cAMP levels.

While Thalidomide was clinically effective, pentoxifylline failed to show clinical efficacy in an open label study with 16 patients with steroid dependent CD.[72] Another target could be the recently cloned TNF convertase which is responsible for the proteolytic cleavage of soluble TNF from the membrane bound form.[73] The efficacy of TNF convertase inhibition remains to be shown, however.

Yet another approach to antagonize TNF represents the molecule CNI-1493 which inhibits MAP-Kinase. This way, the mitogen-activated protein kinase signalling pathways are blocked which leads to the inhibition of TNF gene expression.[74] In a pilot study in patients with CD, a high response rate could be achieved (8 out of 9 patients after 4 weeks).[75] In 3 patients, however, treatment had to be discontinued because of adverse events.

Anti-IL-12 and IL-18 Strategies

Since IL-12 is a key cytokine for Th1-differentiation, it conceptually represents an attractive therapeutic target. And indeed, in an animal model specific anti-IL-12 antibodies abrogated experimental colitis.[45] A recently performed, yet unpublished phase 2 study, however, had negative results. A possible explanation for the discrepancy between findings in experimental models and clinical studies in patients could be an insufficient dose. In addition, there might be a bigger redundancy in the human than in the murine immune system. IL-18 has a synergistic effect to IL-12 in promoting Th1-differentiation of CD4+ T-cells.[76] In an animal model local administration of IL-18 antisense mRNA improved experimental colitis.[38] Anti-IL-18 strategies have not been tested in humans, yet.

IL-10

As a potent antiinflammatory cytokine IL-10 plays a key role in the balance of the intestinal immune system. In fact, inactivation of the IL-10 gene in mice leads to the development of chronic colitis.[77] This observation was the basis for several clinical studies in patients with IBD. In a pilot study recombinant IL-10 was given intravenously on 7 consecutive days[78] During the 3-week-follow-up period 50% of the patients experienced a complete remission versus 23% in the control group. The overall average response, however, was only mild with a decrease in the disease activity score to 179 versus 226. In several consecutive clinical studies, however, comprising a total of 800 patients this effect could not be confirmed.[79,80] A totally novel approach for the local delivery of high doses of a certain factor has been developed by genetic modification of a lactobacillus strain that produces high levels of IL-10.[81] Intragastric administration led to a 50% improvement of experimental colitis in mice.

Interferons

IFN-γ is a key proinflammatory cytokine of a Th1-response, whereas INF-α and IFN-β have antiinflammatory effect. In open-label studies INF-α suggested response rates of 50% in patients with CD and over 90% in patients with UC,[82,83] that could not be confirmed in a controlled study.[84] IFN-β showed effect in UC with response rates of 50% versus 14% in the placebo control group.[85] The efficacy of IFN-β could be confirmed in other studies.[86]

Anti-NFκB Strategies

The key role of NFκB in the pathogenesis of IBD is illustrated by the fact that NFκB is the main target of many established conventional antiinflammatory drugs such as steroids, 5-aminosalicylic acids and others. An innovative experimental anti- NFκB approach has been choosen with the development of antisense oligonucleotides against the p65 subunit of NFκB. These were locally applicated by rectal instillation in mice with experimental colitis and abrogated intestinal inflammation clinically and histologically.[87] This line was pursued further in a controlled pilot study in 11 patients with distal colonic IBD. 5 out of 7 patients treated with antisense oligonucleotides showed a good clinical response and 2 of these developed a long lasting remission.[88]

Inhibition of Activated T-cells

The current paradigm of intestinal inflammation emphasizes the critical pathogenetic role of activated T-cells, and in fact, there are several case reports on patients with IBD and concurrent HIV who experienced a remission of colitis along with the decrease of CD4+ T-helper cells.[89] There have been studies with the anti-CD4 antibody cM-T412, in which high rates of long lasting remissions in CD and UC could be achieved.[90,91] Due to concerns regarding undesired long term side effects the antibody approach has not been pursued further. Of note, promising results have been achieved in patients with UC by elimination of leukocytes via apheresis.[92]

Epidermal growth factor that regulates T-cells differentiation has been tested topically as enema in left sided colitis in a small controlled pilot study in 23 patients and resulted in a remission rate of 82% versus 8% in the placebo group, while the keratinocyte growth factor Repifermin showed no effect in patients with UC.[93,94]

As outlined above, T-cell activation required additional accessory signals, i.e., via B7/CD28 or CD40/CD40-ligand. There is an ongoing study using a monoclonal inhibiting antibody against CD40-ligand in patients with CD.

Anti-Adhesion Molecule Strategies

Proinflammatory adhesion molecules have been the therapeutic target in several experimental models utilizing different approaches. In several experimentel colitis models topically or systemically administered neutralizing antibodies showed clinical effect for ELAM-1 (endothelial leucocyte adhesion molecule-1), ICAM-1 (intercellular adhesion molecule-1) or VCAM-1 (vascular cell adhesion molecule 1).[95,96] Interestingly, the molecule MAdCAM-1 (mucosal addressin cel adhesion molecule 1) which is conceptually interesting because of its mucosa specificity, failed as therapeutic target.

Natalizumab is a recombinant humanized monoclonal IgG-4 antibody against human α4-Integrin which functions as ligand for MAdCAM. Natalizumab was tested successfully in the cotton-top tamarin model of colitis.[97] In 2 placebo controlled clinical trials in patients with CD, Natalizumab showed clinical effect with remission rates of 39% and 46% versus 8% and

27% in the placebo groups.[98,99] In a small study of 10 patients with UC, Natalizumab was less effective.[100]

Alicaforsen (ISIS-2302-CS9) is an antisense phosphorothioate oligondeoxynucleotide to ICAM-1. In a randomized, controlled pilot study 20 patients were assigned to intravenous application of either alicaforsen 13 times every other day or placebo. 47% versus 20% were in remission at the end of the treatment and 5 out of 7 Alicaforsen treated remitters remained in remission for 6 months.[101] These promissing initial results, however, could not be confirmed by larger studies.[102,103] Here, Alicaforsen failed to showed effect beyond placebo.

Others

Most of the components of the intestinal immune system, that are differentially regulated in IBD represent potential therapeutic targets. Thus the number of attempts to downregulate the immune response by targeting these components is high. Many strategies are successful in animal models, but fail in clinical trials in patients.

Human recombinant IL-11 (opreleukin) has been used in clinical practice for some time for its thrombocytopoietic activity in patients with thrombocytopenia caused by intensive cancer chemotherapy. IL-11 was tested successfully in the TNBS model of colitis in the rat but showed little clinical effects in CD patients.[104,105]

In CD the T-cell surface marker CD44, important for T-cell activation and migration could potentially represent a therapeutic target. The importance of CD44v7 could be demonstrated in two animal models and specific antibodies to CD44v7 were able to abrogate TNBS-induced colitis in mice.[106]

The proinflammatory Th1-cytokine IL-2 is target of many conventional immunosuppressive drugs. Dacluzimab is monoclonal antibody against the IL-2 receptor and is being studied in clinical trials.

Human growth hormone reduces intestinal permeability which characterizes the inflamed bowel. In a pilot study in 37 patients growth hormone reduced disease activity.[107] Promising data exist for the androgen hormone Medroxyprogesterone Acetate (CBP-1011)[108] and for the antidiabetic Insulin receptor sensitizing Glitazones.[109]

Concluding Remarks

During the past decade our understanding of the pathogenesis of IBD at the molecular levels has improved dramatically. For CD an experimentally well supported pathogenetic concept has been established based on the predominance of Th1-cytokines that has been reproduced in various animal model systems. This has led to a number of novel therapeutic strategies, of which the anti-TNF-strategy has been pursued most consequently. The story of TNF is a unique example of the success from a close collaboration between clinically oriented basic science and applied biotechnological research. The strength of the anti-TNF approach is the conclusiveness of the pathogenetic concept on the one hand and the availability of appropriate antibodies. New treatments to be successful should fulfill these two prerequisites.

All immunomodulatory strategies should be used cautiously—as any therapy. There is little data available regarding long term effects of these biological drugs. With this in mind, however, the availability of these new therapeutic options allows for a fairly optimistic outlook in providing help for the huge number of patients affected by IBD.

References

1. Duchmann R, Zeitz M. Crohn' disease. In: Ogra P, Strober W, eds. Handbook of mucosal immunology. Academic Press, 1999:1005.
2. Hugot JP, Chamaillard M, Zouali H et al. Association of NOD2 leucine-rich repeat variants with susceptibility to Crohn's disease. Nature 2001; 411:599-603.
3. Ogura Y, Bonen DK, Inohara N et al. A frameshift mutation in NOD2 associated with susceptibility to Crohn's disease. Nature 2001; 411:603-606.
4. Farrell RJ, Peppercorn MA. Ulcerative colitis. The Lancet 2002; 359:331-340.

5. Sandborn WJ, Landers CJ, Tremaine WJ et al. Antineutrophil cytoplasmic antibody correlates with chronic pouchitis after ileal pouch-anal anastomosis. Am J Gastroenterol 1995; 90:740-747.
6. Gordon LK, Eggena M, Targan SR et al. Definition of ocular antigens in ciliary body and retinal ganglion cells by the marker antibody pANCA. Invest Ophthalmol Vis Sci 1999; 40:1250-1255.
7. Gordon LK, Eggena M, Targan SR et al. Mast cell and neuroendocrine cytoplasmic autoantigen(s) detected by monoclonal pANCA antibodies. Clin Immunol 2000; 94:42-50.
8. Cohavy O, Bruckner D, Gordon LK et al. Colonic bacteria express an ulcerative colitis pANCA-related protein epitope. Infect Immun 2000; 68:1542-1548.
9. Vashishtha A, Fischetti VA. Surface-exposed conserved region of the streptococcal M protein induces antibodies cross-reactive with denatured forms of myosin. J Immunol 1993; 150:4693-4701.
10. Qiao L, Schurmann G, Betzler M et al. Activation and signaling status of human lamina propria T lymphocytes. Gastroenterology 1991; 101:1529-1536.
11. Braunstein J, Qiao L, Autschbach F et al. T cells of the human intestinal lamina propria are high producers of interleukin-10. Gut 1997; 41:215-220.
12. Banchereau J, Steinman RM. Dendritic cells and the control of immunity. Nature 1998; 392:245-252.
13. Moser M, Murphy KM. Dendritic cell regulation of TH1-TH2 development. Nat Immunol 2000; 1:199-205.
14. Liu Z, Geboes K, Hellings P et al. B7 interactions with CD28 and CTLA-4 control tolerance or induction of mucosal inflammation in chronic experimental colitis. J Immunol 2001; 167:1830-1838.
15. Kweon MN, Fujihashi K, Wakatsuki Y et al. Mucosally induced systemic T cell unresponsiveness to ovalbumin requires CD40 ligand-CD40 interactions. J Immunol 1999; 162:1904-1909.
16. Higgins LM, McDonald SA, Whittle N et al. Regulation of T cell activation in vitro and in vivo by targeting the OX40-OX40 ligand interaction: Amelioration of ongoing inflammatory bowel disease with an OX40-IgG fusion protein, but not with an OX40 ligand-IgG fusion protein. J Immunol 1999; 162:486-493.
17. Hutloff A, Dittrich AM, Beier KC et al. ICOS is an inducible T-cell costimulator structurally and functionally related to CD28. Nature 1999; 397:263-266.
18. Tafuri A, Shahinian A, Bladt F et al. ICOS is essential for effective T-helper-cell responses. Nature 2001; 409:105-109.
19. Boirivant M, Pica R, DeMaria R et al. Stimulated human lamina propria T cells manifest enhanced Fas-mediated apoptosis. J Clin Invest 1996; 98:2616-2622.
20. Cario E, Rosenberg IM, Brandwein SL et al. Lipopolysaccharide activates distinct signaling pathways in intestinal epithelial cell lines expressing Toll-like receptors. J Immunol 2000; 164:966-972.
21. Cario E, Podolsky DK. Differential alteration in intestinal epithelial cell expression of toll-like receptor 3 (TLR3) and TLR4 in inflammatory bowel disease. Infect Immun 2000; 68:7010-7017.
22. Jackman RM, Moody DB, Porcelli SA. Mechanisms of lipid antigen presentation by CD1. Crit Rev Immunol 1999; 19:49-63.
23. Colgan SP, Hershberg RM, Furuta GT et al. Ligation of intestinal epithelial CD1d induces bioactive IL-10: Critical role of the cytoplasmic tail in autocrine signaling. Proc Natl Acad Sci USA 1999; 96:13938-13943.
24. Watanabe M, Ueno Y, Yajima T et al. Interleukin 7 is produced by human intestinal epithelial cells and regulates the proliferation of intestinal mucosal lymphocytes. J Clin Invest 1995; 95:2945-2953.
25. Schuermann GM, Aber-Bishop AE, Facer P et al. Altered expression of cell adhesion molecules in uninvolved gut in inflammatory bowel disease. Clin Exp Immunol 1993; 94:341-347.
26. Ogra PL, Strober W, Bienenstock Jet al, eds. Strober WSE: Mucosal immunology. San Diego: Academic Press, 1999.
27. Casini-Raggi V, Kam L, Chong YJ et al. Mucosal imbalance of IL-1 and IL-1 receptor antagonist in inflammatory bowel disease. A novel mechanism of chronic intestinal inflammation. J Immunol 1995; 154:2434-2440.
28. Fuss IJ, Neurath M, Boirivant M et al. Disparate CD4+ lamina propria (LP) lymphokine secretion profiles in inflammatory bowel disease. Crohn's disease LP cells manifest increased secretion of IFN-gamma, whereas ulcerative colitis LP cells manifest increased secretion of IL-5. J Immunol 1996; 157:1261-1270.
29. Plevy SE, Landers CJ, Prehn J et al. A role for TNF-alpha and mucosal T helper-1 cytokines in the pathogenesis of Crohn's disease. J Immunol 1997; 159:6276-6282.
30. Boirivant M, Marini M, Di Felice G et al. Lamina propria T cells in Crohn's disease and other gastrointestinal inflammation show defective CD2 pathway-induced apoptosis. Gastroenterology 1999; 116:557-565.
31. Atreya R, Mudter J, Finotto S et al. Blockade of interleukin 6 trans signaling suppresses T-cell resistance against apoptosis in chronic intestinal inflammation: Evidence in crohn disease and experimen-

tal colitis in vivo. Nat Med 2000; 6:583-588. taf/DynaPage.taf?file=/nm/journal/v586/n585/full/nm0500_0583.html taf/DynaPage.taf?file=/nm/journal/v0506/n0505/abs/nm0500_0583.html.
32. Magram J, Sfarra J, Connaughton S et al. IL-12-deficient mice are defective but not devoid of type 1 cytokine responses. Ann NY Acad Sci 1996; 795:60-70.
33. Szabo SJ, Jacobson NG, Dighe AS et al. Developmental commitment to the Th2 lineage by extinction of IL-12 signaling. Immunity 1995; 2:665-675.
34. Simpson SJ, Shah S, Comiskey M et al. T cell-mediated pathology in two models of experimental colitis depends predominantly on the interleukin 12/Signal transducer and activator of transcription (Stat)-4 pathway, but is not conditional on interferon gamma expression by T cells. J Exp Med 1998; 187:1225-1234.
35. Wirtz S, Finotto S, Kanzler S et al. Cutting edge: Chronic intestinal inflammation in STAT-4 transgenic mice: Characterization of disease and adoptive transfer by TNF- plus IFN-gamma-producing CD4+ T cells that respond to bacterial antigens. J Immunol 1999; 162:1884-1888.
36. Oppmann B, Lesley R, Blom B et al. Novel p19 protein engages IL-12p40 to form a cytokine, IL-23, with biological activities similar as well as distinct from IL-12. Immunity 2000; 13:715-725.
37. Wiekowski MT, Leach MW, Evans EW et al. Ubiquitous transgenic expression of the IL-23 subunit p19 induces multiorgan inflammation, runting, infertility, and premature death. J Immunol 2001; 166:7563-7570.
38. Wirtz S, Becker C, Blumberg R et al. Treatment of T cell-dependent experimental colitis in SCID mice by local administration of an adenovirus expressing IL-18 antisense mRNA. J Immunol 2002; 168:411-420.
39. Ten Hove T, Corbaz A, Amitai H et al. Blockade of endogenous IL-18 ameliorates TNBS-induced colitis by decreasing local TNF-alpha production in mice. Gastroenterology 2001; 121:1372-1379.
40. Szabo SJ, Kim ST, Costa GL et al. A novel transcription factor, T-bet, directs Th1 lineage commitment. Cell 2000; 100:655-669.
41. Szabo SJ, Sullivan BM, Stemmann C et al. Distinct effects of T-bet in TH1 lineage commitment and IFN-gamma production in CD4 and CD8 T cells. Science 2002; 295:338-342.
42. Neurath MF, Weigmann B, Finotto S et al. The transcription factor T-bet regulates mucosal T cell activation in experimental colitis and Crohn's disease. J Exp Med 2002; 195:1129-1143.
43. Monteleone G, Biancone L, Marasco R et al. Interleukin 12 is expressed and actively released by Crohn's disease intestinal lamina propria mononuclear cells. Gastroenterology 1997; 112:1169-1178.
44. Christ AD, Stevens AC, Koeppen H et al. An interleukin 12-related cytokine is up-regulated in ulcerative colitis but not in Crohn's disease. Gastroenterology 1998; 115:307-313.
45. Neurath MF, Fuss I, Kelsall BL et al. Antibodies to interleukin 12 abrogate established experimental colitis in mice. J Exp Med 1995; 182:1281-1290.
46. Holtmann M, Schütz M, Galle PR et al. Functional relevance of soluble and transmembrane TNF and TNF-signaltransduction in gastrointestinal diseases with special reference to inflammatory bowel diseases. submitted.
47. Neurath MF, Fuss I, Pasparakis M et al. Predominant pathogenic role of tumor necrosis factor in experimental colitis in mice. Eur J Immunol 1997; 27:1743-1750.
48. Pender SL, Fell JM, Chamow SM et al. A p55 TNF receptor immunoadhesin prevents T cell-mediated intestinal injury by inhibiting matrix metalloproteinase production. J Immunol 1998; 160:4098-4103.
49. Murch SH, Braegger CP, Walker-Smith JA et al. Location of tumour necrosis factor alpha by immunohistochemistry in chronic inflammatory bowel disease. Gut 1993; 34:1705-1709.
50. Beutler B. Tumor necrosis factors: The molecules and their emerging role in medicine. New York: Raven Press, 1992.
51. Grell M, Douni E, Wajant H et al. The transmembrane form of tumor necrosis factor is the prime activating ligand of the 80 kDa tumor necrosis factor receptor. Cell 1995; 83:793-802.
52. Holtmann MH, Douni E, Schütz M et al. Tumor necrosis factor-receptor 2 is upregulated on lamina propria mononuclear cells in Crohn's disease and promotes experimental colitis in vivo. Eur J Immunol in press.
53. Schmid RM, Adler G, Liptay S. Activation of NFκB in inflammatory bowel disease. Gut 1998; 43:587-588.
54. Yin MJ, Yamamoto Y, Gaynor RB. The anti-inflammatory agents aspirin and salicylate inhibit the activity of I(κ)B kinase-beta. Nature 1998; 396:77-80.
55. Auphan N, DiDonato JA, Rosette C et al. Immunosuppression by glucocorticoids: Inhibition of NF-κ B activity through induction of I κ B synthesis. Science 1995; 270:286-290.
56. Schmid RM, Adler G. NF-κB/rel/IκB: Implications in gastrointestinal diseases. Gastroenterology 2000; 118:1208-1228.
57. Knight DM, Trinh H, Le J et al. Construction and initial characterization of a mouse-human chimeric anti-TNF antibody. Mol Immunol 1993; 30:1443-1453.

58. Scallon BJ, Moore MA, Trinh H et al. Chimeric anti-TNF-alpha monoclonal antibody cA2 binds recombinant transmembrane TNF-alpha and activates immune effector functions. Cytokine 1995; 7:251-259.
59. Siegel SA, Shealy DJ, Nakada MT et al. The mouse/human chimeric monoclonal antibody cA2 neutralizes TNF in vitro and protects transgenic mice from cachexia and TNF lethality in vivo. Cytokine 1995; 7:15-25.
60. Targan SR, Hanauer SB, van Deventer SJ et al. A short-term study of chimeric monoclonal antibody cA2 to tumor necrosis factor alpha for Crohn's disease. Crohn's Disease cA2 Study Group. N Engl J Med 1997; 337:1029-1035.
61. Present DH, Rutgeerts P, Targan S et al. Infliximab for the treatment of fistulas in patients with Crohn's disease. N Engl J Med 1999; 340:1398-1405.
62. Hanauer SB, Feagan BG, Lichtenstein GR et al. Maintenance infliximab for Crohn's disease: The ACCENT I randomised trial. Lancet 2002; 359:1541-1549.
63. Lugering A, Schmidt M, Lugering N et al. Infliximab induces apoptosis in monocytes from patients with chronic active crohn's disease by using a caspase-dependent pathway. Gastroenterology 2001; 121:1145-1157.
64. ten Hove T, van Montfrans C, Peppelenbosch MP et al. Infliximab treatment induces apoptosis of lamina propria T lymphocytes in Crohn's disease. Gut 2002; 50:206-211.
65. Keane J, Gershon S, Wise RP et al. Tuberculosis associated with infliximab, a tumor necrosis factor alpha- neutralizing agent. N Engl J Med 2001; 345:1098-1104.
66. Gluck T, Linde HJ, Scholmerich J et al. Anti-tumor necrosis factor therapy and Listeria monocytogenes infection: Report of two cases. Arthritis Rheum 2002; 46:2255-2257; discussion 2257.
67. Stack WA, Mann SD, Roy AJ et al. Randomised controlled trial of CDP571 antibody to tumour necrosis factor-alpha in Crohn's disease. Lancet 1997; 349:521-524.
68. Sandborn WJ, Feagan BG, Hanauer SB et al. An engineered human antibody to TNF (CDP571) for active Crohn's disease: A randomized double-blind placebo-controlled trial. Gastroenterology 2001; 120:1330-1338.
69. Evans RC, Clarke L, Heath P et al. Treatment of ulcerative colitis with an engineered human anti-TNFalpha antibody CDP571. Aliment Pharmacol Ther 1997; 11:1031-1035.
70. Sandborn WJ, Hanauer SB, Katz S et al. Etanercept for active crohn's disease: A randomized, double-blind, placebo-controlled trial. Gastroenterology 2001; 121:1088-1094.
71. Moreira AL, Sampaio EP, Zmuidzinas A et al. Thalidomide exerts its inhibitory action on tumor necrosis factor alpha by enhancing mRNA degradation. J Exp Med 1993; 177:1675-1680.
72. Bauditz J, Haemling J, Ortner M et al. Treatment with tumour necrosis factor inhibitor oxpentifylline does not improve corticosteroid dependent chronic active Crohn's disease. Gut 1997; 40:470-474.
73. Black RA, Rauch CT, Kozlosky CJ et al. A metalloproteinase disintegrin that releases tumour-necrosis factor- alpha from cells. Nature 1997; 385:729-733.
74. Cohen PS, Nakshatri H, Dennis J et al. CNI-1493 inhibits monocyte/macrophage tumor necrosis factor by suppression of translation efficiency. Proc Natl Acad Sci USA 1996; 93:3967-3971.
75. Hommes D, van den Blink B, Plasse T et al. Inhibition of stress-activated MAP kinases induces clinical improvement in moderate to severe Crohn's disease. Gastroenterology 2002; 122:7-14.
76. Barbulescu K, Becker C, Schlaak JF et al. IL-12 and IL-18 differentially regulate the transcriptional activity of the human IFN-gamma promoter in primary CD4+ T lymphocytes. J Immunol 1998; 160:3642-3647.
77. Kuhn R, Lohler J, Rennick D et al. Interleukin-10-deficient mice develop chronic enterocolitis. Cell 1993; 75:263-274.
78. van Deventer SJ, Elson CO, Fedorak RN. Multiple doses of intravenous interleukin 10 in steroid-refractory Crohn's disease. Crohn's Disease Study Group. Gastroenterology 1997; 113:383-389.
79. Fedorak RN, Gangl A, Elson CO et al. Recombinant human interleukin 10 in the treatment of patients with mild to moderately active Crohn's disease. The Interleukin 10 Inflammatory Bowel Disease Cooperative Study Group. Gastroenterology 2000; 119:1473-1482.
80. Schreiber S, Fedorak RN, Nielsen OH et al. Safety and efficacy of recombinant human interleukin 10 in chronic active Crohn's disease. Crohn's Disease IL-10 Cooperative Study Group. Gastroenterology 2000; 119:1461-1472.
81. Steidler L, Hans W, Schotte L et al. Treatment of murine colitis by Lactococcus lactis secreting interleukin- 10. Science 2000; 289:1352-1355.
82. Hanauer SB, Baert F, Robinson M. Interferon treatment in mild to moderate active Crohn's disease: Preliminary results in an open label study. Gastroenterology 1994; 106:A696.
83. Madsen SM, Schlichting P, Davidsen B et al. An open-labeled, randomized study comparing systemic interferon-alpha-2A and prednisolone enemas in the treatement of left-sided ulcerative colitis. Am J Gastroenterol 2001; 96:1807-1815.

84. Gasché C, Schölmerich J, Brynskow J et al. Prospective evaluation of interferon-alpha in treatment of chronic active Crohn's disease. Dig dis Sci 1995; 40:800-804.
85. Nikolaus S, Rutgeerts P, Fedorak RN et al. Recombinant human interferon-beta (IFNbeta-1a) induces remission and is well tolerated in moderately active ulcerative colitis (UC). Gastroenterology 2001; 120:A454.
86. Musch E, Jüntgen C, Witzke O et al. Successful therapy of colitis ulcerosa by beta-interferon. Gut 1995; 37:A141.
87. Neurath MF, Pettersson S, Meyer zum Buschenfelde KH et al. Local administration of antisense phosphorothioate oligonucleotides to the p65 subunit of NF-κ B abrogates established experimental colitis in mice. Nat Med 1996; 2:998-1004.
88. Löfberg R, Neurath M, Ost A et al. Topical NFκ p65 antisense oligonucleotides in patients with active distal colonic IBD. A randomised, controlled pilot trial. Gastroenterology 2002; 122:A60.
89. James SP. Remission of Crohn's disease after human immunodeficiency virus infection. Gastroenterology 1988; 95:1667-1669.
90. Deusch K, Mauthe B, Reiter C et al. CD4-antibody treatment of inflammatory bowel disease: One-year follow-up. Gastroenterology 1993; 104:A691.
91. Stronkhorst A, Radema S, Yong SL et al. CD4 antibody treatment in patients with active Crohn's disease: A phase 1 dose finding study. Gut 1997; 40:320-327.
92. Hanai H, Watanabe F, Matsushida I et al. Granulocyte and monocyte adsorption apheresis in patients with severe corticosteroid unresponsive ulcerative colitis. Gastroenterology 2002; T1198.
93. Sinha A, Nightingale J, West KP et al. Epidermal growth-factor enemas are ffective in the treatment of left-sided ulcerative colitis. Gastroenterology 2001; 120:A11-12.
94. Sandborn WJ, Sands BE, Wolf DC et al. Repifermin (Keratinocyte Growth Factor 2) for the treatment of active ulcerative colitis: A randomized, double-blind, placebo-controlled, dose-escalation trial. Gastroenterology 2002; 122:A61.
95. Hamamoto N, Maemura K, Hirata I et al. Inhibition of dextran sulphate sodium (DSS)-induced colitis in mice by intracolonically administered antibodies against adhesion molecules (endothelial leucocyte adhesion molecule-1 (ELAM-1) or intercellular adhesion molecule-1 (ICAM-1)). Clin Exp Immunol 1999; 117:462-468.
96. Soriano A, Salas A, Sans M et al. VCAM-1, but not ICAM-1 or MAdCAM-1, immunoblockade ameliorates DSS- induced colitis in mice. Lab Invest 2000; 80:1541-1551.
97. Podolsky DK, Lobb R, King N. Attenuation of colitis in the cotton-top tamarin by anti-alpha 4 integrin monoclonal antibody. J clin Invest 1993; 91:372-380.
98. Gordon FH, Lai CW, Hamilton MH et al. A randomized placebo-controlled trial of a humanized monoclonal antibody to alpha4 integrin in active Crohn's disease. Gastroenterology 2001; 121:268-274.
99. Gosh S, Goldin E, Malchow H et al. A randomised, double-blind, placebo-controlled, pan-European study of a recombinant humanised antibody to alpha4 integrin (Antegren TM) in moderate to severely active Crohn's disease. Gastroenterology 2001; 121:A127-128.
100. Gordon FH, Hamilton MI, Donoghue S. Treatment of active ulcerative colitis with a recombinant humanised antibody to alpha4 integrin (Antegren TM). Gastroenterology 1999; 116:A726.
101. Yacyshyn BR, Bowen-Yacyshyn MB, Jewell L et al. A placebo-controlled trial of ICAM- 1 antisense oligonucleotide in the treatment of Crohn's disease. Gastroenterology 1998; 114:1133-1142.
102. Yacyshyn BR, Chey WY, Goff J et al. Double blind, placebo controlled trial of the remisssion inducing and steroid sparing properties of an ICAM-1 antisense oligodeoxynucleotide, alicafersen (ISIS 2302), in active steroid dependent Crohn's disease. Gut 2002; 51:30-36.
103. Schreiber S, Hampe J, Malchow H et al. The german ICAM-1 study group. Absence of efficacy of subcutaneous antisense ICAM-1 treatment of chronic active Crohn's disease. Gastroenterology 2001; 120:1339-1346.
104. Sands BE, Bank S, Sninsky CA et al. Preliminary evaluation of safety and activity of recombinant human interleukin 11 in patients with active Crohn's disease. Gastroenterology 1999; 117:58-64.
105. Sands BE, Winston BD, Salzberg B et al. RHIL-11 crohn's study group. Randomized, controlled trial of recombinant human interleukin-11 in patients wiht active Crohn's disease. Aliment Pharmacol Ther 2002; 16:399-406.
106. Wittig B, Scharzler C, Fohr N et al. Curative treatment of an experimentally induced colitis by a CD44 variant V7-specific antibody. J Immunol 1998; 161:1069-1073.
107. Slonim AE, Bulone L, Damore MB et al. A preliminary study of growth hormone therapy for Crohn's disease. N Engl J Med 2000; 342:1633-1637.
108. Aronchick C, Dalke D, Ionna S et al. Medroxyprogesterone acetate is effective in the treatment of ulcerative colitis. Gastroenterology 2001; 120:A455.
109. Su CG, Wen X, Bailey ST et al. A novel therapy for colitis utilizing PPAE-gamma ligands to inhibit the epithelial inflammation response. J Clin Invest 1999; 104:383-389.

Index

A

Adaptive immunity 162
Adhesion molecule 3, 6, 17, 18, 44, 102, 108, 113, 122, 127, 153, 168, 171-173, 178, 179, 183, 230, 233, 237
Adjuvant 35, 36, 41, 42, 44, 45, 58, 136, 142
$\alpha_E\beta_7^+$ T cell 138, 140
Anaerobic bacteria 35, 36, 38, 41, 63, 153, 154
Anergy 57, 58
Animal model 9, 39, 41, 98, 99, 102, 109, 112, 113, 120, 122, 127, 136, 138, 141, 142, 161, 172, 190, 193, 197, 198, 219, 221, 222, 230-232, 236, 238
Anti-saccharomyces cerevisiae antibody (ASCA) 30, 48, 49, 157, 211, 214
Antibiotics 37-39, 41, 45, 48, 49, 61, 63, 76, 109, 136, 141, 212, 219-223
Antibody 3, 5, 9, 17, 48, 49, 62, 70, 71, 72, 73, 74, 75, 78, 80, 98, 99, 100, 101, 102, 111, 112, 113, 137, 142, 149, 150, 151, 152, 153, 154, 155, 156, 157, 158, 159, 161, 162, 173, 181, 183, 198, 209, 211, 212, 213, 214, 215, 227, 230, 232, 233, 235, 236, 237, 238
Antigen 1-3, 5, 7, 15-20, 35, 36, 41, 42, 44, 46-49, 56-63, 67, 69, 70, 72-80, 84, 87-89, 99, 101, 108, 112, 113, 117-122, 125-127, 134, 136, 137, 139-143, 149-153, 154-158, 160, 162, 168-174, 178, 179, 181, 182, 184, 188, 190, 191, 193, 209, 211-214, 220-223, 227-231
Antigen presenting cell (APC) 1, 3, 15, 17-19, 36, 44, 68, 69, 73, 78-80, 86, 112, 117, 119, 120, 134, 136, 140, 156, 172, 220, 229, 231
Antineutrophil cytoplasmic antibody (ANCA) 49, 156, 157
Antisense 104, 184, 227, 230, 235-238
Appendectomy 47, 101, 157, 158
Appendix 7, 101, 150, 157-159, 187
Association region 27
Autoantibody 84, 101, 137, 154, 155, 158, 209, 213, 214, 229, 231, 236

B

β-adrenergic receptor 191
B cell 3, 7, 8, 15, 17-20, 48, 81, 83, 101, 118-120, 127, 134, 137, 139, 140, 142, 143, 149, 150, 152-155, 157-162, 172, 173, 229, 231
Bacteria 7, 15-18, 24, 29, 35-38, 40-49, 57, 58, 60-65, 68, 76, 81, 89, 99, 100, 108, 109, 112, 113, 119, 120, 125, 127, 136, 141-143, 149, 152-157, 161, 162, 181, 209, 211, 219-223
Bacteroides vulgatus 38, 49, 62, 221
Basal lymphoid aggregate (BLA) 158
BCA-1 158-162
Bifidobacteria 48, 49, 142
Bifidobacterial species 49
Biological drug 238
Bordetella pertussis 119, 142

C

C3H/HeJ Bir mouse 88
CCL25 3, 15, 17, 121, 213
CCR5 3, 120, 121, 142
CCR7 3, 7, 117, 121
CCR9 15, 119, 121, 213, 214
CD1 78, 79, 81, 112, 113
CD1d 19, 78, 80, 111, 113, 137, 141, 162, 230
CD3 5, 86, 99, 102, 133, 134, 139, 140
CD4 3, 5, 6, 16, 19, 20, 38, 46-48, 61, 68, 69, 70, 75, 77, 78, 82, 86, 99, 100, 105, 111, 118-123, 127, 133-141, 143, 159, 169, 221, 222, 229-231, 233, 236, 237
$CD4^+CD25^+$ T cell 123, 138, 140
$CD4^+CD45RB^{hi}$ adoptive transfer model 136
CD8 5, 16, 59, 61, 78, 84, 105, 118, 133, 134, 138-142, 231
$CD8^+CD28^-$ T cell 138, 140
CD25 19, 20, 46, 47, 58, 59, 69-72, 76, 86, 121-123, 125, 127, 135, 138-140, 182
CD40 17, 46, 61, 84, 117, 125, 126, 134, 139, 140, 142, 168, 170, 172-174, 229, 237

CD40/CD40 ligand system 168, 172
CD40/CD40L 134, 172-174
CD46 156
CD55 155
CD59 156
CD89 151
Cell-transfer colitis 69-76
Chemokine 3, 7-9, 15, 17, 18, 20, 44, 64, 68, 74, 102, 108, 110, 111, 117, 119, 121, 122, 125-127, 142, 158-161, 168, 170-174, 179, 181, 182, 184, 188, 211, 213
Citrobacter freundii 40, 47
Clinical trial 109, 111, 219, 221, 237, 238
Colitis 5, 16, 19, 20, 24, 30, 31, 35, 38-42, 45-49, 56-89, 98-105, 108, 109, 111-113, 122, 123, 125, 126, 133, 134, 136-143, 149, 154, 156-158, 162, 169, 172, 177, 183, 186, 195-198, 211-213, 219-221, 227, 228, 230-234, 236-238
Collagen 65, 102, 103, 105, 169, 170, 173
Collagenous colitis 104, 105
Colonize 2, 35, 38, 41, 62, 63, 64, 137, 220
Commensal bacteria 15, 17, 18, 35, 36, 38, 40-49, 57, 89, 119, 120, 122, 125, 127, 143, 149, 152, 156, 162, 209, 211
Commensal fungi 35
Crohn's disease (CD) 24-31, 35, 37, 41, 42, 44, 48, 49, 56, 57, 60, 62, 63, 66-68, 70, 73, 74, 77, 80, 83, 88, 89, 98-104, 108-113, 120, 125-127, 134, 136, 152-155, 157-159, 161, 162, 169-172, 177, 183, 184, 195-197, 209-214, 219-221, 227-238
Cryptopatches (CP) 5, 9, 16, 133
CTLA-4 18-20, 69, 71, 72, 135, 138-140
CXCL13 18, 158, 160
CXCR1 3
CXCR3 3
CXCR5 159-162
Cyclooxygenase-2 (COX-2) 20, 44, 45, 170
Cytokine 1, 3, 8, 9, 16-20, 27, 29, 35, 44, 45, 47, 56-60, 64, 67-81, 85, 89, 99, 100, 102-104, 108-111, 113, 117-119, 121, 122, 125-128, 133-140, 142, 143, 151, 161, 162, 168-173, 178-183, 187-189, 191, 192, 210, 211, 220, 227, 229-238

D

Decay-accelerating factor (DAF) 155
Defensin 42, 44, 110, 112, 211, 212
Dendritic cell 2-4, 6-9, 18, 27, 36, 42, 44, 46, 58, 61, 65, 68, 72, 78, 79, 100, 112, 117-122, 125, 142, 158, 160-162, 168, 172, 186, 220, 227, 229, 231, 232
Dextran sulfate sodium-induced colitis 59
Diagnosis 101, 209-214
Dietary toxin 35
DSS colitis 59, 60, 65, 66, 87, 141

E

E. coli 5, 37, 38, 40-42, 44-48, 57, 61-65, 84, 109, 126, 137, 139, 142, 154, 158, 161, 212, 214, 219-221
E. coli expressing ovalbumin (ECOVA) 57, 61, 84
E. fecalis 40, 46, 47, 63
Endothelial cell 42, 168-173, 178-180, 183, 186, 230, 233
Enteric nervous system (ENS) 184-186, 194, 195, 197, 198
Enterococci 37, 142
Eosinophil (Eo) 3, 57, 82, 151, 152, 177-184, 186-192, 195-198
Epithelial barrier 15, 61, 64, 68, 79, 108, 113, 127, 150, 156
Experimental enterocolitis 37, 49
Experimental model 35, 38, 55, 56, 109, 133, 143, 154, 198, 236, 237
Extracellular matrix (ECM) 103, 159, 170, 171, 178

F

Family history 24
FasL 2
Fcα receptor 126, 151
Feline immunodeficiency virus 7
Fibroblast 1, 102, 103, 168-170, 172, 178, 180, 183, 186
Fibronectin 159, 170, 180
Fluorescent labeled antibody 9
Follicular dendritic cell (FDC) 6, 158-161
Food allergy 186
Food protein 15, 18
Foxp3 20, 59, 70, 86, 135

G

G-protein-linked receptor 58
Gastritis 38, 39, 47, 59, 62, 135, 221
GATA-3 134, 135
Gene 3, 7, 17, 24-31, 41, 64-67, 70, 76, 77, 83, 86, 100, 103, 104, 108, 110, 112, 135, 136, 153, 155, 161, 162, 171, 179, 180, 182, 184, 188, 209-211, 220, 222, 228, 233-236
Genetics 24-27, 30, 31, 35, 36, 38, 41, 42, 48, 55, 56, 65-67, 73, 74, 76, 77, 80, 81, 85, 88, 89, 98, 108, 110, 112, 152, 154, 157, 183, 209, 210, 212-215, 219, 222, 227, 228, 237
Genetic linkage 24-27
Gnotobiotics 38, 41, 44, 46, 49, 59
Granulocyte-macrophage colony stimulating factor (GM-CSF) 42, 119, 126, 170, 171, 179, 182
Granzyme 2, 78
Gut associated lymphoid tissue (GALT) 15-20, 109, 112, 113, 133, 134, 150, 153, 154, 157, 159-161

H

H. hepaticus 40, 46, 47, 61, 72, 83
Hapten-induced colitis 76, 77
Helicobacter bilis 40
Helicobacter pylori 35, 37, 38, 48, 109, 127
High endothelial venule (HEV) 3, 4, 6-9, 171, 178
Histamine 171, 180-183, 185, 189-193, 195, 196
HIV 7, 99, 120, 237
Human Genome Project 24, 25
Human intestinal microvascular endothelial cell (HIMEC) 171-174
Human umbilical vein endothelial cell (HUVEC) 171

I

ICAM-1 44, 170-173, 178, 179, 182, 230, 235, 237, 238
IFN-γ 1, 2, 20, 44, 47, 56, 57, 64, 70, 72-74, 77, 78, 108, 111, 113, 119, 120, 125, 127, 134-137, 139, 141, 142, 171, 173, 182, 183, 211, 229, 231, 232, 237
IgA 3, 8, 15-18, 49, 64, 101, 119, 150, 151, 153, 154, 157, 211, 212, 229
IgD 17, 120
IgG 16, 47, 100-102, 112, 149-158, 162, 211, 212, 235, 237
IgM 8, 17, 149, 150, 154, 160
IκB kinase (IKK) 44, 45, 234
IL-2 1, 38-40, 47, 56, 59, 60, 86, 99, 109, 111, 113, 123, 134, 139-141, 154, 156, 182, 183, 229, 238
IL-4 7, 56-58, 66, 72, 77, 79, 80, 119-121, 134-142, 178, 179, 181, 182, 184, 229, 231
IL-5 1, 3, 7, 17, 77, 99, 111, 134-137, 140, 179-182, 184, 188, 192, 229, 231
IL-6 44, 74, 85, 111, 122, 125-127, 170, 171, 180, 182, 183, 192, 211, 230-235
IL-7 1, 5, 15, 82, 171, 230
IL-8 64, 111, 122, 126, 170, 171, 173, 178-180, 182, 183, 231
IL-10 7, 16, 19, 20, 36, 38-41, 45-48, 56, 58-67, 69, 71, 72, 75, 76, 81, 83, 85, 88, 109, 111, 118-123, 125, 126, 134, 135, 137-143, 162, 170, 182, 219, 221, 229, 230, 235-237
IL-12 7, 19, 20, 44, 46, 56-60, 65-70, 73, 74, 76-79, 81-83, 89, 99, 100, 102, 111, 118-120, 122, 123, 125-127, 134-137, 142, 182, 183, 220, 231, 232, 234-236
IL-13 56, 57, 66, 77-81, 89, 108, 113, 134, 137, 140, 179, 180, 182, 229
IL-15 1, 3, 86, 171, 230
IL-23 44, 58, 70, 74, 83, 89, 118, 122, 123, 125, 142, 231
Immune exclusion 16, 149, 150, 152, 153, 154
Immune villi 5
Immunogenetics 209, 213, 214
Immunomodulation 208
Inflammatory bowel disease (IBD) 9, 15, 20, 24, 26, 27, 29-31, 35, 41, 42, 44, 48, 49, 55, 56, 59-62, 65, 68, 75, 80, 90-94, 96-99, 102-104, 108-113, 117, 122, 123, 125-127, 133-138, 140-143, 149-162, 167-174, 177, 183, 184, 186, 190, 194-198, 209-215, 219, 221, 222, 227-238
Innate immunity 2, 20, 29, 36, 42, 45, 56, 60, 77, 110, 112, 162
Integrin 3, 9, 120, 126, 140, 142, 171, 178, 179, 230, 237

Intestinal barrier 112, 113, 157, 162, 193, 222, 227
Intestinal epithelial cell (IEC) 8, 16, 27, 29, 35, 44, 46, 108-113, 120, 121, 125, 127, 140, 142, 193, 211, 229, 230
Intestinal microbiota 162
Intestinal microflora 70, 156
Intraepithelial lymphocyte (IEL) 1-3, 5, 9, 15, 16, 47, 82, 105, 133, 140
Isolated lymphoid follicle (ILF) 5, 9, 15, 16, 119, 122, 150, 157-159, 162

J

JNK 44

K

Kupfer cell 8

L

L-selectin 8, 171
Lactobacilli 49, 142, 219
Lactobacillus plantarum 64
Lactobacillus salivarius 64
Lamina propria 2-5, 8, 9, 15-18, 20, 41, 42, 44, 46-48, 67-69, 72, 73, 75, 76, 79, 80, 87, 99, 102, 103, 105, 110, 119, 120, 123, 125, 127, 133, 134, 137, 139, 150-153, 155-158, 162, 169, 171, 178, 179, 220, 221, 227, 229-234
Lamina propria lymphocyte (LPL) 3, 44, 82, 133, 139
Lamina propria T cell 20, 48, 105, 127, 133, 134, 229, 231, 233
Laminin 170
Leishmania major 72
Leukotriene 102, 179-183, 192, 193
Linkage studies 24, 26, 27
Listeria monocytogenes 42, 109
Liver 8, 82
Lymph node 3, 7, 8, 15-19, 41, 46, 47, 69, 72, 73, 118, 120, 122, 123, 127, 137, 154, 157-159, 162, 178
Lymphatics 8
Lymphocyte-filled villi (LFV) 4, 5
Lymphocytic colitis 104, 105
Lymphotoxin (LT) 5, 7, 17, 18, 160, 161, 210, 212

M

M cells 5, 7, 9, 15, 112, 121, 125
Macrophages 3, 6, 18, 27, 29, 42, 44, 45, 59, 76, 78, 87, 99, 100, 102, 103, 105, 117, 122, 125-128, 142, 143, 149, 151, 152, 160, 161, 162, 168, 170, 172, 173, 186, 210, 211, 220, 221, 227, 229, 231-235
MAdCAM 3, 6, 17, 18, 159, 178, 183, 230, 237
Major histocompatibility complex (MHC) 16, 18, 27, 30, 44, 65, 70, 78, 84, 112, 113, 119, 120, 122, 134, 140, 141, 168, 169, 171, 172, 182, 183, 212-214, 229, 230
Mast cell (MC) 3, 161, 177-184, 186-198
Matrix metalloproteinase (MMP) 102-104, 127, 173, 231, 233
Membrane cofactor protein (MCP) 111, 127, 156, 179, 182, 184
Memory cell 8, 127, 133, 136
Mesenchymal cell 42, 169, 170
Mesenteric lymph node (MLN) 7, 8, 16, 17, 19, 41, 46-48, 69, 72, 73, 118-122, 125, 127, 137, 154, 157, 158, 162
Metalloproteinase 102, 127, 169, 172, 173, 231
MHC class 2 44
Microbacteria 5
Microbes 15, 108, 110, 112, 126, 142, 219
Migration 7, 9, 16-18, 35, 38, 44, 73, 104, 113, 117, 118, 120, 121, 125, 127, 134, 141, 158, 159, 171, 179, 190, 230, 238
Mitogen-activated protein kinase (MAPK) 44, 45, 236
Monocytes 27, 29, 44, 110, 117-119, 125-127, 151, 152, 161, 173, 211, 227, 228, 230, 236
Mucosal inflammation 9, 41, 42, 55-57, 59-68, 71, 73, 75, 80, 89, 113, 117, 127, 136, 137, 174, 177, 210, 219
Muramyl dipeptide (MDP) 27-29, 31, 43, 44, 67
Muscarinic M2 receptor 182, 192
Muscle cell 168, 169, 186
Mycobacterium 42, 49, 98, 109, 214
Mycobacterium paratuberculosis 42, 98, 109, 212
Mycoplasma pneumoniae 42
Myofibroblast 1, 103, 168-170

N

Neuroimmune interaction 177, 197
Neuropeptide 180, 185, 188-192, 196-198
Neuropeptide Y (NPY) 185, 186, 188-190
Neurotrophin 180, 187-191, 197, 198
NF-κB 27, 29, 44-46, 56, 59, 64, 67, 82, 83, 102, 104, 110, 119, 184, 210, 220, 227, 229, 231, 234, 235, 237
Nitric oxide (NO) 172, 173, 185, 186, 188
NK cell 78, 118, 141, 220
NKT cell 56, 57, 78-81, 126
NOD2/CARD15 24-31, 100, 209
Nonsteroidal antiinflammatory drugs (NSAIDs) 35
Norpeinephrine (NE) 185, 188, 191

O

Oral tolerance 9, 18, 19, 57, 58, 60, 75, 81, 123, 139, 149, 154, 161, 162, 184, 194
Organic cation transporter 29
Organogenesis 9
Osteoprotegerin (OPG) 123
Outer membrane protein 212
Oxazolone colitis 57, 58, 77-81, 126, 137

P

p38 44
P-glycoprotein 211, 212
Paneth cell 29, 42, 44, 112
Pathogen-associated molecular pattern (PAMP) 27, 117, 122, 125
Pattern recognition molecule 108, 110
Pattern recognition receptor (PRR) 27, 43, 44, 117, 122
Permeability 41, 64, 113, 152, 177, 183, 193, 197, 212, 230, 238
Peyer's patch (PP) 3, 5-8, 15-17, 19, 20, 58, 73, 101, 112, 118-123, 137, 150, 159
Pituitary adenylate cyclase activating peptide (PACAP) 186, 188, 189
Plasma cell (PC) 3, 17, 100, 101, 103, 149-151, 153, 156, 187
Platelet-activating factor 179, 180, 182, 193
Pouchitis 37, 41, 109, 155, 220, 221
PPARγ 45, 46
Probiotics 37, 40, 41, 45, 48, 49, 63-65, 109, 112, 142, 143, 219-223
Prostaglandin (PGE_2) 20, 125, 170, 173, 179-183, 193, 230
Protectin 156
Proteinase-activated receptor (PAR) 192
Pseudomembranous colitis 186
Pseudomonas fluorescens 41, 49

R

Regulation 6, 7, 9, 26, 41, 45, 56, 71-73, 75, 113, 117, 121, 126, 127, 134, 140, 142, 162, 177, 178, 180, 181, 184, 185, 189, 191-193, 196, 210, 213, 219, 230-232, 235
Regulatory T cell 19, 20, 36, 46, 47, 49, 58, 59, 64, 69, 71, 74, 75, 77, 85, 86, 88, 89, 103, 117, 119-123, 125, 127, 135, 138, 139, 141, 143, 223
Rodent 6, 7, 35, 38, 46, 49, 122, 154, 161, 178, 181, 187, 193
Rotavirus 150, 219

S

S. typhimurium 29, 64, 126
Saccharomyces 44, 48, 142, 157, 211, 214
Saccharomyces cerevisiae 48, 49, 157, 211
SAMP1/Yit mouse 60, 66, 88
SCID mice 5, 16, 46, 59, 61, 62, 68, 71, 72, 82, 140, 141
Secondary lymphoid tissue chemokine (SLC) 7, 18
Secretory component (SC) 16, 149-151, 153
Secretory immunity 149, 150, 153
Self-antigen 48, 57, 58, 60, 69, 70, 78, 79, 117, 118
Self-tolerance 8, 118
Sensory C fiber 192
Serotonin 186, 191, 193
Severe combined immunodeficient (SCID) 5, 16, 39, 46, 57, 59-62, 68, 69, 71, 72, 77, 82, 84, 85, 122, 136, 141
Single nucleotide polymorphism (SNP) 25, 26, 211
SMAD 20, 76, 103, 104
Smooth muscle cell 168, 169, 186
STAT4 70, 99, 100, 134-136, 231, 232
STAT4 transgenic mice 60, 82, 136, 231

T

T-bet 82, 99, 100, 134, 135, 231, 232
T cell 1, 3-8, 15-20, 36, 38, 41, 46-49, 56-59, 61, 62, 64-66, 68-78, 81-86, 88, 89, 99, 100, 102-105, 108, 111, 113, 117-123, 125-127, 133-143, 157, 159-162, 168-174, 182, 184, 209, 212-214, 220-223, 229-238
TECK 3, 15, 17, 213
Terminal ileitis 99
TGF-β 7, 17, 19, 20, 36, 44, 46, 48, 58, 65, 69-72, 75-77, 79, 81, 82, 86, 99, 103, 104, 111, 119-122, 125-127, 134, 135, 138-140, 142, 143, 184, 188, 192, 229, 230
TH0 1
Th1 1, 19, 36, 38, 40, 44, 46-48, 56-59, 66-70, 72-76, 78, 81-89, 98-100, 111, 118, 119, 125, 126, 134-143, 161, 183, 211, 228, 229, 231-234, 236-238
Th2 1, 19, 36, 48, 56-59, 66, 73, 77-79, 81, 84-89, 99, 118, 119, 134-141, 143, 182, 183, 228, 229, 231, 232
Th3 1, 58, 138, 139, 229, 231
TNBS-colitis 56, 58, 60, 66, 71, 73-77, 81
TNF-α 2, 8, 27, 29, 30, 58, 60, 61, 64, 65, 70, 73, 74, 83, 102-104, 117, 118, 120-123, 125-127, 136, 151, 162, 171, 179-183, 210, 211
Tolerance 7-9, 18, 19, 36, 38, 44, 46-49, 57, 58, 60, 70, 74, 75, 77, 81, 84, 99, 111, 118, 119, 121, 123, 125, 127, 139, 149, 154, 161, 162, 170, 184, 194, 220
Toll-like receptor (TLR) 19, 27, 42-45, 56, 58, 65, 68, 74, 76, 77, 82, 110, 117, 119, 122, 142, 151, 230
Tr1 19, 71, 72, 119, 125, 138, 139, 141, 142, 143
Tumor necrosis factor (TNF) 17, 18, 20, 30, 44, 45, 58, 60, 61, 64, 65, 70, 73, 74, 83, 84, 108, 111, 117, 118, 120, 121, 123, 125-127, 136, 151, 162, 171, 172, 179-183, 210, 211, 227-229, 231-236, 238
Twins 25, 35, 149, 152, 213, 222

U

Ulcerative colitis (UC) 5, 24-27, 29, 30, 37, 41, 42, 47-49, 57, 62, 77, 80-82, 88, 89, 98-103, 108-113, 125, 126, 134, 136, 137, 149, 152, 154-159, 161, 169-172, 177, 183, 195-197, 209-214, 220, 227-232, 236-238

V

VCAM-1 170, 171, 173, 178, 179, 183, 237
VIP 180-182, 186, 188-191, 195, 196

W

Wiskott-Aldrich syndrome protein (WASP) 88

MIX
Papier aus verantwortungsvollen Quellen
Paper from responsible sources
FSC® C105338
FSC
www.fsc.org

If you have any concerns about our products,
you can contact us on
ProductSafety@springernature.com

In case Publisher is established outside the EU,
the EU authorized representative is:
Springer Nature Customer Service Center GmbH
Europaplatz 3, 69115 Heidelberg, Germany

Printed by Libri Plureos GmbH
in Hamburg, Germany